*Nonlinear elliptic boundary value problems
and their applications*

Pitman Monographs and
Surveys in Pure and Applied Mathematics 80

Nonlinear elliptic boundary value problems and their applications

Heinrich G W Begehr

Free University Berlin

and

Guo Chun Wen

Peking University

LONGMAN

Addison Wesley Longman Limited
Edinburgh Gate, Harlow
Essex CM20 2JE, England
and Associated companies throughout the world.

*Published in the United States of America
by Addison Wesley Longman Inc.*

© Addison Wesley Longman Limited 1996

All rights reserved; no part of this publication may be reproduced,
stored in a retrieval system, or transmitted in any form or by any
means, electronic, mechanical, photocopying, recording, or
otherwise, without the prior written permission of the Publishers, or
a licence permitting restricted copying in the United Kingdom
issued by the Copyright Licensing Agency Ltd, 90 Tottenham Court
Road, London, W1P 9HE

First published 1996

AMS Subject Classifications: (Main) 35
                                   (Subsidiary) 32, 30, 73, 76

ISSN 0269-3666

ISBN 0 582 29204 2

British Library Cataloguing in Publication Data

A catalogue record for this book is
available from the British Library

Printed and bound by Bookcraft (Bath) Ltd

Contents

Preface

This volume is a continuation of "Boundary value problems for elliptic equations and systems" (see [140]). Mainly some nonlinear boundary value problems for elliptic equations and systems, various boundary value problems for degenerate and higher order elliptic equations and systems, for functions of several complex variables and functions in Clifford analysis are considered. Finally applications in continuum mechanics are presented.

In Chapter 1, some boundary value problems for elliptic equations and systems under weaker conditions in multiply connected domains are discussed. The authors have studied nonlinear Riemann-Hilbert and Haseman boundary value problems for some classes of quasilinear elliptic complex equations. Moreover, several nonlinear boundary value problems for nonlinear elliptic complex equations of first and second order were discussed by different methods. These subjects can be found in Chapter 2.

In Chapter 3, not only several boundary value problems for degenerate elliptic equations of second order, but also some boundary value problems for degenerate elliptic complex equations of first order are introduced.

Boundary value problems for functions of several complex variables and for functions in Clifford analysis were only recently investigated. In Chapter 4 and Chapter 5 in detail some new results are presented, which cannot be found in other books published.

Similarly to the book [140], the considered complex equations and boundary conditions in this volume are rather general, and several methods are used. There are two characteristics of this book: one is that the research fields involved are very wide and include many deep results, another one is that applications of elliptic boundary value problems to continuum mechanics are considered, which can be seen in Chapter 7 where some free boundary problems in planar filtrations, gas dynamics, elastico-plastic mechanics and some mixed-contact problem for orthotropic elasticity are discussed by using function theoretic methods.

The great majority of the contents originates in investigations of the authors and their cooperative colleagues, postgraduate-students, and many results are published here for the first time. After reading the volume, it can be seen that many questions remain for further investigations.

The preparation of this book was supported by the Deutsche Forschungsgemeinschaft and the National Natural Science Foundation of China, the State Education Commission of the People's Republic of China on the basis of their scientific exchange program as well as through the Freie Universität Berlin and the Peking University. The authors would like to thank Barbara M. Wengel for helping to prepare the index and the camera–ready copy of the manuscript and Dipl. Math. Ute Fuchs for graphing and incorporating the figures .

Berlin and Beijing
August, 1995

Heinrich Begehr   Guo Chun Wen
Free University Berlin   Peking University

I Boundary Value Problems for Nonlinear Elliptic Equations and Systems with Weak Conditions

In this chapter, we discuss some boundary value problems for elliptic systems of first order equations and elliptic equations of second order, in which the boundaries of domains, the boundary conditions and the conditions for the coefficients of the equations are weaker than those in the book [140]. We obtain some results on the solvability of the boundary value problems, which are useful for the following chapters, on the numerical analysis and on some free boundary problems in mechanics and physics.

1 Some Boundary Value Problems for Elliptic Complex Equations with Non-smooth Boundaries

This section mainly deals with the Riemann-Hilbert problem for nonlinear elliptic systems of first order equations in bounded domains with non-smooth boundaries. We first consider the well-posedness of the above boundary value problem, and then discuss its solvability.

1.1 The well-posedness of the Riemann-Hilbert boundary value problem with non-smooth boundary

Let D be an $(N+1)$-connected bounded domain in the complex plane $\mathbb{C}$ with boundary $\Gamma = \cup_{j=0}^{N}\Gamma_j \in C_\alpha(0 < \alpha < 1)$, where $\Gamma_j(j = 1, \cdots, N)$ are located in the domain D_0 bounded by $\Gamma_0 = \Gamma_{N+1}$, and $0 \in D$.

We discuss the nonlinear uniformly elliptic complex equation of first order

$$\begin{cases} w_{\bar{z}} = F(z, w, w_z), \ F = Q_1 w_z + Q_2 \overline{w_z} + A_1 w + A_2 \overline{w} + A_3, \\ Q_j = Q_j(z, w, w_z), \ j = 1, 2, \ A_j = A_j(z, w), \ j = 1, 2, 3, \end{cases} \tag{1.1}$$

which is the complex form of the real nonlinear elliptic system of first order equations

$$\Phi_j(x, y, u, v, u_x, u_y, v_x, v_y) = 0, \ j = 1, 2 \tag{1.2}$$

under certain conditions(see [140]). Suppose that the complex equation (1.1) satisfies the following conditions.

Condition C

1) $Q_j(z, w, U)\,(j = 1, 2)$, $A_j(z, w)\,(j = 1, 2, 3)$ are continuous in $w \in \mathbb{C}$ for almost every $z \in D$, $U \in \mathbb{C}$, and $Q_j = 0\,(j = 1, 2)$, $A_j = 0\,(j = 1, 2, 3)$ for $z \notin D$.

2) The above functions are measurable in $z \in D$ for all continuous functions $w(z)$ on $\bar{D}$ and all measurable functions $U(z) \in L_{p_0}(\tilde{D})$, and satisfy

$$L_p[A_j, \bar{D}] \leq k_0, \ j = 1, 2, \ L_p[A_3, \bar{D}] \leq k_1, \tag{1.3}$$

where $\tilde{D}$ is any closed subset in D, $p_0, p\,(2 < p_0 \leq p)$, k_0, k_1 are nonnegative constants.

3) The complex equation (1.1) satisfies the uniform ellipticity condition

$$\mid F(z, w, U_1) - F(z, w, U_2) \mid \leq q_0 \mid U_1 - U_2 \mid, \tag{1.4}$$

in which $U_1, U_2 \in \mathbb{C}$ and $q_0(< 1)$ is a nonnegative constant.

The Riemann–Hilbert boundary value problem for equation (1.1) may be formulated as follows:

Problem A   Find a continuous solution $w(z)$ of (1.1) on $\bar{D}$ satisfying the boundary condition

$$\mathrm{Re}[\overline{\lambda(z)}w(z)] = r(z), \ z \in \Gamma, \tag{1.5}$$

where $\mid \lambda(z) \mid = 1$, $z \in \Gamma$, and $\lambda(z)$, $r(z)$ satisfy the conditions

$$C_\alpha[\lambda(z), \Gamma] \leq k_0, \ C_\alpha[r(z), \Gamma] \leq k_2, \tag{1.6}$$

in which $\alpha\,(0 < \alpha < 1), k_2$ are nonnegative constants.

This boundary value problem for (1.1) with $A_3(z, w) = 0$, $z \in D, w \in \mathbb{C}$ and $r(z) = 0, z \in \Gamma$ will be called Problem A_0. The integer $K = \frac{1}{2\pi}\Delta_\Gamma \arg \lambda(z)$ is the index of Problem A and Problem A_0. When the index $K < 0$, Problem A may not be solvable, when $K \geq 0$, the solution of Problem A is not necessarily unique. Hence we consider the well posedness of Problem A with modified boundary conditions.

Problem B   Find a continuous solution $w(z)$ of the complex equation (1.1) satisfying the boundary condition

$$\mathrm{Re}[\overline{\lambda(z)}w(z)] = r(z) + h(z), \ z \in \Gamma, \tag{1.7}$$

where

$$h(z) := \begin{cases} 0, z \in \Gamma, & \text{if } K \geq N, \\ \left.\begin{array}{l} h_j, z \in \Gamma_j, j = 1, \cdots, N - K, \\ 0, z \in \Gamma_j, j = N - K + 1, \cdots, N + 1, \end{array}\right\} & \text{if } 0 \leq K < N, \\ \left.\begin{array}{l} h_j, z \in \Gamma_j, j = 1, \cdots, N, \\ h_0 + \mathrm{Re} \sum_{m=1}^{-K-1}(h_m^+ + ih_m^-)z^m, z \in \Gamma_0, \end{array}\right\} & \text{if } K < 0, \end{cases} \tag{1.8}$$

in which $h_j \, (j = 0, 1, \cdots, N)$, $h_m^{\pm} \, (m = 1, \cdots, -K-1, K < 0)$ are unknown real constants to be determined appropriately. In addition, for $K \geq 0$ the solution $w(z)$ is assumed to satisfy the point conditions

$$\mathrm{Im}[\overline{\lambda(a_j)}w(a_j)] = b_j, \ j \in J = \begin{cases} 1, \cdots, 2K-N+1, & \text{if } K \geq N, \\ N-K+1, \cdots, N+1, & \text{if } 0 \leq K < N, \end{cases} \tag{1.9}$$

where $a_j \in \Gamma_j \, (j = 1, \cdots, N)$, $a_j \in \Gamma_0 \, (j = N+1, \cdots, 2K-N+1, K \geq N)$ are distinct points, and $b_j (j \in J)$ are all real constants satisfying the conditions

$$\mid b_j \mid \leq k_3, \ j \in J, \tag{1.10}$$

herein k_3 is a nonnegative constant.

Problem B' If $\Gamma_j \, (j = 0, 1, \cdots, N)$ are rectifiable curves, the point condition (1.9) are replaced by the integral conditions

$$L_j(\bar{\lambda}w) = \begin{cases} \mathrm{Im} \int_{\Gamma_j} \overline{\lambda(z)}w(z)ds = B_j, \ j = \begin{cases} N-K+1, \cdots, N+1 \text{ if } 0 \leq K < N, \\ 1, \cdots, N+1 \quad \text{if } K \geq N, \end{cases} \\ \mathrm{Re} \int_{\Gamma_0} z^{j-N-1}\overline{\lambda(z)}w(z)ds = B_j, j = N+2, \cdots, K+1 \\ \mathrm{Im} \int_{\Gamma_0} z^{j-K-1}\overline{\lambda(z)}w(z)ds = B_j, j = K+2, \cdots, 2K-N+1 \end{cases} \Big\} \text{ if } K \geq N, \tag{1.11}$$

where $B_j (j \in J)$ are real constants satisfying the conditions

$$\mid B_j \mid \leq k_3, \ j \in J. \tag{1.12}$$

For convenience, we sometimes will subsume the integral conditions or the point conditions under boundary conditions.

In the following, we first give a priori estimates of solutions for Problem B for (1.1) with Condition C, afterwards by using the foregoing estimates and the Schauder fixed–point theorem, the solvability of Problem B for (1.1) is proved. Moreover, the solvability conditions of Problem A for (1.1) are derived. In order to verify the uniqueness of solutions for Problem B, we need to add the following condition: For any continuous functions $w_1(z), w_2(z)$ on $\bar{D}$ and $U(z) \in L_{p_0}(\bar{D})$, there is

$$F(z, w_1, U) - F(z, w_2, U) = A(z, w_1, w_2, U)(w_1 - w_2), \tag{1.13}$$

where $L_{p_0}[A, \bar{D}] < \infty$ and $p_0(2 < p_0 \leq p)$ is a constant. In particular, when (1.1) is a linear equation, the condition (1.13) obviously is satisfied.

1.2   A priori estimates for solutions of Problem B and Problem B'

First of all, we give a representation theorem for Problem B and for Problem B'.

Theorem 1.1 *Suppose that the complex equation (1.1) satisfies Condition C, and $w(z)$ is a solution of Problem B (or Problem B') for (1.1). Then $w(z)$ is representable by*

$$w(z) = \Phi[\zeta(z)]e^{\phi(z)} + \psi(z), \tag{1.14}$$

where $\zeta(z)$ is a homeomorphism on $\bar{D}$, which quasiconformally maps D onto an $(N+1)$–connected circular domain G with boundary $L = \cup_{j=0}^{N}L_j$, where the $L_j = \zeta(\Gamma_j)$ $(j = 1, \cdots, N)$ are located in the unit disk bounded by $L_0 = \zeta(\Gamma_0) = \{| \zeta |= 1\}$, and $\zeta(0) = 0, \Phi(\zeta)$ is an analytic function in G, $\psi(z), \phi(z), \zeta(z)$ and its inverse function $z(\zeta)$ satisfy the estimates

$$C_\beta[\tau, \bar{D}] \le k_4, \ \tau = \psi(z), \varphi(z), \zeta(z), \ C_\beta[z(\zeta), \bar{G}] \le k_4, \tag{1.15}$$

$$L_{p_0}[| \psi_{\bar{z}} | + | \psi_z |, \bar{D}] \le k_4, \ L_{p_0}[| \phi_{\bar{\zeta}} | + | \phi_\zeta |, \bar{G}] \le k_4, \tag{1.16}$$

$$L_{p_0}[| \zeta_{\bar{z}} | + | \zeta_z |, \tilde{D}] \le k_5, \ L_{p_0}[| z_{\bar{\zeta}} | + | z_\zeta |, \tilde{G}] \le k_5, \tag{1.17}$$

in which $\beta = \min(\alpha, 1 - 2/p_0), p_0(2 < p_0 \le p), k_4, k_5$ are nonnegative constants, $k_4 = k_4(q_0, p_0, k_0, k_1, D)$, $k_5 = k_5(q_0, p_0, k_0, k_1, \tilde{D})$, $\tilde{D} = \{z \mid z \in D, \ \text{dist}(z, \Gamma) \ge \varepsilon > 0\}$, ε is a constant and $\tilde{G} = \zeta(\tilde{D})$.

Proof   Similarly to Theorem 2.4, Chapter 2 in [140], we substitute the solution $w(z)$ of Problem B(or Problem B') into the coefficients of the equation (1.1) and consider the following system

$$\psi_{\bar{z}} = Q\psi_z + A_1\psi + A_2\bar{\psi} + A_3, Q = \begin{cases} Q_1 + Q_2\overline{w_z}/w_z \ \text{for} \ w_z \ne 0, \\ 0 \ \text{for} \ w_z = 0 \ \text{or} \ z \notin G, \end{cases} \tag{1.18}$$

$$\phi_{\bar{z}} = Q\phi_z + A, A = \begin{cases} A_1 + A_2\bar{w}/w \ \text{for} \ w(z) \ne 0, \\ 0 \ \text{for} \ w(z) = 0 \ \text{or} \ z \notin D, \end{cases} \tag{1.19}$$

$$W_{\bar{z}} = QW_z, \ \ W(z) = \Phi[\zeta(z)]. \tag{1.20}$$

By using the continuity method and the principle of contracting mappings, we can find the solutions $\psi(z) = Tf = -\frac{1}{\pi} \iint_D[f(\zeta)/(\zeta - z)]d\sigma_\zeta, \ \phi = Tg, \ \zeta(z) = \Psi[\chi(z)], \ \chi(z) = z + Th$, where $f(z), g(z), h(z) \in L_{p_0}(\bar{D}), 2 < p_0 \le p$, $\chi(z)$ is a homeomorphism on $\bar{D}, \Psi(\chi)$ is a univalent analytic function, which conformally maps $E = \chi(D)$ onto an $(N+1) - -$connected circular domain G, and $\Phi(\zeta)$ is an analytic function in G. We can verify that $\psi(z), \phi(z), \zeta(z)$ satisfy the estimates (1.15) and (1.16). It remains to prove that $\zeta(z)$ and its inverse function $z = z(\zeta)$ satisfy the estimate (1.17). In fact, we first find a univalent analytic function $Z = Z(z)$, which maps D onto an $(N+1) - -$connected circular domain Δ similar to D, and denote by $z = z(Z)$ the inverse function of $Z = Z(z)$. Thus the complex equation (1.20) is reduced to

$$W_{\bar{Z}} = [Q\overline{z'(Z)}/z'(Z)]W_Z, \ | Q\overline{z'(Z)}/z'(Z) |\le q_0 < 1. \tag{1.21}$$

As is known, the complex equation (1.21) has a homeomorphic solution of the form $V(Z) = Z + TH(H(Z) \in L_{p_0}(\bar{\Delta}))(\text{see } [131]1')$. Next, we find a univalent analytic function $\zeta = \Omega(V)$, which maps $V(\Delta)$ onto G, hence $\zeta = \zeta(z) = \Omega\{V[Z(z)]\}$. By the result on conformal mappings (see [137],[139]17)), applying the method of Lemma 2.1, Chapter 2 in [140], we can prove that (1.17) is true.

Secondly, we shall give a priori estimates of solutions for Problem B and for Problem B'.

Theorem 1.2 *Under the same conditions as in Theorem 1.1, any solution $w(z)$ of Problem B (or Problem B') for (1.1) satisfies*

$$C_\gamma[w(z), \bar{D}] \leq M_1 = M_1(q_0, p_0, k, \alpha, K, D), \tag{1.22}$$

$$L_{p_0}[| w_{\bar{z}} | + | w_z |, \tilde{D}] \leq M_2 = M_2(q_0, p_0, k, \alpha, K, \tilde{D}), \tag{1.23}$$

where $k = k(k_0, k_1, k_2, k_3)$, $\gamma = \alpha\beta^2$, $2 < p_0 \leq p$, $\tilde{D} = \{z \mid z \in D, \text{dist}\,(z, \Gamma) \geq \varepsilon > 0\}$.

Proof On the basis of Theorem 1.1, the solution $w(z)$ of Problem B can be expressed as in (1.14), hence the boundary conditions (1.7)–(1.9) can be transformed into the boundary conditions of Problem $\tilde{B}$:

$$\text{Re}[\overline{\Lambda(\zeta)}\Phi(\zeta)] = R(\zeta) + H(\zeta), \; \zeta \in L, \tag{1.24}$$

$$H(\zeta) = \begin{cases} 0, \zeta \in L, \; \text{if} \; K \geq N, \\ \left.\begin{array}{l} h_j, \zeta \in L_j, j = 1, \cdots, N - K, \\ 0, \zeta \in L_j, j = N - K + 1, \cdots, N + 1 \end{array}\right\} \; \text{if} \; 0 \leq K < N, \\ \left.\begin{array}{l} h_j, \zeta \in L_j, j = 1, \cdots, N \\ h_0 + \text{Re} \sum_{m=1}^{-K-1}(h_m^+ + ih_m^-)[z(\zeta)]^m, \zeta \in L_0 \end{array}\right\} \; \text{if} \; K < 0, \end{cases} \tag{1.25}$$

$$\text{Im}[\overline{\Lambda(a_j')}\Phi(a_j')] = b_j', \; j \in J, \tag{1.26}$$

where

$$\overline{\Lambda(\zeta)} = \overline{\lambda[z(\zeta)]}e^{\phi[z(\zeta)]}, \; R(\zeta) = r[z(\zeta)] - \text{Re}\{\overline{\lambda[z(\zeta)]}\psi[z(\zeta)]\},$$

$$a_j' = \zeta(a_j), \; b_j' = b_j - \text{Im}[\overline{\lambda(a_j)}\psi(a_j)], \; j \in J.$$

By (1.15), it can be seen that $\Lambda(\zeta), R(\zeta), b_j' \, (j \in J)$ satisfy the conditions

$$C_{\alpha\beta}[\Lambda(\zeta), L] \leq M_3, \; C_{\alpha\beta}[R(\zeta), L] \leq M_3, \; | \, b_j' \, | \leq M_3, \; j \in J, \tag{1.27}$$

where $M_3 = M_3(q_0, p_0, k, \alpha, K, D)$. If we can prove that the solution $\Phi(\zeta)$ of Problem $\tilde{B}$ satisfies the estimate

$$C_{\alpha\beta}[\Phi(\zeta), \bar{G}] \leq M_4, \; C_{\alpha\beta}[\Phi'(\zeta), \tilde{G}] \leq M_5, \tag{1.28}$$

in which $\tilde{G} = \zeta(\tilde{D})$, $M_4 = M_4(q_0, p_0, k, \alpha, K, D)$, $M_5 = M_5(q_0, p_0, k, \alpha, K, \tilde{D})$, then from the representation (1.14) of the solution $w(z)$ of Problem B and the estimates (1.15),(1.16) and (1.17), the estimates (1.22) and (1.23) can be derived.

It remains to prove that (1.28) holds. For this, we first verify the boundness of $\Phi(\zeta)$, i.e.

$$C[\Phi(\zeta), \bar{G}] \leq M_6 = M_6(q_0, p_0, k, \alpha, K, D). \tag{1.29}$$

Suppose that (1.29) is not true. Then there exist sequences of functions $\{\Lambda_n(\zeta)\}$, $\{R_n(\zeta)\}$, $\{b'_{jn}\}$ satisfying the same conditions as $\Lambda(\zeta), R(\zeta), b'_j$, which uniformly converge to $\Lambda_0(\zeta), R_0(\zeta), b'_{j0}$ $(j \in J)$ on L respectively. For the solution $\Phi_n(\zeta)$ of the boundary value problem (Problem B_n) corresponding to $\Lambda_n(\zeta), R_n(\zeta), b'_{jn}$ $(j \in J)$, we have $I_n = C[\Phi_n(\zeta), \bar{G}] \to \infty$ as $n \to \infty$. There is no harm in assuming that $I_n \geq 1$, $n = 1, 2, \cdots$. Obviously $\tilde{\Phi}_n(\zeta) := \Phi_n(\zeta)/I_n$ satisfies the boundary conditions

$$\mathrm{Re}[\overline{\Lambda_n(\zeta)}\tilde{\Phi}_n(\zeta)] = [R_n(\zeta) + H(\zeta)]/I_n, \zeta \in L, \tag{1.30}$$

$$\mathrm{Im}[\overline{\Lambda_n(a'_n)}\tilde{\Phi}_n(a'_n)] = b'_{jn}/I_n, j \in J. \tag{1.31}$$

Applying the Schwarz formula, the Cauchy formula and the method of symmetric extension (see Theorem 1.4, Chapter 1, [140]), the estimates

$$C_{\alpha\beta}[\tilde{\Phi}_n(\zeta), \bar{G}] \leq M_7, \ C[\tilde{\Phi}'_n(\zeta), \tilde{G}] \leq M_8 \tag{1.32}$$

can be obtained, where $M_7 = M_7(q_0, p_0, k, \alpha, K, D)$, $M_8 = M_8(q_0, p_0, k, \alpha, K, \tilde{D})$. Thus we can select a subsequence of $\{\tilde{\Phi}_n(\zeta)\}$, which uniformly converge to an analytic function $\tilde{\Phi}_0(\zeta)$ in $\bar{G}$, and $\tilde{\Phi}_0(\zeta)$ satisfies the homogeneous boundary value conditions

$$\mathrm{Re}[\overline{\Lambda_0(\zeta)}\tilde{\Phi}_0(\zeta)] = H(\zeta), \ \zeta \in L, \tag{1.33}$$

$$\mathrm{Im}[\overline{\Lambda_0(a'_j)}\tilde{\Phi}_0(a'_j)] = 0, \ j \in J. \tag{1.34}$$

On the basis of the uniqueness theorem (see Theorem 1.4), we conclude that $\tilde{\Phi}_0(\zeta) = 0, \zeta \in \bar{G}$. However, from $C[\tilde{\Phi}_n(\zeta), \bar{G}] = 1$, it follows that there exists a point $\zeta_* \in \bar{G}$, such that $|\tilde{\Phi}_0(\zeta_*)| = 1$. This contradiction proves that (1.29) holds. Afterwards using the method which leads from $C[\tilde{\Phi}_n(\zeta), \bar{G}] = 1$ to (1.32), the estimate (1.28) can be derived.

Similarly, we can verify that any solution $w(z)$ of Problem B' satisfies the estimates (1.22) and (1.23).

Theorem 1.3   *Under the same conditions as in Theorem 1.1, any solution $w(z)$ of Problem B (or Problem B') for (1.1) satisfies*

$$C_\gamma[w(z), \bar{D}] \leq M_9 k_*, \ L_{p_0}[|w_{\bar{z}}| + |w_z|, \tilde{D}] \leq M_{10} k_*, \tag{1.35}$$

where $\gamma, p_0, \tilde{D}$ are as stated in Theorem 1.2, $k_* = k_1 + k_2 + k_3$, $M_9 = M_9(q_0, p_0, k, \alpha, K, D)$, $M_{10} = M_{10}(q_0, p_0, k, \alpha, K, \tilde{D})$.

Proof   If $k_* = 0$, i.e. $k_1 = k_2 = k_3 = 0$, from Theorem 1.4, it follows that $w(z) = 0, z \in D$. If $k_* > 0$, it is easy to see that $W(z) = w(z)/k_*$ satisfies the complex equation and boundary conditions

$$W_{\bar{z}} - Q_1 W_z - Q_2 \overline{W_z} - A_1 W - A_2 \overline{W} = A_3/k_*, z \in D, \tag{1.36}$$

$$\mathrm{Re}[\overline{\lambda(z)}W(z)] = [r(z) + h(z)]/k_*, z \in \Gamma, \tag{1.37}$$

$$\mathrm{Im}[\overline{\lambda(a_j)}W(a_j)] = b_j/k_*, j \in J. \tag{1.38}$$

Noting that $L_p[A_3/k_*, \bar{D}] \le 1$, $C_\alpha[r(z)/k_*, \Gamma] \le 1$, $\mid b_j/k_* \mid \le 1, j \in J$ and according to the proof of Theorem 1.2, we have

$$C_\gamma[W(z), \bar{D}] \le M_9, \ L_{p_0}[\mid W_{\bar{z}} \mid + \mid W_z \mid, \tilde{D}] \le M_{10}. \tag{1.39}$$

From the above estimates, it immediately follows that (1.35) holds.

We mention that when $w(z)$ is any solution of Problem B', the point conditions (1.38) shall be replaced by the integral conditions.

1.3 Uniqueness and solvability for Problem B and Problem B'

Theorem 1.4 *If Condition C and (1.13) hold, then the solution of Problem B (or Problem B') for (1.1) is unique.*

Proof Let $w_1(z), w_2(z)$ be two solutions of Problem B for (1.1). By Condition C and (1.13), we see that $w(z) = w_1(z) - w_2(z)$ satisfies the equation and boundary conditions

$$w_{\bar{z}} - \tilde{Q}w_z = \tilde{A}w, \ z \in D, \tag{1.40}$$

$$\mathrm{Re}[\overline{\lambda(z)}w(z)] = h(z), \ z \in \Gamma, \tag{1.41}$$

$$\mathrm{Im}[\overline{\lambda(a_j)}w(a_j)] = 0, \ j \in J, \tag{1.42}$$

where

$$\tilde{Q} = \begin{cases} [F(z, w_1, w_{1z}) - F(z, w_1, w_{2z})]/(w_1 - w_2)_z & \text{for } w_{1z} \neq w_{2z}, \\ 0 \text{ for } w_{1z} = w_{2z}, \ z \in D, \end{cases}$$

$$\tilde{A} = \begin{cases} [F(z, w_1, w_{2z}) - F(z, w_2, w_{2z})]/(w_1 - w_2) & \text{for } w_1(z) \neq w_2(z), \\ 0 \text{ for } w_1(z) = w_2(z), \ z \in D, \end{cases}$$

and $\mid \tilde{Q} \mid \le q_0 < 1, z \in D, L_{p_0}(\tilde{A}, \bar{D}) < \infty$. According to the representation (1.14), we have

$$w(z) = \Phi[\zeta(z)]e^{\phi(z)}, \tag{1.43}$$

where $\phi(z), \zeta(z), \Phi(\zeta)$ are as stated in Theorem 1.1. It can be seen that the analytic function $\Phi(z)$ satisfies the boundary conditions of Problem B_0 :

$$\mathrm{Re}[\overline{\Lambda(\zeta)}\Phi(\zeta)] = H(\zeta), \ \zeta \in L = \zeta(\Gamma), \tag{1.44}$$

$$\mathrm{Im}[\overline{\Lambda(a'_j)}\Phi(a'_j)] = 0, \ j \in J, \tag{1.45}$$

where $\Lambda(\zeta), H(\zeta) \, (\zeta \in L), a'_j \, (j \in J)$ are as stated in (1.24)–(1.26). In accordance with Theorem 4.1, Chapter 2 in [140], we can see that when $K \geq 0$, there is $\Phi(\zeta) = 0, \ \zeta \in G = \zeta(D)$. Hence, $w(z) = \Phi[\zeta(z)]e^{\phi(z)} = 0$, i.e. $w_1(z) = w_2(z), \ z \in D$.

Next, we discuss the case of $K < 0$. According to Theorem 4.12 of Chapter 4 in [131]1), the necessary and sufficient condition for solvability of Problem B_0 is that

$$\int_L [\overline{\Lambda(\zeta)}]^{-1}\Psi_n(\zeta)H(\zeta)\zeta'(s)ds = 0, \ n = 1, \cdots, N - 2K - 1, \tag{1.46}$$

where $\Psi_n(\zeta)(n = 1, \cdots, N - 2K - 1)$ is a complete system of linearly indepedent solutions for the boundary value problem B'_0 for analytic functions with the boundary conditions

$$\mathrm{Re}\{[\overline{\Lambda(\zeta)}]^{-1}\Psi_n(\zeta)\zeta'(s)\} = 0, \ \zeta \in L, \ n = 1, \cdots, N - 2K - 1. \tag{1.47}$$

Taking into account the definition of $h(z)$ in (1.8), the conditions (1.46) can be rewritten as

$$\begin{aligned}
&\textstyle\sum_{j=0}^{N} h_j \int_{L_j} \mathrm{Im}[\overline{\Lambda(\zeta)}]^{-1}\Psi_n(\zeta)\zeta'(s)ds+ \\
&+ \textstyle\sum_{m=1}^{-K-1}\{h_m^+ \int_{L_0} \mathrm{Im}[\overline{\Lambda(\zeta)}]^{-1}\Psi_n(\zeta)\zeta'(s) \cos[m \arg z(\zeta)]ds \\
&- h_m^- \int_{L_0} \mathrm{Im}[\overline{\Lambda(\zeta)}]^{-1}\Psi_n(\zeta)\zeta'(s) \sin[m \arg z(\zeta)]ds\} = 0, \\
&\hspace{3cm} n = 1, \cdots, N - 2K - 1.
\end{aligned} \tag{1.48}$$

Suppose that $\sum_{j=0}^{N} \mid h_j \mid + \sum_{m=1}^{-K-1}[\mid h_m^+ \mid + \mid h_m^- \mid] \neq 0$, then the coefficients determinant of the algebraic system (1.48) certainly equals zero. Hence we can find real constants $c_1, \cdots, c_{N-2K-1}$, which are not all equal to zero, such that

$$\begin{aligned}
&\textstyle\int_{L_j} \mathrm{Im}[\overline{\Lambda(\zeta}]^{-1}\Psi(\zeta)\zeta'(s)ds = 0, \ j = 0, \cdots, N, \\
&\textstyle\int_{L_0} \mathrm{Im}[\overline{\Lambda(\zeta)}]^{-1}\Psi(\zeta)\zeta'(s) \cos[m \arg z(\zeta)]ds = 0, \\
&\textstyle\int_{L_0} \mathrm{Im}[\overline{\Lambda(\zeta)}]^{-1}\Psi(\zeta)\zeta'(s) \sin[m \arg z(\zeta)]ds = 0, \\
&\hspace{3cm} m = 1, \cdots, -K - 1,
\end{aligned} \tag{1.49}$$

in which $\Psi(\zeta) = \sum_{j=1}^{N-2K-1} c_n\Psi_n(\zeta)$ is a solution of Problem B'_0 and $\Psi(\zeta) \not\equiv 0$ in $G = \zeta(D)$. From (1.49) and Theorem 4.7 from Chapter 4 in [131]1), it can be derived that $\Psi(\zeta)$ has at least $2N - 2K - 1$ zero points on L. Noting that $N - K - 1 = \frac{1}{2\pi}\Delta_\Gamma \arg[\Lambda(\zeta)]^{-1}\overline{\zeta'(s)}$ and according to the formula (4.19) from Chapter 4 in [131]1), we get the absurd inequality

$$2N - 2K - 1 \leq 2N_D + N_\Gamma = 2N - 2K - 2.$$

This contradiction proves that

$$h_j = 0, \; j = 0, 1, \cdots, N, \quad h_m^{\pm} = 0, \; m = 1, \cdots, -K - 1,$$

i.e. $H(\zeta) = 0, \zeta \in L$, and so $\Phi(\zeta) = 0, \zeta \in \bar{G}$. Consequently $w(z) = 0$ in D, i.e. $w_1(z) = w_2(z), \; z \in D$.

Applying a similar method, we can also prove that the solution of Problem B' is unique.

Theorem 1.5 *Suppose that Condition C holds. Then Problem B and Problem B' are solvable.*

Proof Here we only discuss Problem B for (1.1). Let us introduce a closed, convex and bounded subset B_1 in the Banach space $B = L_{p_0}(\bar{D}) \times L_{p_0}(\bar{D}) \times L_{p_0}(\bar{D})$, whose elements are systems of functions $q = [Q(z), f(z), g(z)]$ with norms $\| q \| = L_{p_0}(Q, \bar{D}) + L_{p_0}(f, \bar{D}) + L_{p_0}(g, \bar{D})$, which satisfies the condition

$$| Q(z) | \leq q_0 < 1 \; (z \in D), \; L_{p_0}[f(z), \bar{D}] \leq k_4, \; L_{p_0}[g(z), \bar{D}] \leq k_4, \tag{1.50}$$

where q_0, k_4 are nonnegative constants as stated in (1.4), (1.15) and (1.16). Moreover introduce a closed and bounded subset B_2 in B, the elements of which are systems of functions $\omega = [f(z), g(z), h(z)]$ satisfying the condition

$$L_{p_0}[f(z), \bar{D}] \leq k_4, \; L_{p_0}[g(z), \bar{D}] \leq k_4, \; | h(z) | \leq q_0 | 1 + \Pi h |, \tag{1.51}$$

where $\Pi h = -\frac{1}{\pi} \iint_D [h(\zeta)/(\zeta - z)^2] d\sigma_\zeta$.

We arbitrarily select $q = [Q(z), f(z), g(z)] \in B_1$, and using the principle of contracting mappings, a unique solution $h(z) \in L_{p_0}(\bar{D})$ of the integral equation

$$h(z) = Q(z)[1 + \Pi h] \tag{1.52}$$

is found, which satisfies the third estimate in (1.51). Moreover, $\chi(z) := z + Th$ is a homeomorphism on $\bar{D}$. Now, we find a univalent analytic function $\zeta = \Phi(\chi)$, which maps $\chi(D)$ onto a circular domain G as stated in Theorem 1.1. Afterwards, we find an analytic function $\Phi(\zeta)$ in G satisfying the boundary conditions

$$\mathrm{Re}[\overline{\Lambda(\zeta)}\Phi(\zeta)] = R(\zeta) + H(\zeta), \; \zeta \in L = \zeta(\Gamma), \tag{1.53}$$

$$\mathrm{Im}[\overline{\Lambda(a_j')}\Phi(a_j')] = b_j', \; j \in J, \tag{1.54}$$

in which $\zeta(z) = \Psi[\chi(z)]$, $z(\zeta)$ is its inverse function, $\psi(z) = Tf$, $\phi(z) = Tg$, $\Lambda(\zeta)$, $R(\zeta)$, $H(\zeta)$, a_j', $b_j'(j \in J)$ are as stated in $(1.24) - (1.26)$. In the following, we only consider the case of $K < 0$. It is sufficient to verify that the following conditions hold (see Chapter 4 in [131]1))

$$\int_L [\overline{\Lambda(\zeta)}]^{-1}\Psi_n(\zeta)[R(\zeta) + H(\zeta)]\zeta'(s)ds = 0, \; n = 1, \cdots, N - 2K - 1, \tag{1.55}$$

where $\Psi_n(\zeta)(n = 1, \cdots, N - 2K - 1)$ are as stated in (1.47). It can be proved that the coefficients determinant of $h_j(j = 1, \cdots, N - 2K - 1)$ in (1.55) is not equal to zero. Hence from (1.55), we may find a unique system $h_j(j = 1, \cdots, N - 2K - 1)$. For the system $h_j(j = 1, \cdots, N - 2K - 1)$, the boundary value problem (1.53), (1.54) possesses a unique solution $\Phi(\zeta)$. Thus the function $w(z) = \Phi[\zeta(z)]e^{\phi(z)} + \psi(z)$ is determined.

Finally, we find out the solution $[f^*(z), g^*(z), h^*(z), Q^*(z)]$ of the system of integral equations

$$f^*(z) = F(z, w, \Pi f^*) - F(z, w, 0) + A_1(z, w)Tf^* + A_2(z, w)\overline{Tf^*} + A_3(z, w), \quad (1.56)$$

$$Wg^*(z) = F(z, w, W\Pi g^* + \Pi f^*) - F(z, w, \Pi f^*) + A_1(z, w)W + A_2(z, w)\overline{W}, \quad (1.57)$$

$$S'(\chi)h^*(z)e^{\phi(z)} = F[z, w, S'(\chi)(1 + \Pi h^*)e^{\phi(z)} + W\Pi g^* + \Pi f^*]$$
$$- F(z, w, W\Pi g^* + \Pi f^*), \quad (1.58)$$

$$Q^*(z) = h^*(z)/(1 + \Pi h^*), \ S'(\chi) = [\Phi(\Psi(\chi))]_\chi, \quad (1.59)$$

and denote by $q^* = E(q)$ the mapping from $q = (Q, f, g)$ to $q^* = (Q^*, f^*, g^*)$. According to Theorem 2.1 from Chapter 4 in [140], we can prove that $q^* = E(q)$ continuously maps B_1 onto a compact subset in B_1. On the basis of the Schauder fixed–point theorem (see [117]), there exists a system $q = (Q, f, g) \in B_1$, such that $q = E(q)$. Applying the above method, from $q = (Q, f, g)$, we can construct a function $w(z) = \Phi[\zeta(z)]e^{\phi(z)} + \psi(z)$, which is just a solution of Problem B for (1.1).

In a similar method as in the proof of Theorem 4.8, Chapter 2 in [140], from Theorem 1.5 the following result can be derived.

Theorem 1.6   *Under the same conditions as in Theorem 1.5, the following statements hold.*

(1) *If the index $K \geq N$, then Problem A for (1.1) is solvable.*

(2) *If $0 \leq K < N$, then the total number of solvability conditions for Problem A does not exceed $N - K$.*

(3) *If $K < 0$, then Problem A has $N - 2K - 1$ solvability conditions.*

Besides, we can also treat the Poincaré boundary value problem for nonlinear elliptic equations of second order with non–smooth boundary (see [155],[151]2)).

2   Compound Boundary Value Problem for Some Nonlinear Elliptic Complex Equations of First Order

Here, we discuss the compound boundary value problem for more general nonlinear elliptic complex equations of first order in multiply connected domains, which includes the Riemann–Hilbert problem as its special case.

2.1 Formulation of the compound boundary value problem

Let D be an $(N+1)$-connected domain with the boundary $\Gamma = \cup_{j=0}^{N}\Gamma_j \in C_{\alpha}^1$ as stated in Section 1, $L_j\,(j = 1,\cdots,m)$ be m mutually disjoint contours, and $L = \cup_{j=1}^{m}L_j \in C_{\alpha}^1$, where the positive constant α satisfies $1/2 < \alpha < 1$. Denote

$$D^+ = D\backslash D^-, \ D^- = \cup_{j=0}^{m}D_j^-, \ D_j^- = D \cap d_j, \ j = 1,\cdots,m,$$

where d_j is the bounded domain surrounded by $L_j, j = 1,\cdots,m$. Without loss of generality, we assume that $0 \in D^+$.

We consider the nonlinear uniformly elliptic complex equation

$$w_{\bar{z}} = F(z,w,w_z) + f(z,w), \ z \in \tilde{D} = D\backslash L, \tag{2.1}$$

in which $F(z,w,U)$, $f(z,w)$ satisfy the following conditions.

Condition C' For any functions $w(z), w_j(z) \in C(\bar{D}\backslash L)$, $U(z), U_j(z) \in L_{p_0}(\bar{D})\,(j = 1,2)$, $F(z,w,U)$, $f(z,w)$ are measurable in D, and satisfy the conditions

$$\begin{cases} \mid F(z,w,U_1) - F(z,w,U_2) \mid \leq q_0 \mid U_1 - U_2 \mid, \\ \mid F(z,w_1,U) - F(z,w_2,U) \mid \leq A_0(z) \mid w_1 - w_2 \mid, \end{cases} \tag{2.2}$$

$$\mid f(z,w) \mid \leq B(z) \mid w \mid^{\sigma}, \sigma > 0, \tag{2.3}$$

where $L_p(A_0,\bar{D}) \leq k_0$, $L_p(B,\bar{D}) \leq k_0$, $L_p(A_3,\bar{D}) \leq k_1$, $A_3 = F(z,0,0)$, and $q_0, p_0\,(2 < p_0 < p), k_0, \sigma(> 0)$ are nonnegative constants. Besides, we assume that $f(z,w)$ is continuous in $w \in \mathbb{C}$ for almost every point $z \in D$.

Problem F The compound boundary value problem for (2.1) is to find a piecewise continuous solution

$$w(z) = \begin{cases} w^+(z), & z \in \overline{D^+}, \\ w^-(z), & z \in \overline{D^-}, \end{cases} \tag{2.4}$$

satisfying the boundary conditions:

$$\mathrm{Re}[\overline{\lambda(z)}w(z)] = r(z), \ z \in \Gamma, \tag{2.5}$$

$$w^+(z) = G(z)w^-(z) + g(z), \ z \in L, \tag{2.6}$$

where $\lambda(z), r(z)$ are as stated in (1.5) and (1.6), and $G(z), g(z)$ satisfy

$$C_{\alpha}[G(z), L] \leq k_0, \ G(z) \neq 0, \ z \in L, \ C_{\alpha}[g(z), L] \leq k_2, \tag{2.7}$$

herein $\alpha\,(0 < \alpha < 1), k_0, k_2$ are nonnegative constants. Problem F with $A_3(z) = 0$ in D, $r(z) = 0$ on Γ and $g(z) = 0$ on L is called Problem F_0. The index of Problem F and Problem F_0 is defined as

$$K = \frac{1}{2\pi}[\Delta_{\Gamma}\arg\lambda(z) + \Delta_L\arg G(z)]. \tag{2.8}$$

Because Problem F with $K < N$ may not be solvable, similarly to [140], we modify the boundary conditions so that the problem becomes well–posed.

Problem G   Find a piecewise continuous solution $w(z)$ in the form (2.4) satisfying the modified boundary conditions

$$\text{Re}[\overline{\lambda(z)}w(z)] = r(z) + h(z), \tag{2.9}$$

$$h(z) = \begin{cases} 0, z \in \Gamma, \ K \geq N, \\ h_j, z \in \Gamma_j, j = 1, \cdots, N - K \\ 0, z \in \Gamma_j, j = N - K + 1, \cdots, N + 1 \end{cases} \left. \begin{array}{c} \\ \\ \end{array} \right\} 0 \leq K < N, \\ \begin{cases} h_j, z \in \Gamma_j, j = 1, \cdots, N, \\ h_0 + \sum_{m=1}^{-K-1} \text{Re}(h_m^+ + ih_m^-)z^m, z \in \Gamma_0 \end{array} \left. \begin{array}{c} \\ \\ \end{array} \right\} K < 0, \tag{2.10}$$

where h_j $(j = 0, 1, \cdots, N)$ and $h_m^\pm$ $(m = 1, \cdots, -K - 1, K < 0)$ are appropriate real constants. Moreover, we can require that the solution $w(z)$ satisfies the point conditions

$$\text{Im}[\overline{\lambda(a_j)}w(a_j)] = b_j, \ j \in J = \begin{cases} 1, \cdots, 2K - N + 1, \ K \geq N, \\ N - K + 1, \cdots, N + 1, 0 \leq K < N, \\ \phi, K < 0, \end{cases} \tag{2.11}$$

in which a_j, b_j are as stated in (1.9) and (1.10). Problem G with $A_3(z) = 0$ in D, $r(z) = 0$ on Γ and $g(z) = 0$ on L is called Problem G_0.

Besides, we denote

$$X(z) = \begin{cases} X^+(z) = e^{\Gamma^+(z)/\Pi(z)}, \ z \in D^+, \\ X^-(z) = e^{\Gamma^-(z)}, \ z \in D^-, \end{cases} \tag{2.12}$$

where $\Pi(z) = \Pi_{j=1}^m (z - s_j)^{\kappa_j}$, $s_j \in d_j$, $\kappa_j = \frac{1}{2\pi}\Delta_{L_j} \arg G(z)$, $j = 1, \cdots, m$, and

$$\Gamma(z) = \frac{1}{2\pi i} \int_L \frac{\ln[G(t)\Pi(t)]}{t - z} dt.$$

In [36], the Riemann problem, the Riemann–Hilbert problem, and the compound boundary value problem in a simply connected domain D are considered for some nonlinear elliptic complex equation (2.1). In this section, we shall prove that Problem G for the complex equation (2.1) with Condition C' has a piecewise continuous solution for $0 < \sigma < 1$, and for $\sigma > 1$ also a piecewise continuous solution provided that the positive constants k_1, k_2, k_3 are small enough. Moreover, we shall also give a uniqueness result and some examples. In order to show that Problem G is uniquely solvable, we need to add some condition, namely the function $f(z, w)$ has to satisfy

$$\mid f(z, w_1) - f(z, w_2) \mid \leq B_0(z, w_1, w_2) \mid w_1 - w_2 \mid \tag{2.13}$$

with $B_0(z, w_1, w_2) \in L_p(\bar{D})$, if $\mid w_1(z) \mid + \mid w_2(z) \mid$ is bounded in $\bar{D}$.

2.2 Solvability of the compound boundary value problem

First of all, we state and prove some lemmas.

Lemma 2.1 *If $f(z,w)$ satisfies the condition stated in Condition C', then the nonlinear mapping $f: \tilde{C}(\bar{D}) \to L_p(\bar{D})$ defined by $f = f[z, w(z)]$ is continuous and bounded:*

$$L_p[f(z, w(z)), \bar{D}] \leq L_p[B, \bar{D}][\tilde{C}(w, \bar{D})]^\sigma, \tag{2.14}$$

where $\tilde{C}[w, \bar{D}] = \max[C(w^+, \overline{D^+}), C(w^-, \overline{D^-})].$

Proof Suppose that f is not continuous, then there exist a positive number ε and a sequence of functions $\{w_n(z)\}$, $w_n(z) \in \tilde{C}(\bar{D})$, such that $\tilde{C}[w_n - w_0, \bar{D}] \to 0$ as $n \to \infty$, but

$$L_p[f(z, w_n) - f(z, w_0), \bar{D}] \geq \varepsilon, \ n \geq 1. \tag{2.15}$$

Denote

$$I_n(z) = f(z, w_n), \ J_n(z) = B(z) \mid w_n(z) \mid^\sigma, \ n = 0, 1, 2, \cdots. \tag{2.16}$$

It is clear that the functions $I_n(z), J_n(z)$ converge to $I_0(z), J_0(z)$ for almost every point $z \in D$ respectively, and

$$\mid I_n(z) \mid \leq J_n(z), \ n = 0, 1, 2, \cdots. \tag{2.17}$$

Because of

$$\iint_D \mid J_n - J_0 \mid^p d\sigma_z = \iint_D \mid B(\mid w_n \mid^\sigma - \mid w_0 \mid^\sigma) \mid^p d\sigma_z$$

$$\leq \sup_{z \in D} \mid\mid w_n \mid^\sigma - \mid w_0 \mid^\sigma\mid^p \iint_D \mid B \mid^p d\sigma_z \to 0,$$

and due to $\tilde{C}[w_n - w_0, \bar{D}] \to 0$ as $n \to \infty$, we obtain

$$\mid L_p[J_n, \bar{D}] - L_p[J_0, \bar{D}] \mid \leq L_p[J_n - J_0, \bar{D}] \to 0 \ \text{ as } \ n \to \infty,$$

i.e.

$$L_p[J_n, \bar{D}] \to L_p[J_0, \bar{D}] \ \text{ as } \ n \to \infty. \tag{2.18}$$

Now from the inequality

$$\mid I_n - I_0 \mid^p \leq 2^p(\mid I_n \mid^p + \mid I_0 \mid^p) \leq 2^p(J_n^p + J_0^p)$$

and Fatou's lemma, it follows that

$$\iint_D \varliminf_{n \to \infty} [2^p(J_n^p + J_0^p) - \mid I_n - I_0 \mid^p] d\sigma_z \leq$$

$$\leq \varliminf_{n \to \infty} \iint_D [2^p(J_n^p + J_0^p) - \mid I_n - I_0 \mid^p] d\sigma_z.$$

Moreover by (2.18), we have

$$2^{p+1} \iint_D \mid J_0 \mid^p d\sigma_z \leq 2^{p+1} \iint_D \mid J_0 \mid^p d\sigma_z - \overline{\lim_{n \to \infty}} \iint_D \mid I_n - I_0 \mid^p d\sigma_z,$$

and then

$$\lim_{n \to \infty} \iint_D \mid I_n - I_0 \mid^p d\sigma_z = 0. \tag{2.19}$$

This contradicts with (2.15). Hence the operator f is continuous.

The inequality (2.14) is obvious.

Lemma 2.2 *If the functions* $w^+(z) \in W_p^1(D^+)$, $w^-(z) \in W_p^1(D^-)$, $p > 2$, *and the function* $w(z)$ *in the form (2.4) is continuous in* D, *then* $w(z) \in W_p^1(D)$.

Proof It is sufficient to verify $w_{\bar{z}}, w_z \in L_p(\bar{D})$. For $\phi(z) \in D_\infty^0(D)$, we have by Green's formula

$$\iint_D w\phi_{\bar{z}}d\sigma_z = \iint_{D^+} w^+\phi_{\bar{z}}d\sigma_z + \iint_{D^-} w^-\phi_{\bar{z}}d\sigma_z$$

$$= -\iint_{D^+} w_{\bar{z}}^+\phi d\sigma_z + \frac{1}{2i}\int_{\Gamma \cap \bar{D}^+} w^+\phi dz - \frac{1}{2i}\int_L w^+\phi dz -$$

$$- \iint_{D^-} w_{\bar{z}}^-\phi d\sigma_z + \frac{1}{2i}\int_L w^-\phi dz - \frac{1}{2i}\int_{\Gamma \cap \bar{D}^-} w^-\phi dz =$$

$$= -\iint_{D^+} w_{\bar{z}}^+\phi d\sigma_z - \iint_{D^-} w_{\bar{z}}^-\phi d\sigma_z - \frac{1}{2i}\int_L (w^+ - w^-)\phi dz$$

$$= -\iint_{D^+} w_{\bar{z}}^+\phi d\sigma_z - \iint_{D^-} w_{\bar{z}}^-\phi d\sigma_z.$$

Thus $w_{\bar{z}} \in L_p(\bar{D})$. Similarly, we can prove $w_z \in L_p(\bar{D})$. Hence $w(z) \in W_p^1(D)$.

Next, we give an a priori estimate for solutions of Problem G for the complex equation

$$w_{\bar{z}} = F(z, w, w_z) + f(z) \quad \text{in} \quad D, \tag{2.20}$$

where $f(z)$ is a measurable function in D satisfying the condition

$$L_p[f(z), \bar{D}] \leq k_4, \tag{2.21}$$

herein $p(> 2)$ and k_4 are nonnegative constants.

Theorem 2.3 *Let the complex equation (2.20) satisfy Condition* C' *and (2.21). Then Problem* G *for (2.20) has a unique solution* $w(z) \in \tilde{W}_{p_0}^1(D)$ *and* $w(z)$ *satisfies the estimate*

$$\| w \|_{\tilde{W}_{p_0}^1(D)} = \tilde{C}[w, \bar{D}] + L_{p_0}[|w_z| + |w_{\bar{z}}|, \overline{D^+}] + L_{p_0}[|w_z| + |w_{\bar{z}}|, \overline{D^-}] \leq$$

$$\leq M_1\{L_p[A_3, \bar{D}] + L_p[f, \bar{D}] + C_\alpha[r, \Gamma] + C_\alpha[g, L] + \sum_{j \in J} \mid b_j \mid\} \tag{2.22}$$

$$\leq M_1\{k_1 + 2k_2 + k_4 + \max(2K - N + 1, K + 1)k_3\},$$

where p_0 $(2 < p_0 < p)$, k_j $(j = 1, 2, 3, 4)$ are nonnegative constants as stated before, and $M_1 = M_1(q_0, p_0, k_0, \alpha, K, D^{\pm})$.

Proof By the method of elimination (see [97]1),[140]), this theorem can be proved. In fact, introducing a transformation of the function $w(z)$ by

$$w(z) = \Phi(z) + X(z)W(z), \ \Phi(z) = X(z)\Psi(z), \tag{2.23}$$

where $X(z)$ is as stated in (2.12) and

$$\Psi(z) = \frac{1}{2\pi i} \int_L \frac{g(t)}{X^+(t)} \frac{dt}{t - z},$$

it is not difficult to see that Problem G for (2.20) is reduced to the Riemann–Hilbert boundary value problem(Problem B):

$$W_{\bar{z}} = F_0(z, W, W_z) + f(z)/X(z), \ z \in D \backslash L, \tag{2.24}$$

$$\mathrm{Re}[\overline{\Lambda(z)}W(z)] = R(z) + h(z), \ z \in \Gamma, \tag{2.25}$$

$$W^+(z) = W^-(z), \ z \in L, \tag{2.26}$$

$$\mathrm{Im}[\overline{\Lambda(a_j)}W(a_j)] = B_j, \ j \in J, \tag{2.27}$$

in which

$$F_0 = F(z, \Phi + XW, \Phi' + X'W + XW_z)/X, \ z \in D,$$

$$\Lambda(z) = \lambda(z)\overline{X(z)}, \ z \in \Gamma,$$

$$R(z) = r(z) - \mathrm{Re}[\overline{\lambda(z)}\Phi(z)], \ z \in \Gamma,$$

$$B_j = b_j - \mathrm{Im}[\overline{\lambda(a_j)}\Phi(a_j)], \ j \in J.$$

Noting that (2.24)–(2.27) satisfy conditions similar to Condition C in Section 1 and (2.5)–(2.7),(2.11), according to the method of proofs as used in Section 1, it can be verified that the above Problem G has a unique solution $W(z)$. Moreover, because the functions $X(z), \Psi(z)$ and $\Phi(z)$ in (2.12) and (2.23) satisfy the estimates

$$C_\alpha[Y(z), \overline{D^{\pm}}] \leq M_2, \ L_{p_0}[Y'(z), \overline{D^{\pm}}] \leq M_3, \tag{2.28}$$

where by $Y(z)$ any of the functions $X(z), \Psi(z)$ and $\Phi(z)$ is defined, α and p_0 $(2 < p_0 \leq \min(p, 1/(1 - \alpha))$ are positive constants as stated before, and $M_j = M_j(\alpha, k, K, D^{\pm})$, $j = 2, 3$, $k = (k_0, k_1, k_2, k_3)$. Hence the function $f(z)/X(z), \Lambda(z), R(z)$ and the constants B_j $(j \in J)$ in (2.24),(2.25) and (2.27) satisfy the conditions

$$\begin{cases} L_{p_0}[f(z)/X(z), \bar{D}] \leq M_4, \ C_\alpha[\Lambda(z), \Gamma] \leq M_5, \\ C_\alpha[R(z), \Gamma] \leq M_6, \ | B_j | \leq M_7 (j \in J), \end{cases} \tag{2.29}$$

in which $M_j = M_j(\alpha, k, K, D^{\pm}), j = 4, \cdots, 7$. On the basis of the result in [140] and Lemma 2.2, we can obtain the estimate of $W(z)$

$$\| W \|_{W_{p_0}^1 (D)} = C[W, \bar{D}] + L_{p_0}[W_{\bar{z}}, \bar{D}] + L_{p_0}[W_z, \bar{D}]$$

$$\leq M_8\{L_p[A_3(z)/X(z), \bar{D}] + L_p[f(z)/X(z), \bar{D}] + \tag{2.30}$$

$$+ C_\alpha[R(z), \Gamma] + \sum_{j \in J} | B_j |\}$$

with $M_8 = M_8(q_0, p_0, \alpha, k, K, D^{\pm})$. Combining (2.28)–(2.30), the estimate (2.22) can be derived.

Now we prove the solvability of Problem G for the complex equation (2.1) satisfying Condition C'.

Theorem 2.4   *Let the complex equation (2.1) satisfy Condition C'.*

(1) When $0 < \sigma < 1$, Problem G for (2.1) has a solution $w(z)$ in $\tilde{W}_{p_0}^1(D)$, where $p_0\,(2 < p_0 < p)$ is as stated before.

(2) When $\sigma > 1$, Problem G for (2.1) has a solution $w(z) \in \tilde{W}_{p_0}^1(D)$, provided that

$$M_9 = L_p[A_3, \bar{D}] + C_\alpha[r, \Gamma] + C_\alpha[g, L] + \sum_{j \in J} | b_j | \tag{2.31}$$

is sufficiently small.

(3) If $f(z, w)$ satisfies the condition (2.13). Then the above solution is unique.

Proof   Consider the algebraic equation for t :

$$M_1\{L_p[A_3, \bar{D}] + L_p[B, \bar{D}]t^\sigma + L_\alpha[r, \Gamma] + C_\alpha[g, \Gamma] + \sum_{j \in J} | b_j |\} = t. \tag{2.32}$$

Because $0 < \sigma < 1$, the equation (2.32) has a unique solution $t = M_{10} > 0$. Now, we introduce a closed and convex subset B_0 of the Banach space $\tilde{C}(\bar{D})$, whose elements are of the form $w(z) = \Phi(z) + X(z)W(z)$, $w(z)$ satisfying the condition

$$w(z) \in \tilde{C}(\bar{D}), \quad \tilde{C}[w(z), \bar{D}] \leq M_{10}. \tag{2.33}$$

We choose an arbitrary function $\tilde{w}(z) \in B_0$ and substitute it in the position of w in $f(z, w)$. By Theorem 2.3, a solution of Problem G for the complex equation

$$w_{\bar{z}} = F(z, w, w_z) + f(z, \tilde{w}). \tag{2.34}$$

can be found. Noting that $f[z, \tilde{w}(z)] \in L_p(\bar{D})$, the above solution of Problem G for (2.34) is unique. Denote by $w(z) = T[\tilde{w}(z)]$ the mapping from $\tilde{w}(z)$ to $w(z)$. From

Lemma 2.1 and Theorem 2.3, we know that

$$\| w \|_{\tilde{W}^1_{p_0}(D)} \leq M_1\{L_{p_0}[A_3, \bar{D}] + C_\alpha[r, \Gamma] + C_\alpha[g, L] +$$

$$+ \sum_{j \in J} | b_j | + L_{p_0}[f, \bar{D}]\} \leq M_1\{M_9 + L_{p_0}[B, \bar{D}]\tilde{C}[|w|^\sigma, \bar{D}]\} \tag{2.35}$$

$$\leq M_1\{M_9 + L_{p_0}[B, \bar{D}]M_{10}^\sigma\} = M_{10}.$$

This shows that T maps B_0 onto a compact subset in B_0. Next, we verify that T in B_0 is a continuous operator. In fact, arbitrarily select a sequence $\{\tilde{w}_n(z)\}$ in B_0, such that $\tilde{C}(w_n - w_0, \bar{D}) \to 0$ as $n \to \infty$. By Lemma 2.1, it can be seen that $L_p[f(z, \tilde{w}_n) - f(z, \tilde{w}_0), \bar{D}] \to 0$ as $n \to \infty$. Moreover, from $w_n = T[\tilde{w}_n]$, $w_0 = T[\tilde{w}_0]$, it is clear that $w_n - w_0$ is a solution of Problem G_0 for the following complex equation

$$(w_n - w_0)_{\bar{z}} = F(z, w_n, w_{nz}) - F(z, w_0, w_{0z}) + f(z, \tilde{w}_n) - f(z, \tilde{w}_0) \quad \text{in} \quad D. \tag{2.36}$$

By means of Theorem 2.3 and the method in [140], we can obtain the estimate

$$\| w_n - w_0 \|_{\tilde{W}^1_{p_0}(D)} \leq M_1 L_{p_0}[f(z, w_n(z)) - f(z, w_0(z)), \bar{D}]. \tag{2.37}$$

Hence $\tilde{C}[w_n - w_0, \bar{D}] \to 0$ as $n \to \infty$. On the basis of the Schauder fixed–point theorem, there exists a function $w(z) \in \hat{C}(D)$, such that $w(z) = T[w(z)]$, and from Theorem 2.3, it is seen that $w(z) \in \tilde{W}^1_{p_0}(D)$, and $w(z)$ is a solution of Problem G for the complex equation (2.1) with $0 < \sigma < 1$.

Secondly, we discuss the case: $\sigma > 1$. In this case, (2.32) has the solution $t = M_{10}$ provided that M_9 in (2.31) is small enough. Now we consider a closed and convex subset B_1 in the Banach space $\tilde{C}(\bar{D})$, i.e.

$$B_1 = \{w(z) | \tilde{C}[w(z), \bar{D}] \leq M_{10}\}, \tag{2.38}$$

and apply a method similar as before. It can be proved that there exists a solution $w(z) \in \tilde{W}^1_{p_0}(D)$ of Problem G for (2.1) for the constant $\sigma > 1$.

When $f(z, w)$ satisfies condition (2.13), we can verify the uniqueness of solutions in Theorem 2.4 by using a similar method as in the proof of Theorem 2.3.

From the above theorem, the next result can be derived.

Theorem 2.5 *Under the same conditions as in Theorem 2.4, the following statements hold.*

(1) *When the index $K > N$, Problem F for (2.1) is solvable.*

(2) *When $0 \leq K < N$, Problem F for (2.1) is solvable, if $N - K$ solvability conditions are satisfied.*

(3) *When $K < 0$, Problem F for (2.1) is solvable under $N - 2K - 1$ conditions.*

2.3   Some examples

Finally, we give some examples to illustrate the above results.

Example 1   The function $f(z,w) = |w| \sin(|w|^2)$ satisfies condition (2.3), but does not satisfy condition (2.2). Moreover, the function

$$f(z,w) = B(z) |w|^{\rho_1} \sin(|w|^{\rho_2}), \quad \rho_1, \rho_2 \geq 0 \tag{2.39}$$

satisfies (2.3). Problem G for (2.1) and (2.39) has a solution $w(z)$ in $\tilde{W}^1_{p_0}(D)$ for

   (1) $\rho_1 \leq 1$ by virtue of Theorem 2.4 (1).

   (2) $\rho_1 > 1$ and M_9 being small enough by virtue of Theorem 2.4 (2).

   (3) If $\rho_1 \leq 1, \rho_1 + \rho_2 - 1 \geq 0$, the solution of Problem G is unique (see [11]4)).

Example 2   Let $D = \{|z| < 1\}$. The Dirichlet boundary value problem

$$\begin{cases} w_{\bar{z}} = -w |w|^{-3/4} \ \text{ in } \ D \ \text{ with } \ \sigma = \tfrac{1}{4} < 1, \\[2mm] \operatorname{Re} w(z) = -(\dfrac{3}{2})^{4/3} \cdot \dfrac{1}{2}(1-x)^{4/3} \ \text{ on } \ \Gamma = \{|z| = 1\}, \\[2mm] \operatorname{Im} w(1) = 0 \end{cases}$$

has two smooth solutions

$$\begin{cases} w_1(z) = -(\dfrac{3}{2})^{4/3} e^{i\frac{\pi}{3}} (1-x)^{4/3}, \\[3mm] w_2(z) = -(\dfrac{3}{2})^{4/3} e^{-i\frac{\pi}{3}} (1-x)^{4/3}. \end{cases}$$

This shows that to ensure uniqueness of solutions, we need a condition like the assumption (2.13).

Example 3   Let D be a simply connected domain with $0 \in D$ and n be a positive integer. The boundary value problem

$$\begin{cases} w_{\bar{z}} = -\dfrac{z}{|z|} w |w|^{1/n} \ \text{ in } \ D, \\[3mm] \operatorname{Re} w(z) = \operatorname{Re}(\dfrac{n}{\bar{z}})^n \ \text{ on } \ \Gamma, \\[3mm] \operatorname{Im} w(t_0) = \operatorname{Im}(\dfrac{n}{t_0})^n, \ t_0 \in \Gamma \end{cases}$$

has a non–smooth solution

$$w(z) = (\dfrac{n}{\bar{z}})^n,$$

where $\sigma = 1 + 1/n > 1$(see [36]).

3 Discontinuous Riemann–Hilbert Problem for Nonlinear Elliptic Complex Equations of First Order

In [139]10), the discontinuous Riemann–Hilbert boundary value problem for nonlinear uniformly elliptic complex equations of first order in a multiply connected domain was discussed, but there the solution for the modified boundary value problem with index $K < -1$ has a pole of order $[-K + 1/2] - 1$ in the domain. In this section, we give a new form of the well–posed discontinuous Riemann–Hilbert problem for elliptic complex equations of first order, so that its solution will be continuous in the domain.

3.1 Formulation of the discontinuous Riemann–Hilbert problem and its well–posedness

Let D be an $(N + 1)$–connected domain in the complex plane $\mathbb{C}$ with the boundary Γ as stated in Section 1, but here $\Gamma \in C^1_\alpha(0 < \alpha < 1)$. We may assume that D is a circular domain in the unit disk $\mid z \mid < 1$, which is bounded by $N + 1$ circles $\Gamma_j = \{\mid z - z_j \mid\} = \sigma_j, j = 1, \cdots, N$ and $\Gamma_0 = \Gamma_{N+1} = \{|z| = 1\}$, and $0 \in D$.

We consider the nonlinear elliptic complex equation (1.1) in D satisfying Condition C, see Section 1.

Problem A^* The discontinuous Riemann–Hilbert boundary value problem for the complex equation (1.1) is to find a continuous solution $w(z)$ of (1.1) in $\bar{D}$, except possibly in $m(0 < m < \infty)$ points $t_1, \cdots, t_m \in \Gamma$, satisfying the boundary condition

$$\mathrm{Re}[\overline{\lambda(z)}w(z)] = r(z), \quad z \in \Gamma^* = \Gamma \backslash T, \tag{3.1}$$

where $T = \{t_1, \cdots, t_m\}$, $\lambda(z)(\mid \lambda(z) \mid = 1)$, $r(z) = \Pi_{j=1}^m \mid \zeta(z) - \zeta(t_j) \mid^{-\beta_j} r_0(z)$ are continuous on Γ^* and satisfy the condition

$$C_\alpha[\lambda(z(\zeta)), \zeta(\Gamma^j)] \le k_0, \ C_\alpha[r_0(z(\zeta)), \zeta(\Gamma)] \le k_2, \tag{3.2}$$

in which $\Gamma^j(j = 1, \cdots, m)$ are all arcs on Γ^* with end points in T. Moreover, if $\lambda(z)$ is continuous on $\Gamma_j(0 \le j \le N)$, we shall add arbitrarily chosen points $t_j \in \Gamma_j$ to T relabelling this set again as T. $\zeta(z)$ is assumed to be any homeomorphism of the Beltrami equation

$$\zeta_{\bar{z}} = Q\zeta_z, \ \mid Q \mid \le q_0, \ z \in D, \tag{3.3}$$

which quasiconformally maps D onto a circular domain G similar to D, such that $0 = \zeta(0)$ and $L_0 = \zeta(\Gamma_0) = \{\mid \zeta \mid = 1\}$. $z(\zeta)$ is the inverse function of $\zeta(z)$, $\alpha \, (0 < \alpha < 1)$, $\beta_j \, (0 \le \beta_j < 1, j = 1, \cdots, m)$, k_0, k_2 are nonnegative constants. The index of Problem A^* is defined by

$$K = \frac{1}{2}(K_1 + \cdots + K_m) \tag{3.4}$$

where

$$K_j = \left[\frac{\phi_j}{\pi}\right] + J_j, \ J_j = 0 \text{ or } 1, \ e^{i\phi_j} = \frac{\lambda(t_j - 0)}{\lambda(t_j + 0)}, \ \gamma_j = \frac{\phi_j}{\pi} - K_j, \ j = 1, \cdots, m. \quad (3.5)$$

It is easy to see that the index is not unique and may not be an integer.

In order to discuss the solvability of Problem A^*, we modify the boundary condition so that the problem becomes well posed.

Problem B^*   Find a continuous solution $w(z)$ of (1.1) in $\bar{D}\backslash T$ satisfying the modified boundary condition

$$\text{Re}[\overline{\lambda(z)}w(z)] = r(z) + \mid Y(z) \mid h(z), \ z \in \Gamma^*, \quad (3.6)$$

where

$$h(z) = \begin{cases} 0, \ z \in \Gamma, \ \text{if } K > N - 1, \\ h_j, \ z \in \Gamma_j, \ j = 1, \cdots, N - K', \\ 0, \ z \in \Gamma_j, \ j = N - K' + 1, \cdots, N + 1 \\ h_j, \ z \in \Gamma_j, \ j = 1, \cdots, N, \\ h_j\eta_j(z), \ z \in \Gamma_0^j, \ j = N + 1, \cdots, N - 2K' - 1 \end{cases} \left.\begin{array}{c} \\ \\ \end{array}\right\} \text{if } 0 \le K \le N - 1, \quad (3.7)$$

and

$$Y(z) = \Pi_{j=1}^{m_1}[\zeta(z) - \zeta(t_j)]^{\gamma_j} \Pi_{j=m_1+1}^{m_2}\left[\frac{\zeta(z) - \zeta(t_j)}{\zeta - \zeta_j}\right]^{\gamma_j} \cdots \Pi_{j=m_N+1}^{m}\left[\frac{\zeta(z) - \zeta(t_j)}{\zeta - \zeta_N}\right]^{\gamma_j}, \quad (3.8)$$

in which $\gamma_j(j = 1, \cdots, m)$ and $\zeta(z)$ are stated as before, ζ_j is the centre of the circle $L_j = \zeta(\Gamma_j), \ j = 0, 1, \cdots, N, \ h_j \in J'(J' = \phi \text{ if } K > N - 1; \ J' = 1, \cdots, N - K', \text{ if } 0 \le K \le N - 1; \ J' = 1, \cdots, N - 2K' - 1, \text{ if } K < 0; \ K' = [\mid K \mid +1/2])$ are unknown real constants to be determined appropriately, herein $h_{N+1} = 0$, if $2K$ is odd, and $\Gamma_0^j(j = N + 1, \cdots, N - 2K' - 1)$ are nondegenerate, mutually disjoint arcs on Γ_0, $\eta_j(z)$ is a positive continuous function on the interior point set of Γ_0^j, such that $\eta_j(z) = 0$ on $\overline{\Gamma\backslash\Gamma_0^j}$ and $C_\alpha[\eta_j(z(\zeta)), L_j] \le k_0, \ j = N + 1, \cdots, N + 1, N - 2K' - 1$, where we assume that the signs of $\lambda(z)\overline{Y(z)}$ has been adjusted similarly to those in the proof of Theorem 4.3, Chapter 4 in [140]. If $-1 \le K < 0$, we choose $\eta_j(z) = 1, z \in \Gamma_0$. Moreover, we may require that the solution $w(z)$ satisfies the point conditions

$$\text{Im}[\overline{\lambda(a_j)}w(a_j)] = \mid Y(a_j) \mid b_j, \ j \in J^*, \quad (3.9)$$

where

$$J^* = \begin{cases} 1, \cdots, 2K - N + 1, \text{ if } K > N - 1, \\ N_0 + 1, \cdots, N_0 + [K] + 1, \ 0 \le K' \le N - N_0, \\ N - K' + 1, \cdots, N - K' + [K] + 1, \\ N - N_0 < K' \le N - 1, \text{ if } 0 \le K \le N - 1, \end{cases}$$

and $a_j(\notin T) \in \Gamma_j$, $j = 1, \cdots, N_0$, $a_j(\notin T \cup \overline{\Gamma_0^j}) \in \Gamma_0$, $j = N+1, \cdots, \max(2K - N + 1$, $N - K' + [K] + 1)$, $b_j(j \in J^*)$ are all real constants satisfying the condition $\mid b_j \mid \leq k_3 < \infty$ $(j \in J^*)$. There is no harm in assuming that $\kappa_j = \mathrm{Ind}\,\lambda(z) \mid_{\Gamma_j}(j = 1, \cdots, N_0)$ are integers and $\kappa_j = \mathrm{Ind}\,\lambda(z) \mid_{\Gamma_j}(j = N_0 + 1, \cdots, N)$ are not integers. We mention that $\zeta(z) = z$ in (3.8), if (1.1) is $w_{\bar{z}} = 0$ in D.

In the following, we first discuss the existence and uniqueness of solutions of Problem B^* for analytic functions, and then discuss a priori estimates of solutions for (1.1), moreover by using the above estimates of solutions and the Schauder fixed–point theorem, the solvability of Problem B^* for (1.1) can be proved.

3.2 Existence and uniqueness of solutions to Problem B^* for analytic functions

Theorem 3.1 *If the conditions γ_j $(j = 1, \cdots, m)$ in (3.5) are fixed, then the solution of Problem B^* for analytic functions is unique.*

Proof Let $\Phi_1(z)$ and $\Phi_2(z)$ be any two arbitrary solutions of Problem B^* for analytic functions, and denote $\Phi(z) = \Phi_1(z) - \Phi_2(z)$. Then the function $\Psi(z) = \Phi(z)/Y(z)$ satisfies the boundary and side conditions:

$$\mathrm{Re}[\overline{\Lambda(z)}\Psi(z)] = h(z), \Lambda(z) = \lambda(z)\overline{Y(z)}/\mid Y(z) \mid, \; z \in \Gamma^*, \tag{3.10}$$

$$\mathrm{Im}[\overline{\Lambda(a_j)}\Psi(a_j)] = 0, \; j \in J^*. \tag{3.11}$$

Suppose that $\Psi(z) \not\equiv 0, z \in D$. By an argument similar to those in Chapter 5, [139]17), or Chapter 1, [140], we can obtain the inequality

$$2N_G + N_{\Gamma^*} \leq 2K + I, \; I = \begin{cases} 0, \text{ if } K \geq 0, \\ 2K' - 1, \text{ if } K < 0 \text{ and } 2K \text{ is odd}, \\ 2K' - 2, \text{ if } K < 0 \text{ and } 2K \text{ is even}, \end{cases} \tag{3.12}$$

where N_D and N_{Γ^*} are the numbers of zero points of $\Psi(z)$ in D and on Γ^* respectively, and the number of zero points on $\Gamma_j(j = 1 \leq j \leq N)$ is odd or even depending whether $2\kappa_j$ is odd or even. Thus, the following absurd inequalities would be derived:

$$\begin{cases} 2K + 1 = 2N_0 + N - 2N_0 + 2K - N + 1 \leq 2N_D + N_{\Gamma^*} \leq 2K, \text{ if } K > N - 1, \\ 2K + 1 \leq K' + [K] + 1 \leq 2N_D + N_{\Gamma^*} \leq 2K \text{ for } 0 \leq K' \leq N - N_0, \\ 2K + 1 \leq 2N_D + N_{\Gamma^*} \leq 2K \text{ for } N - N_0 \leq K \leq N - 1, \text{ if } 0 \leq K \leq N - 1, \\ 0 \leq 2N_D + N_{\Gamma^*} \leq 2K + 2K' - 2 < 0, \text{ if } K < 0 \text{ and } K \text{ is even}, \\ 1 \leq 2N_D + N_\Gamma \leq 2K + 2K' - 1, \text{ if } K < 0 \text{ and } K \text{ is odd}. \end{cases}$$

$$\tag{3.13}$$

This contradiction proves $\Psi(z) = 0$, hence $\Phi(z) = 0$, i.e. $\Phi_1(z) = \Phi_2(z)$ in D.

Theorem 3.2 *Problem B^* for analytic functions has a solution.*

Proof　In the case of $K \geq -1$, the solvability of Problem B^* for analytic functions has been proved in Chapter 5 of [139]17). Now, we consider the case of $K < -1$, and let $\Phi_0(z)$ be a solution of the following boundary value problem

$$\mathrm{Re}[\overline{\Lambda(z)}\Phi_0(z)] = r(z) + h(z), \ \Lambda(z) = \lambda(z)\bar{z}^{1-K'}, \ z \in \Gamma, \tag{3.14}$$

where

$$K' = \left[\,|\,K\,| + \frac{1}{2}\right], \ \frac{1}{2\pi}\Delta_\Gamma \arg \Lambda(z) = \begin{cases} -1/2 & \text{if } 2K \text{ is odd,} \\ -1 & \text{if } 2K \text{ is even.} \end{cases}$$

If $\Phi_0(z)$ has a zero of order $K' - 1$ at $z = 0$, then $\Phi(z) = \Phi_0(z)z^{1-K'}$ is the solution of Problem B^* with index $K < -1$. Otherwise, we can find $\kappa = 2(K' - 1)$ analytic functions $\Phi_1(z), \cdots, \Phi_\kappa(z)$, which satisfy the boundary conditions

$$\mathrm{Re}[\overline{\Lambda(z)}\Phi_k(z)] = h(z) + \eta_{1+N+k}(z), \ z \in \Gamma, \ k = 1, \cdots, \kappa, \tag{3.15}$$

respectively. It is clear that $\Phi_1(z), \cdots, \Phi_\kappa(z)$ are linearly independent, and we proceed to verify that

$$\Delta = \begin{vmatrix} \mathrm{Re}\,\Phi_1(0) & \cdots & \mathrm{Re}\,\Phi_\kappa(0) \\ \mathrm{Im}\,\Phi_1(0) & \cdots & \mathrm{Im}\,\Phi_\kappa(0) \\ \vdots & \ddots & \vdots \\ \mathrm{Re}\,\Phi_1^{(l)}(0) & \cdots & \mathrm{Re}\,\Phi_\kappa^{(l)}(0) \\ \mathrm{Im}\,\Phi_1^{(l)}(0) & \cdots & \mathrm{Im}\,\Phi_\kappa^{(l)}(0) \end{vmatrix} \neq 0, \tag{3.16}$$

herein $l = K' - 2$. In fact, suppose $\Delta = 0$, then there exist κ real constants $c_1, \cdots, c_\kappa$, which are not all equal to zero, such that

$$c_1\Phi_1^{(j)}(0) + \cdots + c_\kappa\Phi_\kappa^{(j)}(0) = 0, \ j = 0, 1, \cdots, l,$$

and according to (3.12), again the following absurd inequality would be derived:

$$\kappa = 2K' - 2 \leq 2N_D + N_{\Gamma^*} \leq 2K' - 3,$$

where N_D, N_{Γ^*} denote the number of zeros for the analytic function $\sum_{k=1}^{\kappa} c_k\Phi_k(z)(\not\equiv 0)$ in D and on Γ^* respectively. Hence $\Delta \neq 0$, and then there exist κ real constants $c_1, \cdots, c_\kappa$, which satisfy the following algebraic system

$$c_1\Phi_1^{(j)}(0) + \cdots + c_\kappa\Phi_\kappa^{(j)}(0) = \Phi_0^{(j)}(0), \ j = 0, 1, \cdots, l. \tag{3.17}$$

Thus

$$\tilde{\Phi}(z) = \Phi_0(z) - \sum_{k=1}^{\kappa} c_k\Phi_k(z)$$

has a zero of order $\kappa = K' - 1$ at $z = 0$, and $\Phi(z) = \tilde{\Phi}(z)z^{1-K'}$ is just the solution of Problem B^* for analytic functions with $K < -1$.

Finally, we examine the behaviour of the solution of Problem B^* for analytic functions in the neighborhoods $D_j^* \subset \bar{D}$ of the discontinuous points $t_j\,(j = 1, \cdots, m)$ of $\lambda(z)$ and $r(z)$. We may require that the solution $\Phi(z)$ of Problem B^* satisfies

$$\Phi(z) = O(|z - t_j|)^{-\tau_j}) \text{ in } D_j^*, j = 1, \cdots, m, \tag{3.18}$$

where

$$\tau_j = \begin{cases} \beta_j + \delta & \text{for } \gamma_j \geq 0 \text{ or } \gamma_j < 0, \beta_j > |\gamma_j|, \\ |\gamma_j| + \delta & \text{for } \gamma_j < 0 \text{ and } \beta_j \leq |\gamma_j|, \end{cases} \quad j = 1, \cdots, m,$$

in which δ is an arbitrarily given small positive number. Moreover, we shall give a priori estimates of any solution $\Phi(z)$ of Problem B^* for analytic functions as stated below.

3.3   A priori estimates of Problem B^* for the complex equation (1.1)

Theorem 3.3   *Let $\Phi(z)$ be any solution of Problem B^* for analytic functions. Then $\Phi(z)$ satisfies the estimates*

$$C_{\beta_0}[\Phi Z, \bar{D}] \leq M_1, \ C[\Phi', D_n] \leq M_2, \tag{3.19}$$

where $Z(z) = \Pi_{j=1}^m (z - t_j)^{\tau_j}$, $\tau_j (j = 1, \cdots, m)$ are as stated in (3.18), $D_n = \{z \mid |z - t_j| \geq 1/n, n > 0, j = 1, \cdots, m, z \in \bar{G}\}$, $M_1 = M_1(\lambda, r, D, \delta, \beta_0)$, $0 < \beta_0 < \delta$, $M_2 = M_2(\lambda, r, D, \delta, \beta_0, D_n)$.

Proof   Similarly as for Lemma 1.2 and Theorem 1.7, Chapter 4 in [140], it is not difficult to see that $\Phi(z)Z(z) \in C_{\beta_0}(\bar{D})$. Suppose that the estimate

$$C[\Phi Z, \bar{D}] \leq M_3 = M_3(\lambda, r, D, \delta, \beta_0) \tag{3.20}$$

is not true. Then there exist sequences of functions $\{\lambda^{(n)}(z)\}$, $\{r_0^{(n)}(z)\}$ and $\{b_j^{(n)}\}$, which satisfy the same conditions as $\lambda(z)$, $r_0(z)$ and b_j in (3.2) and (3.9), so that when $n \to \infty$, $\{\lambda^{(n)}(z)\}$, $\{r_0^{(n)}(z)\}$, $\{b_j^{(n)}\}$ uniformly converge respectively to $\lambda^{(0)}(z)$, $r_0^{(0)}(z)$, $b_j^{(0)}$ on Γ^j, $j = 1, \cdots, m$. Furthermore there exists a sequence of solutions: $\{\Phi_n(z)\}$ of Problem B^* for analytic functions satisfying the boundary conditions

$$\text{Re}[\overline{\lambda^{(n)}(z)}\Phi_n(z)] = r^{(n)}(z) + |Y(z)| h(z), \tag{3.21}$$

$$\text{Im}[\overline{\lambda^{(n)}(a_j)}\Phi_n(a_j)] = |Y(a_j)| b_j^{(n)}, \ j \in J^*, \tag{3.22}$$

so that $C[\Phi_n Z, \bar{D}] = H_n \to \infty$. There is no harm assuming that $H_n \geq 1$, $n = 1, 2, \cdots$. By setting $\Psi_n(z) = \Phi_n(z)/H_n$, it is clear that $\Psi_n(z)$ satisfies the boundary conditions

$$\text{Re}[\overline{\lambda^{(n)}(z)}\Psi_n(z)] = [r^{(n)}(z) + |Y(z)| h(z)]/H_n, \tag{3.23}$$

$$\text{Im}[\overline{\lambda^{(n)}(a_j)}\Psi_n(a_j)] = \mid Y(z) \mid b_j^{(n)}/H_n, \; j \in J^*. \tag{3.24}$$

By using the method in the proofs of Theorem 1.4, Chapter 1, [140] and Theorem 6.6, Chapter 5, [139]17), we can obtain the estimate

$$C_{\beta_0}[\Psi_n Z, \bar{D}] \leq M_4 = M_4(\lambda, r, D, \delta, \beta_0). \tag{3.25}$$

Hence, we may choose a subsequence $\{\Psi_{n_k}(z)Z(z)\}$ from $\{\Psi_n(z)Z(z)\}$ on any closed set D^* in $D\backslash T$, and the limited function $\Psi_0(z)$ of $\{\Psi_{n_k}(z)\}$ satisfies the boundary conditions

$$\text{Re}[\overline{\lambda^{(0)}(z)}\Psi_0(z)] = \mid Y(z) \mid h(z), \; z \in \Gamma^*, \tag{3.26}$$

$$\text{Im}[\overline{\lambda^0(a_j)}\Psi_0(a_j)] = 0, \; j \in J^*. \tag{3.27}$$

By Theorem 3.1, $\Psi_0(z) = 0, z \in D$. However, from $C[\Psi_n Z, \bar{D}] = 1$, it follows that $C[\Psi_0(z)Z(z), \bar{D}] = 1$. This contradiction shows that (3.20) is true. Hence the first estimate in (3.19) can be derived, and then the second estimate in (3.19) is easily obtained.

Theorem 3.4   *Let the nonlinear elliptic complex equation* (1.1) *satisfy Condition C. Then any solution w(z) of Problem B^* for* (1.1) *satisfies the estimates*

$$L(w) = C_{\beta\beta_0}[wX, \bar{D}] \leq M_3, \tag{3.28}$$

$$S(w) = L_{p_0}[\mid w_{\bar{z}} \mid + \mid w_z \mid, D_n] \leq M_4, \tag{3.29}$$

where

$$X(z) = \Pi_{j=1}^m (z - t_j)^{\eta_j} \text{ in } D_j^*, \; j = 1, \cdots, m, \tag{3.30}$$

herein

$$\eta_j = \begin{cases} (\beta_j + \delta)/\beta \text{ for } \gamma_j \geq 0 \text{ and } \gamma_j < 0, \beta \mid \gamma_j \mid \leq \beta_j, \\ (\mid \gamma_j \mid + \delta)/\beta \text{ for } \gamma_j < 0, \beta \mid \gamma_j \mid > \beta_j, \end{cases} \quad j = 1, \cdots, m,$$

$\beta = 1 - 2/p_0 \leq \alpha, 2 < p_0 \leq p, \; \beta_0 \text{ and } \delta(0 < \beta_0 < \delta)$ *are small positive constants,* D_n *is as stated in* (3.19), $M_3 = M_3(q_0, p_0, k, \lambda, r, D, \delta, \beta_0), M_4 = M_4(q_0, p_0, k, \lambda, r, D, \delta, \beta_0, D_n), \; k = (k_0, k_1, k_2, k_3).$

Proof   According to Theorem 1.1, the solution $w(z)$ of Problem B^* for the complex equation (1.1) can be expressed as

$$w(z) = \Phi[\zeta(z)]e^{\phi(z)} + \psi(z), \tag{3.31}$$

where $\phi(z), \psi(z), \zeta(z)$ and its inverse function $z(\zeta)$ satisfy the estimates

$$C_\beta[\sigma(z), \bar{D}] + L_{p_0}[\mid \sigma_{\bar{z}} \mid + \mid \sigma_z \mid, \bar{D}] \leq M_5, \tag{3.32}$$

$$C_\beta[z(\zeta), \bar{G}] + L_{p_0}[\mid z_{\bar{\zeta}} \mid + \mid z_\zeta \mid, \bar{G}] \leq M_6, \tag{3.33}$$

in which $\sigma(z) = \{\phi(z), \psi(z), \zeta(z)\}, M_j = M_j(q_0, p_0, k, D), j = 5, 6$. By (3.31), the solution $w(z)$ of Problem B^* for (1.1) can be transformed into the solution $\Phi(z)$ of Problem B' for analytic functions satisfying the boundary conditions

$$\mathrm{Re}\,[\overline{\Lambda(\zeta)}\Phi(\zeta)] = R(\zeta) + |\,Y(z(\zeta))\,|\,H(\zeta),\ \zeta \in L = \zeta(\Gamma), \tag{3.34}$$

$$\mathrm{Im}\,[\overline{\Lambda(a_j')}\Phi(a_j')] = b_j' = |\,Y(a_j)\,|\,b_j - \mathrm{Im}\,[\overline{\lambda(a_j)}\psi(a_j)], j \in J^*, \tag{3.35}$$

where $R(\zeta) = r[z(\zeta)] - \mathrm{Re}[\overline{\lambda(z(\zeta))}\psi(z(\zeta))]$, $\overline{\Lambda(\zeta)} = \overline{\lambda[z(\zeta)]}e^{\phi[z(\zeta)]}$, $H(\zeta) = h[z(\zeta)]$, and $\Lambda(\zeta), R(\zeta)$ are discontinuous at the points $t_j' = \zeta(t_j)\,(j = 1, \cdots, m)$ on $L = \zeta(\Gamma)$, $R(\zeta) = \Pi_{j=1}^m (\zeta - t_j')^{-\beta_j} R_0(\zeta)$, $\Lambda(\zeta)$ and $R_0(\zeta) = r_0[z(\zeta)]$ satisfy

$$C_\alpha[\lambda, L^j] \le k_4,\ C_\alpha[R_0, L^j] \le k_4, \tag{3.36}$$

in which $L^j = \zeta(\Gamma^j)(j = 1, \cdots, m)$, $k_4 = k_4(q_0, p_0, k, \lambda, r, D)$. According to Theorem 3.3, the solution $\Phi(\zeta)$ of Problem B' for analytic functions satisfies the estimates

$$C_{\beta_0}[\Phi(\zeta)X(\zeta), \bar{G}] \le M_7,\ C[\Phi'(\zeta), G_n] \le M_8, \tag{3.37}$$

where $G_n = \zeta(D_n), 0 < \beta_0 < \delta, M_7 = M_7(q_0, p_0, k, \lambda, r, D, \delta, \beta_0), M_8 = M_8(q_0, p_0, k, \lambda, r, D, D_n)$. Combining (3.31)-(3.33),(3.37), the estimate (3.28) and (3.29) can be obtained.

3.4 Solvability of Problem B^* for the nonlinear complex equations (1.1)

We first prove that Problem B^* for (1.1) is solvable, and then derive the solvability conditions for Problem A^*.

Theorem 3.5 *Suppose that the nonlinear complex equation* (1.1) *satisfies Condition C. Then Problem B^* for* (1.1) *has a solution $w(z)$.*

Proof Similarly to the proof of Theorem 1.5, we introduce two bounded and closed subsets B_1 and B_2 in the Banach space $B = L_{p_0}(\bar{D}) \times L_{p_0}(\bar{D}) \times L_{p_0}(\bar{D}), 2 < p_0 \le p$. B_1 consists of all triplets of measurable functions: $q = [Q(z), f(z), g(z)]$ satisfying the condition:

$$|Q(z)| \le q_0 < 1, z \in D, L_{p_0}[f(z), \bar{D}] \le M_5, L_{p_0}[g(z), \bar{D}] \le M_5, \tag{3.38}$$

and B_2 consists of all triplets of measurable functions: $\omega = [f(z), g(z), h(z)]$ satisfying the condition:

$$L_{p_0}[f(z), \bar{D}] \le M_5, L_{p_0}[g(z), \bar{D}] \le M_5, h(z) \le q_0|1 + \Pi h|, \tag{3.39}$$

where q_0, M_5 are constants as stated in (1.4) and (3.32). We introduce a system of integral equations as follows

$$h(z) = Q(z)(1 + \Pi h), \tag{3.40}$$

$$f^* = F(z, w, \Pi f^*) - F(z, w, 0)$$
$$+ A_1(z, w)Tf^* + A_2(z, w)\overline{Tf^*} + A_3(z, w), \tag{3.41}$$

$$Wg^* = F(z, w, W\Pi g^* + \Pi f^*) - F(z, w, \Pi f^*)$$
$$+ A_1(z, w)W + A_2(z, w)\overline{W}, \tag{3.42}$$

$$S'(\chi)h^* e^\phi = F[z, w, S'(\chi)(1 + \Pi h^*)e^\phi + W\Pi g^* + \Pi f^*]$$
$$- F(z, w, W\Pi g^* + \Pi f^*), \tag{3.43}$$

$$Q^* = h^*(z)/(1 + \Pi h^*), \tag{3.44}$$

in which $q = [Q(z), f(z), g(z)] \in B_1$. By the principle of contracting mappings, equation (3.40) has a unique solution $h(z) \in L_{p_0}(\bar{D})$. Set $\chi(z) = z + Th$, $\psi(z) = Tf$, $\phi(z) = Tg$, and introduce $\zeta(z) = \Psi[\chi(z)]$, $S(\chi) = \Phi[\Psi(\chi)]$, $w(z) = \Phi[\zeta(z)]e^{\phi(z)} + \psi(z) = W(z) + \psi(z)$, where $\zeta = \Psi(\chi)$ is a conformal mapping which maps the domain $\chi(D)$ onto an $(N + 1)$–connected circular domain G similar to D, such that $\zeta(0) = 0$, and $f(z), g(z), h(z), \psi(z), \phi(z), \zeta(z)$ and its inverse function $z(\zeta)$ satisfy the inequalities (3.38),(3.39),(3.32) and (3.33), and the analytic function $\Phi(\zeta)$ satisfies the boundary condition as in (3.34) and (3.35). Substitute $w(z), W(z), S'(\chi), \phi(z)$ into the corresponding positions in (3.41)-(3.43), by using the continuity method and the principle of contracting mappings, we can find the solution $f^*(z), g^*(z), h^*(z), Q^*(z) \in L_{p_0}(\bar{D})$ of the system of integral equations (3.41)-(3.44) successively. Let $q^* = [Q^*(z), f^*(z), g^*(z)]$ and denote by $q^* = S_1(q)$ and $w(z) = S_2(q)$ the mappings from $q = [Q(z), f(z), g(z)]$ onto q^* and $w(z)$ respectively. We can prove $q^* = S_1(q)$ continuously maps B_1 onto a compact set in B_1. By the Schauder fixed–point theorem, there exists a system of functions $q = [Q(z), f(z), g(z)] \in B_1$ satisfying $q = S_1(q)$ and $w(z) = S_2(q) = \Phi[\zeta(z)]e^{\phi(z)} + \psi(z)$ is just a solution of Problem B^* for (1.1).

Now we can state and prove a solvability result of Problem A^* as follows.

Theorem 3.6 *Suppose that the same conditions as in Theorem 3.5 hold. Then*

(1) *If $K \geq N - 1/2$, Problem A^* for (1.1) is solvable and its general solution $w(z)$ is dependent on $2K - N + 1$ arbitrary real constants.*

(2) *If $0 \leq K \leq N - 1$, the total number of solvability conditions of Problem A^* does not exceed $N - K' = N - [K + 1/2]$. When those conditions are satisfied, then the general solution of Problem A^* includes $[K] + 1$ arbitrary real constants.*

(3) *If $K < 0$, Problem A^* has $N - 2K - 1$ solvability conditions.*

Proof Let $w(z)$ be any solution of Problem B^* for (1.1) and substitute it into the boundary condition (3.6). If $h(z) = 0$, i.e. $h_j = 0, j \in J'$, then $w(z)$ is also a solution of Problem A^*. Hence Problem A^* has as many solvability conditions as stated before. Noting that $b_j \, (j \in J^*)$ are arbitrary, this shows that the general solution $w(z)$ of Problem A^* depends on (J^*) arbitrary real constants (see [152]).

4 Mixed Boundary Value Problem for Elliptic Equations of Second Order

The results in this section will be used in latter chapters.

4.1 The mixed problem for elliptic equations of second order

As in Section 3, let D be an $(N+1)$–connected domain with the boundary $\Gamma = \cup_{j=0}^{N}\Gamma_j$, but here $\Gamma \in C_\alpha^2 (0 < \alpha < 1)$. We consider the second order equation

$$\begin{cases} u_{z\bar{z}} = F(z, u, u_z, u_{zz}), \ F = \mathrm{Re}\,[Qu_{zz} + A_1 u_z] + A_2 u + A_3, \\ Q = Q(z, u, u_z, u_{zz}), \ A_j = A_j(z, u, u_z), \ j = 1, 2, 3, \end{cases} \tag{4.1}$$

satisfying the following conditions.

Condition C^*

1) $Q(z, u, u_z, U), A_j(z, u, u_z)(j = 1, 2, 3)$ are continuous in $u \in R$, $u_z \in \mathbb{C}$ for almost every $z \in D$, $U \in \mathbb{C}$, and $Q = 0$, $A_j = 0 \, (j = 1, 2, 3)$ for $z \notin D$.

2) The above functions are measurable in $z \in D$ for all continuously differentiable functions $u(z)$ and all measurable functions $U(z) \in L_{p_0}(\tilde{D})$, and satisfy

$$L_p[A_j(z, u, u_z), \bar{D}] \leq k_0, \ j = 1, 2, 3, \tag{4.2}$$

in which $\tilde{D}$ is any closed subset in D, $p_0, p \, (2 < p_0 \leq p), k_0$ are nonnegative constants.

3) The equation (4.1) satisfies the uniform ellipticity condition

$$|F(z, u, u_z, U_1) - F(z, u, u_z, U_2)| \leq q_0|U_1 - U_2|, \tag{4.3}$$

for almost every point $z \in D$, any continuously differentiable function $u(z)$ in D and $U_1, U_2 \in \mathbb{C}$, where $q_0(< 1), p(> 2), k_0$ are all nonnegative constants.

Problem M In the domain D, find a solution $u(z)$ of equation (4.1), which is continuous on $\bar{D}$, and satisfies the boundary condition

$$a_1(z)\frac{\partial u}{\partial \nu} + a_2(z)u(z) = a_3(z) + h, \ z \in \Gamma, \ u(0) = u_0, \tag{4.4}$$

where $\nu(= \nu_1 + i\nu_2)$ can be arbitrary provided that $\cos(\nu, n)$ has a positive lower bound on $\Gamma = \partial D$, n being the outward normal vector, ν_1 and ν_2 are the two components of the vector ν, $a_1(z)$ and $a_2(z)$ are nonnegative functions satisfying

$$a_1 + a_2 \geq 1, \ C_\alpha^1[a_j, \Gamma] \leq k_0, \ C_\alpha[\nu_j, \Gamma] \leq k_0, \ |u_0| \leq k_0, \tag{4.5}$$

in which $\alpha\,(1/2 < \alpha < 1)$ is a constant and h is an unknown constant to be determined appropriately.

If $a_2(z) \not\equiv 0$ for $z \in \Gamma$, we do not assume $u(0) = u_0$, but let $h = 0$. When $a_1 > 0$, $z \in \Gamma$, Problem M is the third boundary value problem (Problem III).

4.2   A priori estimates of solutions for the mixed problem

First of all, we give the estimate of boundedness for solutions of (4.1).

Lemma 4.1   *Let $u(z)$ be a solution of Problem M for (4.1). Then we have*

$$\frac{\partial u}{\partial \nu} \text{ or } u(z) \geq -M_1, \ \frac{\partial u}{\partial \nu} \text{ or } u(z) \leq M_1, \ z \in \Gamma, \tag{4.6}$$

where $M_1 = M_1(q_0, p_0, \alpha, k_0, D) > 0$, $2 < p_0 < 1/(1 - \alpha)$.

Proof   Without loss of generality, we only discuss the linear equation

$$Lu = u_{z\bar{z}} - \text{Re}[Q u_{zz} + A_1 u_z] - A_2 u = A_3 \tag{4.7}$$

satisfying Condition C^*. Denote the solutions for the inhomogeneous equation (4.7) and the homogeneous equation $Lu = 0$ in the disk $D_1 = \{|z| < 1\}$ by $\psi(z)$ and $\Psi(z)$ respectively, which satisfies the boundary condition $\psi(z) = 0$ and $\Psi(z) = 1$ for $z \in \Gamma_0 = \{|z| = 1\}$. It can be seen that

$$U(z) = u(z) - \psi(z) - [u_0 - \psi(0)]\Psi(z)/\Psi(0) \tag{4.8}$$

is a solution of Problem M for the equation

$$LU = U_{z\bar{z}} - \text{Re}[Q U_{zz} + A_1 U_z] - A_2 U = 0 \tag{4.9}$$

satisfying the boundary condition

$$l\,U = a_1(z)\frac{\partial U}{\partial \nu} + a_2(z)U(z) = r(z) + h, \ z \in \Gamma, \ U(0) = 0, \tag{4.10}$$

where

$$r(z) = a_3(z) - a_1(z)\frac{\partial \psi}{\partial \nu} - a_2(z)\psi(z) - \frac{u_0 - \psi(0)}{\Psi(0)}[a_1(z)\frac{\partial \Psi}{\partial \nu} + a_2(z)\Psi(z)].$$

By using Theorem 2.5 and Theorem 2.9 of Chapter 3 in [140], we can obtain

$$\frac{\partial U}{\partial \nu} \text{ or } U(z) \geq -M_2, \ \frac{\partial U}{\partial \nu} \text{ or } U(z) \leq M_2, \ z \in \Gamma, \tag{4.11}$$

where $M_2 = M_2(q_0, p_0, \alpha, k_0, D) > 0$, and so (4.6) holds.

Theorem 4.2 *If $u(z)$ is any solution of Problem M for the nonlinear equation (4.1) with Condition C^*, then $u(z)$ satisfies*

$$|u(z)| \le C[u(z), \bar{D}] \le M_3 = M_3(q_0, p_0, \alpha, k_0, D). \tag{4.12}$$

Proof Similarly to the proof of Lemma 4.1, we only discuss the solution $U(z)$ of the homogeneous equation (4.9) which satisfies the boundary condition (4.10).

Let $U(z) \not\equiv$ constant and so $U(z) \not\equiv 0$ for $z \in D$. Denote $M = \max_{z \in \bar{D}} U(z)$, then the function $M\Psi(z) - U(z)$ is a positive solution of (4.9) in D. By means of the Harnack inequality (see Theorem 3.9, Chapter 3 in [140]), we have

$$\frac{1}{q(r)} \le \frac{M\Psi(z) - U(z)}{M\Psi(0)} \quad \text{or} \quad M_4 = \max_{|z|=\gamma_0^*} U(z) \le M(1 - \frac{M_5}{2}), \tag{4.13}$$

where γ_0^* is a sufficiently small positive constant, and M_5 $(0 < M_5 < 1)$ is a positive constant.

Moreover, we choose the harmonic measure $\eta(z)$ on $D_0 = D \cap \{|z| > \gamma_0^*\}$ satisfying the boundary condition

$$\eta(z) = \begin{cases} 0, & z \in \Gamma, \\ 1, & z \in \Gamma_{N+1} = \{|z| = \gamma_0^* = \gamma_{N+1}\}. \end{cases} \tag{4.14}$$

It is clear that $\eta(z) > 0$, $z \in D_0$, and

$$\gamma_k \frac{\partial \eta}{\partial s} = 2\mathrm{Re}[i(z - z_k)\eta_k] = 0, \quad \gamma_k \frac{\partial \eta}{\partial n} = 2(-1)^m \mathrm{Re}[(z - z_k)\eta_z] < 0, \quad z \in \Gamma_k,$$

where $m = 0$ for $k = 0$ and $m = 1$ for $k = 1, \cdots N$, and so $\eta_z \ne 0$, $z \in \Gamma \cup \Gamma_{N+1}$. According to the property of the solution η_z for the Riemann–Hilbert problem with the boundary condition

$$\mathrm{Re}[i(z - z_k)\eta_z] = 0, \quad z \in \Gamma_k, \quad k = 0, 1, \cdots, N + 1,$$

we conclude that η_z possesses N zero points $c_1, \cdots, c_N$ in the neighborhoods $|z - c_k| < \gamma_k^*$, $k = 1, \cdots, N$, herein γ_k^* $(k = 1, \cdots, N)$ are sufficiently small positive constants. By using again the Harnack inequality, it is seen that

$$M_6 = \max_{1 \le k \le N} \max_{|z - c_k| = \gamma_k^*} U(z) \le M(1 - \frac{M_5}{2}). \tag{4.15}$$

Let $\phi(z) = e^{c\eta(z)}$, where c is a positive constant chosen such that

$$\frac{\partial \phi}{\partial \nu} = c\phi[\cos(\nu, n)\frac{\partial \eta}{\partial n} + \cos(\nu, s)\frac{\partial \eta}{\partial s}] = c\phi \cos(\nu, n)\frac{\partial \eta}{\partial n} < -1, \quad z \in \Gamma, \tag{4.16}$$

then

$$(L + A_2)\phi \geq 0 \text{ for } z \in D^* = \bar{D} \cap \{z| \geq \gamma_0^*\} \cap \cdots \cap \{|z - c_N| \geq \gamma_N^*\} \qquad (4.17)$$

holds.

Next, we consider the function $V(z) = M_2\phi(z) + U(z)$, and can derive

$$(L + A_2)V \geq A_2 U \text{ in } D^*. \qquad (4.18)$$

It is obvious that $U(z)$ is bounded on $\Gamma^* = \{z \mid z \in \Gamma, a_1(z) = 0\}$, and $\partial V/\partial \nu < 0$ or $U(z) < M_2$, $z \in \Gamma\backslash\Gamma^*$. Because $V(z)$ cannot attain its maximum at the points $z \in \Gamma\backslash\Gamma^*$ for $\partial V/\partial \nu < 0$, the maximum of $V(z)$ on $\overline{D}^*$ will be attained in one of the following three cases:

1) $V(z)$ reaches its maximum at the point $z = b_2 \in \Gamma$ and $U(b_2) \leq M_2$.

2) $V(z)$ reaches its maximum at the point $z = b_3 \in D^*$. In this case, we can prove $U(b_3) \leq 0$ and obtain

$$M = U(b_1) = \max_{z \in \bar{D}} U(z) \leq M_2\phi(b_3) - M_2\phi(b_1) = M_2[\phi(b_3) - 1]. \qquad (4.19)$$

3) $V(z)$ reaches its maximum at the point $z = b_4 \in \sum_{k=0}^{N}\{|z - c_k| = \gamma_k^*\}$. Similarly to 2), it can be derived that

$$M \leq M_2[\phi(b_4) - 1] + M_6 \leq M_2[\phi(b_4) - 1] + M(1 - \frac{M_5}{2}). \qquad (4.20)$$

Combining $(4.19), (4.20)$ and $U(b_2) \leq M_2$, we obtain

$$M = \max_{z \in \bar{D}} U(z) \leq \max\{M_2, M_2[\phi(b_3) - 1], [\phi(b_4) - 1]\frac{2M_2}{M_5}\}. \qquad (4.21)$$

In order to obtain a lower bound estimate of $U(z)$, we only assume that $U^*(z) = -U(z)$ and use the above proved result for $U^*(z)$.

Furthermore, we shall give some interior estimates for the bounded solution of (4.1).

Theorem 4.3   *Under the hypothesis of Theorem 4.2, any solution $u(z)$ of Problem M for (4.1) satisfies the following estimate on any closed subset D_m in D:*

$$C_\beta^1[u, D_m] \leq M_7, \ L_{p_0}[|u_{z\bar{z}}| + |u_{zz}|, D_m] \leq M_8, \qquad (4.22)$$

where $\beta = 1 - 2/p_0$, $M_j = M_j(q_0, p_0, \alpha, k_0, D_m), j = 7, 8$.

Proof   Let $2R$ denote the distance between D_m and the boundary Γ of D. Choose any point $z_0 \in D_m$ and assume $\Delta = \{|z - z_0| \leq R\}$. Similarly to Lemma 4.1, it is sufficient to consider the linear equation (4.7), and then using the relation

$$U(z) = [u(z) - \psi(z)]/\Psi(z), \qquad (4.23)$$

the solution $u(z)$ of (4.7) can be transformed into the solution $U(z)$ of the homogeneous equation

$$U_{z\bar{z}} - \text{Re}[QU_{zz} + AU_z] = 0, \tag{4.24}$$

where Q is a bounded measurable function as in (4.7), and

$$A = -2(\ln \Psi)_{\bar{z}} + 2Q(\ln \Psi)_z + A_1, \ L_{p_0}[A, \Delta] \le M_9,$$

herein $M_9 = M_9(q_0, p_0, \alpha, k_0, D)$. Now, we discuss the sequence of equations

$$U_{z\bar{z}} - \text{Re}[Q_nU_{zz} + AU_z] = 0, \ z \in \Delta, n = 4, 5, \cdots \tag{4.25}$$

in which

$$Q_n(z) = \sigma_n(z)Q(z), \ A_n = \sigma_n(z)A(z),$$

$$\sigma_n = \begin{cases} 1, & |z - z_0| \le (1 - 1/n)R, \\ 0, & (1 - 1/n)R < |z - z_0| \le R, \end{cases} \ n = 4, 5, \cdots.$$

Denote the solution of (4.25) with the boundary value $U(z)$ by $U_n(z)$, then we can prove that $\{U_n(z)\}$ and $\{U_{nz}\}$ uniformly converge to $U(z)$ and U_z on Δ respectively. Setting $\zeta = (z - z_0)/R$ and $\tilde{U}_n(\zeta) = U_n(z_0 + R\zeta)$, the equation (4.25) can be reduced to the equation

$$\tilde{U}_{\zeta\bar{\zeta}} - \text{Re}[Q_n\tilde{U}_{\zeta\zeta} + A_n\tilde{U}_\zeta] = 0, \ n = 4, 5, \cdots. \tag{4.26}$$

Now we apply the formulae (3.40) and (3.42) of Chapter 3 in [140], i.e.

$$\tilde{U}_n(\zeta) = \int_0^{2\pi} P(\zeta, e^{i\omega})\tilde{U}_n(e^{i\omega})d\omega, \ |P_\zeta| \le M_{10}\frac{P(0)}{|1 - \zeta|^{2/\beta}}, \tag{4.27}$$

where $M_{10} = M_{10}(q_0, p_0, k_0), \beta = 1 - p_0/2$. By (4.12), it can be assumed that $\tilde{U}_n(\zeta) \ge 0$ in $\{|\zeta| \le 1\}$(otherwise, we may only add a positive constant to $\tilde{U}_n(\zeta)$), and so

$$\tilde{U}_{n\zeta} = \int_0^{2\pi} P_\zeta\tilde{U}_n(e^{i\omega})d\omega, \ |\zeta| < 1, \tag{4.28}$$

holds. Moreover, we have

$$|\tilde{U}_{n\zeta}| \le \frac{M_{10}}{|1 - \zeta|^{2/\beta}} \int_0^{2\pi} P(0)\tilde{U}_n(e^{i\omega})d\omega = \frac{M_{10}}{2^{-2/\beta}}\tilde{U}_n(0) \ \text{in} \ |\zeta| \le \frac{1}{2}, \tag{4.29}$$

and

$$|U_{nz}| \le M_{11} = M_{11}(q_0, p_0, \alpha, k_0, D_m), \ z \in D_m. \tag{4.30}$$

It is easy to see that $W_n(z) = U_{nz}$ is a bounded solution of the uniformly elliptic complex equation

$$W_{\bar{z}} - \text{Re}[Q_nW_z + A_nW] = 0, \ z \in D_m, \tag{4.31}$$

and may be expressed as

$$W_n(z) = \Phi_n[\zeta_n(z)]e^{\phi_n(z)}, \; z \in D. \tag{4.32}$$

By this expression, the estimates with respect to Hölder continuity of $W_n(z)$ and the boundness of $L_{p_0}[|W_{n\bar{z}}| + |W_{nz}|, D_m]$ can be derived. Hence we obtain the estimate (4.22).

Theorem 4.4 *Under the hypothesis of Theorem 4.3, any solution $u(z)$ of Problem M for (4.1) satisfies the estimate*

$$C_\beta^1[u, D_*] \le M_{12}, \; L_{p_0}[|u_{z\bar{z}}| + |u_{zz}|, D_*] \le M_{13}, \tag{4.33}$$

where $D_* = \bar{D} \cap \{\cap_{z^* \in \partial\Gamma^*}(|z - z^*| \ge \delta > 0)\}$, $\Gamma^* = \{z | z \in \Gamma, a_1(z) = 0\}$, $\partial\Gamma^*$ is a set of end points of Γ^* and $2 < p_0 < \min(p, 1/(1 - \alpha))$, $\beta = 1 - 2/p_0$, $M_j = M_j(q_0, p_0, \alpha, k_0, D, \Gamma^*, \delta)$, $j = 12, 13$.

Proof By Theorem 4.3, it remains only to show that the solution $u(z)$ of Problem M satisfies the estimate (4.33) in the neighborhood of $\Gamma \backslash \partial\Gamma^*$. Choose an open arc $\tilde{\Gamma}$ in $\Gamma^* \backslash \partial\Gamma^*$ such that the length of $\tilde{\Gamma}$ is not equal to zero. Without loss of generality, we may assume that $\tilde{\Gamma}$ lies on the unit circle with its middle point at $z = 1$. We find out a solution $u_0(z)$ of (4.1) or (4.7) with the boundary value $r(z) = [a_3(z) + h]/a_2(z)$ on $\tilde{\Gamma}$, and thus $\tilde{u}(z) = u(z) - u_0(z)$ is a solution of the homogeneous equation (4.9) which satisfies the homogeneous boundary condition of (4.4) on $\tilde{\Gamma}$. Hence such a solution $\tilde{u}(z)$ can be extended continuously across $\tilde{\Gamma}$ from D to $\tilde{D}$(the reflected domain of D). To do this, we just assume that

$$u^*(z) = \begin{cases} \tilde{u}(z), \; z \in D \cup \tilde{\Gamma}, \\ -\tilde{u}(1/\bar{z}), \; z \in \tilde{D}. \end{cases} \tag{4.34}$$

It is not difficult to verify the continuity of $u^*(z)$ on $\tilde{\Gamma}$. Noting that

$$u_{z\bar{z}}^* = -|z|^{-4}\tilde{u}_{\zeta\bar{\zeta}}, \; u_{\bar{z}\bar{z}}^* = -\bar{z}^{-4}\tilde{u}_{\zeta\zeta} - 2\bar{z}^{-3}\tilde{u}_\zeta, \; z \in \tilde{D} \cup \tilde{\Gamma},$$

$u^*(z)$ is a solution of the equation

$$u_{z\bar{z}} - \mathrm{Re}[Q^* u_{zz} + A_1^* u_z] - A_2^* u - A_3^* = 0, \tag{4.35}$$

where

$$Q^* = \begin{cases} Q_1(z), \\ \overline{Q_1(1/\bar{z})}(z/\bar{z})^2, \end{cases} \quad A_1^* = \begin{cases} A_1(z), \\ -\overline{A_1(1/\bar{z})}/z^2 + 2z\overline{Q(1/\bar{z})}/z^2, \end{cases}$$

$$A_2^* = \begin{cases} A_2(z), \\ \overline{A_2(1/\bar{z})}/|z|^4, \end{cases} \quad A_3^* = \begin{cases} A_3(z), & z \in D \cup \tilde{\Gamma}, \\ -\overline{A_3(1/\bar{z})}/|z|^4, & z \in \tilde{D}, \end{cases}$$

it is not difficult to see that equation (4.35) satisfies a condition similar to Condition C^* in the neighborhood of $\tilde{\Gamma}$. With a similar method as in the proof of Theorem 4.3, we obtain estimates for $u^*(z)$ on any closed subset of $D \cup \tilde{\Gamma} \cup \tilde{D}$ and then $u(z)$ satisfies the estimates

$$C_\beta^1[u, \check{D}] \leq M_{14}, \quad L_{p_0}[|u_{z\bar{z}}| + |u_{zz}|, \check{D}] \leq M_{15}, \tag{4.36}$$

where $\check{D}$ is any closed subset of $D \cup \tilde{\Gamma}$ and $M_j = M_j(q_0, p_0, \alpha, k_0, D, \Gamma^*, \delta, \check{D})$, $j = 14, 15$.

Next, if there exists an arc $\hat{\Gamma} = \{z | z \in \Gamma, a_1(z) > 0, \nu = n\}$, there is no harm assuming $\hat{\Gamma} \subset \Gamma_0$, then we can find a harmonic function $\sigma(z)$ in D, which satisfies the boundary condition

$$\frac{\partial \sigma}{\partial n} = \frac{a_2(z)}{a_1(z)} \text{ on } \hat{\Gamma}.$$

Thus

$$V(z) = u(z)e^{\sigma(z)}$$

satisfies the boundary condition

$$\frac{\partial V}{\partial n} = \frac{a_3 + h}{a_1}e^\sigma, \text{ i.e. } a_1\frac{\partial u}{\partial n} + a_2 u = a_3 + h \text{ on } \hat{\Gamma}. \tag{4.37}$$

Noting that

$$u_z = e^{-\sigma}[V_z - \sigma_z V], \quad u_{z\bar{z}} = e^{-\sigma}[V_{z\bar{z}} - 2\text{Re}(\sigma_{\bar{z}}V_z) + |\sigma_{\bar{z}}|^2 V],$$

$$u_{zz} = e^{-\sigma}[V_{zz} - 2\sigma_z V_z - (\sigma_{zz} - \sigma_z^2)V],$$

it is easy to see that the function $V(z)$ satisfies the equation

$$V_{z\bar{z}} - \text{Re}[\hat{Q}V_{zz} + \hat{A}_1 V_z] - \hat{A}_2 V - \hat{A}_3 = 0 \text{ in } D,$$

the coefficients of which satisfy a condition similar to Condition C^*. Now, we find a harmonic function $V_0(z)$ in D satisfying the boundary condition (4.37). Moreover, the function

$$\hat{V}(z) = \begin{cases} V(z) - V_0(z), & z \in D \cup \hat{\Gamma}, \\ V(1/\bar{z}) - V_0(1/\bar{z}), & z \in \tilde{D}, \end{cases}$$

satisfies the equation in the form (4.35), where the coefficients satisfy the condition similar to Condition C^*. Hence by using a method as stated before, a priori estimates of $\hat{V}(z)$ can be derived, and $u(z)$ can be shown to satisfy the estimates

$$C_\beta^1[u, \hat{D}] \leq M_{16}, \quad \| u \|_{W_{p_0(D)}^2} \leq M_{17}, \tag{4.38}$$

where $\hat{D} = \{z | z \in D \cup \hat{\Gamma}, \text{dist}(z, \Gamma \backslash \hat{\Gamma}) \geq 2\delta > 0\}$, $M_j = M_j(q_0, p_0, \alpha, k_0, D, \Gamma^*, \delta, \hat{D})$, $j = 16, 17$.

Finally, we discuss the arc $\Gamma_* = \{z | z \in \Gamma, a_1(z) > 0, \nu \not\equiv n\}$. Applying a similar method as before, we can transform the boundary condition (4.4) into a homogeneous boundary condition

$$\frac{\partial V_*}{\partial \nu} = 0, \ \ z \in \Gamma_*,$$

where the function $V_*(z)$ satisfies the equation in the form (4.35). Without loss of generality, we may assume that D in the half-plane $\operatorname{Im} z < 0$ and $0 \in \Gamma_*$, because through a conformal mapping this requirement can be realized. Setting $b_1 = \cos(\nu, x)$, $b_2 = \cos(\nu, y)$, and making the transformation

$$z = \frac{1}{2}(1 + b_1 + ib_2)\zeta + \frac{1}{2}(-1 + b_1 + ib_2)\bar{\zeta}, \ \zeta = \xi + i\eta, \tag{4.39}$$

it is obvious that (4.39) is a homeomorphism $\zeta = \zeta(z)$ in a neighborhood of $\zeta = 0$, which maps the arc Γ_* in the ζ-plane onto a segment L_* in $x = 0$. Denote by $z = z(\zeta)$ the inverse function of $\zeta = \zeta(z)$, thus the function $V^* = V_*[z(\zeta)]$ satisfies the following equation and the boundary condition

$$V^*_{\zeta\bar\zeta} - \operatorname{Re}[Q^* V^*_{\zeta\zeta} * A^* V^*_\zeta] - A^*_1 V^* = A^*_3 \ \text{in} \ G^* = \zeta(D), \tag{4.40}$$

$$\frac{\partial V^*}{\partial n} = 0 \ \text{on} \ L^* = \zeta(\Gamma_*). \tag{4.41}$$

Furthermore, by applying the method of deriving the estimate (4.36), we can obtain estimates of $V^*(z)$, $V_*(z)$ and $u(z)$, namely

$$C^1_\beta[u, D_*] \leq M_{18}, \ \| u \|_{W^2_{p_0}(\dot{D})} \leq M_{19}, \tag{4.42}$$

where $\dot{D} = \{z | z \in D \cup \Gamma_*, \operatorname{dist}(z, \Gamma \backslash \Gamma_*) \geq 2\delta > 0\}$, $M_j = M_j(q_0, p_0, \alpha, k_0, D, \Gamma^*, \delta, \dot{D})$, $j = 18, 19$. Combining (4.22),(4.36),(4.38) and (4.42), the estimates in (4.33) are derived.

4.3 Uniqueness and existence of solutions for the mixed problem

Finally, we prove the existence and uniqueness of solutions for Problem M of (4.1).

Theorem 4.5 *Let the equation (4.1) satisfy Condition C^* and the following condition:*

$$F(z, u_1, u_{1z}, U) - F(z, u_2, u_{2z}, U) = \operatorname{Re}[B_1(u_1 - u_2)_z] + B_2(u_1 - u_2), \tag{4.43}$$

where $B_j = B_j(z, u_1, u_2)$, $L_{p_0}[B_j, \bar{D}] \leq k_1$, $j = 1, 2$, $B_2 \geq 0$, k_1 is a constant. Then the solution of Problem M for (4.1) is unique.

Proof Let $u_1(z), u_2(z)$ be two solutions of Problem M for (4.1) and denote $u(z) = u_1(z) - u_2(z)$. From Condition C^* and (4.43), it is easily seen that $u(z)$ is a solution of the following uniformly elliptic equation

$$u_{z\bar{z}} = \text{Re}[Q(z)u_{zz} + B_1 u_z] - B_2 u, \ |Q| \leq q_0 < 1, \tag{4.44}$$

and satisfies the boundary condition

$$a_1(z)\frac{\partial u}{\partial \nu} + a_2(z)u(z) = h, \ z \in \Gamma, \ u(0) = 0. \tag{4.45}$$

In the same way used in the proof of Theorem 4.5, Chapter 3 of [140], we can derive that $u(z) = 0$, i.e. $u_1(z) = u_2(z)$ in D.

Theorem 4.6 *Suppose that the nonlinear equation (4.1) satisfies Condition C^*, then its Problem M has at least one solution.*

Proof Choosing an arbitrary positive integer n, we consider the third boundary value problem III for (4.1) with the boundary condition

$$l_n u = [a_1(z) + \frac{1}{n}]\frac{\partial u}{\partial \nu} + a_2(z)u(z) = a_3(z) + h, \ z \in \Gamma, \ u(0) = u_0. \tag{4.46}$$

By Theorem 4.9 of Chapter 3 in [140] and Theorem 4.5, Problem III has a solution $u_n(z)$ satisfying the estimate (4.33). Because of the compactness of $\{u_n(z)\}$, it can be selected a subsequence of $\{u_n(z)\}$ which uniformly converges to a solution $u_0(z)$ for (4.1) in any closed subset of $\bar{D}\backslash\Gamma^*$, and $u_0(z)$ satisfies the boundary condition (4.4) on $\Gamma\backslash\Gamma^*$. It remains to show that $u_0(z)$ is continuous on $\bar{D}$ and satisfies the boundary condition (4.4) on Γ^*.

Choose any point $z^* \in \Gamma^*$ and let Γ_δ denote the point set $\{(|z - z^*| < \delta) \cap \Gamma\}$, where δ is a sufficiently small positive constant. We construct a real continuously differentiable function $f(z)$ as follows;

$$f(z) = \begin{cases} M_{20} + 1, \ z \in \Gamma\backslash\Gamma_{\delta/2}, \\ \eta > 0, \ z \in \Gamma_{\delta/4}, \end{cases} \ \eta \leq f(z) \leq M_{20} + 1, \ z \in \Gamma_{\delta/2}\backslash\Gamma_{\delta/4}, \tag{4.47}$$

where M_{20} is a constant to be determined appropriately. $f(z)$ satisfies the estimate

$$C^1_{\alpha-\varepsilon}[f, \Gamma] \leq \frac{M_{21}}{\delta^{1+\alpha-\varepsilon}}, \ \frac{1}{2} < \alpha - \varepsilon < 1, \tag{4.48}$$

where $M_{21} = M_{21}(M_{20}, \Gamma)$. Let $\hat{u}_n(z)$ denote the solution of the homogeneous equation corresponding to (4.1), i.e. $Lu = 0$ on $\bar{D}$ with the boundary value $\hat{u}_n(z) = f(z), z \in \Gamma$. We can prove that $\hat{u}_n(z)$ satisfies the estimate

$$C^1[\hat{u}_n, \bar{D}] \leq \frac{M_{22}}{\delta^{1+\alpha-\varepsilon}}, M_{22} = M_{22}(M_{20}, \Gamma). \tag{4.49}$$

Now extend $a_2(z)$ from Γ^* to Γ such that the new function $a_2^*(z) \in C_\alpha^1(\Gamma)$, $a_2^*(z) > 0$ and $a_2^*(z) = a_2(z)$ for $z \in \Gamma_{\delta/2}$. Then we can find a solution $u_n^*(z)$ of $Lu = A_3$ with the boundary value $u_n^*(z) = [a_3^*(z) + h]/a_2^*(z)$, $z \in \Gamma$. Set

$$\tilde{u}_\pm(z) = \pm\hat{u}_n(z) - u_n(z) + u_n^*(z).$$

It is easy to see that $\tilde{u}_\pm(z)$ are the solutions of $Lu = 0$. Moreover, we can prove that

$$\tilde{u}_+(z) \geq 0, \ \tilde{u}_-(z) \leq 0, \ z \in \bar{D}. \tag{4.50}$$

In fact, obviously $\tilde{u}_+(z) \geq M_{20} + 1 - M_{20} > 0$ on $\Gamma\backslash\Gamma_{\delta/2}$, where

$$M_{20} = M_3 + \max_\Gamma \frac{a_3(z) + h}{a_2^*(z)} > 0, \ M_3 = \max_{\bar{D}} |u_n(z)|.$$

If $\tilde{u}_+(z)$ takes a negative minimum in $\bar{D}$, then there exists a point $z' \in D_{\delta/2}$, such that $\tilde{u}_+(z') \leq \min_D \tilde{u}_+(z)$. However, we have

$$[a_1(z) + \frac{1}{n}]\frac{\partial\tilde{u}_+}{\partial\nu} + a_2(z)\tilde{u}_+(z) = [a_1(z) + \frac{1}{n}]\frac{\partial\hat{u}_n}{\partial\nu}$$

$$+a_2(z)\hat{u}_n(z) + [a_1(z) + \frac{1}{n}]\frac{u_n^*}{\partial\nu} > M_{23}\eta - \max_{\Gamma_{\delta/2}} a_1(z)\frac{M_{24}}{\delta^{1+\alpha-\varepsilon}}$$

$$-\frac{M_{24}}{n\delta^{1+\alpha-\varepsilon}} + [a_1(z) + \frac{1}{n}]\frac{\partial u_n^*}{\partial\nu} \text{ on } \Gamma_{\delta/2},$$

where $M_{23} = \min_{\Gamma_{\delta/2}} a_2(z)$, $M_{24} = M_{24}(M_{22}, \Gamma)$. Due to $a_1(z) \leq M_{25}\delta^{1+\alpha}$ on $\Gamma_{\delta/2}$, herein M_{25} is a positive constant, and because we can first choose δ small enough, and then select n large enough, such that $\delta^\varepsilon M_{24}M_{25}$, $M_{24}/n\delta^{1+\alpha-\varepsilon}$, $|a_1(z)\partial u_n^*/\partial\nu|$, and $|1/n\partial u_n^*/\partial\nu|$ on $\Gamma_{\delta/2}$ are less than $\frac{1}{4}M_{23}\eta$, we have

$$[a_1(z) + \frac{1}{n}]\frac{\partial\tilde{u}_+}{\partial\nu} + a_2(z)\tilde{u}_+ > 0 \text{ on } \Gamma_{\delta/2}.$$

This shows that $\tilde{u}_+(z)$ cannot take a negative minimum on $\Gamma_{\delta/2}$. On the basis of the maximum principle of solutions to $Lu_n = 0$,

$$\tilde{u}_+(z) = \hat{u}_n(z) - u_n(z) + u_n^*(z) \geq 0, \text{ i.e. } u_n(z) - u_n^*(z) \leq \hat{u}_n(z) \text{ in } \bar{D}$$

can be obtained. By the same reason, we have

$$\tilde{u}_-(z) \leq 0, \text{ i.e. } u_n(z) - u_n^*(z) \geq -\hat{u}_n(z) \text{ in } \bar{D}.$$

From $|\hat{u}_n(z)| \leq \eta$ on $\Gamma_{\delta/4}$, it follows that

$$|u_n(z) - u_n^*(z)| < \eta \text{ on } \Gamma_{\delta/4}.$$

By the equicontinuity of $\{\hat{u}_n(z)\}$ in $\bar{D}$, it is seen that there is

$$|u_n(z) - u_n^*(z)| \leq |\hat{u}_n(z)| \leq 2\eta$$

in a neighborhood of z^* in $\bar{D}$. Denote by $\tilde{u}_0(z)$ the limit function of a subsequence of $\{u_n(z) - u_n^*(z)\}$ in $\bar{D}$. It is clear that $|\tilde{u}_0(z)| \leq 2\eta$. Noting that η is an arbitrary positive number, it is easy to see that $\tilde{u}_0(z) = u_0(z) - u_0^*(z)$ is continuous at $z^* \in \Gamma^*$ and $\tilde{u}_0(z^*) = 0$, where $u_0^*(z)$ is the limit of a subsequence of $\{u_n^*(z)\}$. Therefore $u_0(z) = \tilde{u}_0(z) + u_0^*(z)$ is continuous at $z^* \in \Gamma^*$ (see [139]6)).

II Nonlinear Boundary Value Problems for Elliptic Complex Equations and Systems

Nonlinear boundary value problems for analytic functions and generalized analytic functions have been considered by several authors. In this chapter, we discuss several nonlinear boundary value problems for more general elliptic complex equations and systems of first and second order in simply connected and in multiply connected domains, which include the Riemann–Hilbert problem, the Haseman problem and the irregular oblique derivative problem. The boundary condition of the Riemann–Hilbert problem may be piecewise Hölder continuous. The greater part of the contents in this chapter can be found in Refs. [11]1),2),3), [147]3) and [139]10),20).

1 Nonlinear Boundary Value Problems for Quasilinear Elliptic Complex Equations of First Order

In this section, we mainly discuss the Riemann–Hilbert problem and the Haseman problem for quasilinear elliptic complex equations of first order in a disk.

1.1 Formulation of the nonlinear Riemann–Hilbert problem in a disk

We consider a class of quasilinear elliptic complex equation

$$w_{\bar{z}} = F(z, w), \; F = A_1 w + A_2 \bar{w} + A_3, \; A_j = A_j(z, w), \; j = 1, 2, 3 \qquad (1.1)$$

in the unit disk D and suppose that (1.1) satisfies the following conditions.

Condition C For any function $w(z) \in C_\alpha(\bar{D})$, the coefficients $A_j(z, w)(j = 1, 2, 3)$ and the function $F(z, w, w_z)$ in (1.1) satisfy

$$L_p[A_j(z, w), \bar{D}] \le k_0, \; j = 1, 2, \; L_p[A_3(z, w), \bar{D}] \le k_1, \qquad (1.2)$$

$$|F(z, w_1) - F(z, w_2)| \le k_0|w_1 - w_2| \text{ for } w_1, w_2 \in \mathbb{C}, \qquad (1.3)$$

where $p(> 0), \alpha(0 < \alpha < 1), k_0$ and k_1 are nonnegative constants.

Problem A_1 Find a solution $w(z) \in C_\alpha(\bar{D})$ for the complex equation (1.1) satisfying the nonlinear Riemann–Hilbert boundary condition

$$\text{Re}[\overline{\lambda(z)}w(z)] = r(z, w) + h(z) \text{ on } \Gamma = \{|z| = 1\} \qquad (1.4)$$

and if $K \ge 0$ the nonlinear point conditions

$$\text{Im}[\overline{\lambda(a_j)}w(a_j)] = b_j(w), \; j \in J = \{1, \cdots, 2K + 1\}, \qquad (1.5)$$

in which $|\lambda(z)| = 1$, $\lambda(z) \in C_\alpha(\Gamma)$, $a_j(j \in J)$ are distinct fixed points on Γ, $r[z, w(z)] \in C_\alpha(\Gamma)$ and $b_j(w) \in C_\alpha(\Gamma)$ for all $w(z) \in C_\alpha(\Gamma)$, and satisfy

$$\begin{cases} C_\alpha[r(z, w_1(z)) - r(z, w_2(z)), \Gamma] \leq \varepsilon C_\alpha[w_1 - w_2, \Gamma], \\ C_\alpha[\lambda(z), \Gamma] \leq k_0, \ C_\alpha[r(z, 0), \Gamma] \leq k_2, \ b_j = b_j(0), \ |b_j| \leq k_3, \\ |b_j(w_1) - b_j(w_2)| \leq \varepsilon |w_1 - w_2|, \ j \in J, \end{cases} \qquad (1.6)$$

for all $w_1(z), w_2(z) \in C_\alpha(\Gamma)$, $K = \frac{1}{2\pi}\Delta_\Gamma \arg\lambda(z)$ is the index of the problem, and $k_2, k_3, \varepsilon(> 0)$ are nonnegative constants. Moreover,

$$h(z) = \begin{cases} 0, \ \text{for } K \geq 0, \\ h_0 + \text{Re} \sum_{m=1}^{-K-1}(h_m^+ + h_m^-)z^m, \ z \in \Gamma, \ \text{for } K < 0, \end{cases} \qquad (1.7)$$

where $h_0, h_m^\pm \ (m = 1, \cdots, |K| - 1)$ are unknown real constants to be determined appropriately. Problem A_1 with the condition $r(z, w) = 0$ and $b_j(w) = 0 \ (j \in J)$ is called Problem A_0. In the following, we shall give a priori estimates of solutions for Problem A_1, and using these estimates and the method of parameter extension, the result of solvability for Problem A_1 will be obtained. For the sake of convenience, we assume $\alpha \leq 1 - 2/p$ in this chapter.

1.2   A priori estimates of solutions for Problem A_1

First of all, we give an a priori estimate of solutions to the linear boundary value problem (Problem A_1^*) for the quasilinear complex equation (1.1), in which the boundary condition and point conditions are

$$\text{Re}[\overline{\lambda(z)}w(z)] = R(z) + h(z) \text{ on } \Gamma, \ C_\alpha[R(z), \Gamma] \leq k_2, \qquad (1.8)$$

$$\text{Im}[\overline{\lambda(a_j)}w(a_j)] = b_j, \ |b_j| \leq k_3, \ j \in J, \qquad (1.9)$$

where $\alpha, k_2, a_j, b_j, k_3, h(z)$ are as stated in (1.4)-(1.7).

Lemma 1.1   *Let the coefficients of the complex equation (1.1) satisfy Condition C. Then the solution $w(z)$ of Problem A_1^* for (1.1) satisfies the estimate*

$$C_\alpha[w, \bar{D}] \leq M_1(k_1 + k_2 + k_3), \qquad (1.10)$$

where the constant $M_1 = M_1(\alpha, k_0, K)$ depends only on α, k_0, K.

Proof   First we assume that the constant $k := k_1 + k_2 + k_3 > 0$ and denote $W(z) = w(z)/k$, in which $w(z)$ is a solution of Problem A_1^* for (1.1). It is easily seen that $W(z)$ is a solution of the following boundary value problem

$$W_{\bar{z}} = A_1(z, w)W + A_2\overline{W} + A_3(z, w)/k, \qquad (1.11)$$

$$\mathrm{Re}[\overline{\lambda(z)}W(z)] = [R(z) + h(z)]/k, \ z \in \Gamma, \tag{1.12}$$

$$\mathrm{Im}[\overline{\lambda(a_j)}W(a_j)] = b_j/k, \ j \in J, \tag{1.13}$$

with the conditions

$$L_p[A_3(z,w)/k, \bar{D}] \leq 1, \ C_\alpha[R(z)/k, \Gamma] \leq 1, \ |b_j/k| \leq 1, \ j \in J. \tag{1.14}$$

From Theorem 4.3, Chapter 2 in [140], the estimate

$$C_\alpha[W, \bar{D}] \leq M_1 = M_1(\alpha, k_0, K) \tag{1.15}$$

can be derived. Hence the estimate (1.10) holds for $w(z)$.

If $k_1 + k_2 + k_3 = 0$, we replace $k_1 + k_2 + k_3 = 0$ by $k > 0$, then there is the estimate (1.10) where the constant $k_1 + k_2 + k_3$ is represented by $k > 0$. Afterwards, letting k tend to 0, the estimate (1.10) with $k_1 + k_2 + k_3 = 0$ is derived.

Theorem 1.2 *Suppose that (1.1) satisfies Condition C and the positive constant ε in (1.6) is small enough. Then any solution $w(z)$ of Problem A_1 for (1.1) satisfies the estimate*

$$C_\alpha[w, \bar{D}] \leq M_2, \tag{1.16}$$

where $M_2 = M_2(\alpha, k, \varepsilon, K)$ is a nonnegative constant, herein $k = (k_0, k_1, k_2, k_3)$.

Proof Let the solution $w(z)$ of Problem A_1 be substituted into the complex equation (1.1), the boundary condition (1.4) and the point condition (1.5). There is no harm in assuming that $w(z) \not\equiv 0$ in D. Taking condition (1.6) into account, we have

$$C_\alpha[r(z,w), \Gamma] \leq C_\alpha[r(z,w) - r(z,0), \Gamma] + C_\alpha[r(z,0), \Gamma] \leq$$
$$\leq \varepsilon C_\alpha[w, \Gamma] + k_2 \leq \varepsilon C_\alpha[w, \bar{D}] + k_2, \tag{1.17}$$

and

$$|b_j(w)| \leq |b_j(w) - b_j(0)| + |b_j(0)| \leq \varepsilon C_\alpha[w, \Gamma] + k_3 \leq$$
$$\leq \varepsilon C_\alpha[w, \bar{D}] + k_3, \ j \in J. \tag{1.18}$$

The constants $k_1, \varepsilon C_\alpha[w, \bar{G}] + k_2, \varepsilon C_\alpha[w, \bar{G}] + k_3$ correspond to k_1, k_2, k_3 in Lemma 1.1 respectively. According to the estimate (1.10), it can be seen that

$$C_\alpha[w, \bar{G}] \leq M_1[k_1 + k_2 + k_3 + 2\varepsilon C_\alpha(w, \bar{D})]. \tag{1.19}$$

We choose ε so small that $2\varepsilon M_1 < 1$. From (1.19), it follows that

$$C_\alpha[w, \bar{G}] \leq M_1(k_1 + k_2 + k_3)/(1 - 2\varepsilon M_1) = M_2. \tag{1.20}$$

1.3 Solvability of Problem A_1 for the complex equation (1.1)

We first prove the uniqueness of solutions for Problem A_1.

Theorem 1.3 *Under the assumption of Condition C and (1.6), if the constant ε is sufficiently small, then the solution of Problem A_1 for (1.1) is unique.*

Proof Let $w_1(z)$ and $w_2(z)$ be two solutions of Problem A_1. Denoting $w(z) = w_1(z) - w_2(z)$ and taking Condition C and (1.3),(1.6) into account, it is clear that the function $w(z)$ is a solution of the complex equation

$$w_{\bar{z}} = A(z, w_1, w_2)w, \quad A := \begin{cases} [F(z, w_1) - F(z, w_2)]/w, & \text{for } w(z) \neq 0, \\ 0, & \text{for } w(z) = 0, \ z \in D \end{cases} \tag{1.21}$$

and satisfies the following boundary and point conditions

$$\mathrm{Re}[\overline{\lambda(z)}w(z)] = r(z, w_1) - r(z, w_2) + h(z), \ z \in \Gamma, \tag{1.22}$$

$$\mathrm{Im}[\overline{\lambda(a_j)}w(a_j)] = b_j(w_1) - b_j(w_2), \ j \in J, \tag{1.23}$$

where

$$\begin{cases} L_p[A(z, w_1, w_2), \bar{D}] \leq k_0, \\ C_\alpha[r(z, w_1) - r(z, w_2), \Gamma] \leq \varepsilon C_\alpha[w, \bar{D}], \\ |b_j(w_1) - b_j(w_2)| \leq \varepsilon C_\alpha[w, \bar{D}], \ j \in J. \end{cases} \tag{1.24}$$

The constants $0, \varepsilon C_\alpha(w, \bar{D}), \varepsilon C_\alpha(w, \bar{D})$ correspond to k_1, k_2, k_3 in Lemma 1.1 respectively. Hence we have

$$C_\alpha[w, \bar{D}] \leq 2\varepsilon M_1 C_\alpha[w, \bar{D}]. \tag{1.25}$$

Let the constant ε be so small that $2\varepsilon M_1 < 1$. It follows from (1.25) that $C_\alpha[w, \bar{D}] = 0$, i.e. $w_1(z) \equiv w_2(z), \ z \in D$.

Next, we use the estimate (1.10) and the method of parameter extension to prove the existence of solutions of Problem A_1 for (1.1).

Theorem 1.4 *Under the same hypotheses as in Theorem 1.3, Problem A_1 for the complex equation (1.1) has a solution $w(z) \in C_\alpha(\bar{D})$.*

Proof We introduce the boundary value problem B_1 with a parameter $t(0 \leq t \leq 1)$,

$$\begin{cases} w_{\bar{z}} - t[F(z, w) - F(z, 0)] = A(z), \ A(z) \in L_p(\bar{D}), \\ \mathrm{Re}[\overline{\lambda(z)}w(z)] = t[r(z, w) - r(z, 0)] + R(z) + h(z) \ z \in \Gamma, \\ \mathrm{Im}[\overline{\lambda(a_j)}w(a_j)] = t[b_j(w) - b_j(0)] + B_j, \ j \in J, \end{cases} \tag{1.26}$$

for any $A(z) \in L_p(\bar{G}), R(z) \in C_\alpha(\Gamma)$ and any real constants $B_j, j \in J$. It is evident that when $t = 1, A(z) = F(z, 0), R(z) = r(z, 0), B_j = b_j(0)$, Problem B_1 is just Problem A_1.

When $t = 0$, the complex equation (1.1) is $w_{\bar{z}} = A(z)$ and Problem B_1 is a special case of Problem B in Section 1 of Chapter 1, and so there exists a unique solution $w(z) \in C_\alpha(\bar{D})$. Suppose that when $t = t_0 \, (0 \leq t_0 < 1)$, Problem B_1 is solvable, i.e. Problem B_1 for (1.26) has a unique solution $w_0(z) = w(z, t_0) \in C_\alpha(\bar{D})$. We can find a neighborhood $T_\delta = \{|t - t_0| < \delta, 0 \leq t \leq 1\}$ of t_0 such that for every $t \in T_\delta$, Problem B_1 is solvable. In fact, Problem B_1 can be written in the form

$$
\begin{cases}
w_{\bar{z}} - t_0[F(z,w) - F(z,0)] \\
\quad = (t - t_0)[F(z,w) - F(z,0)] + A(z), \ z \in D, \\
\mathrm{Re}[\overline{\lambda(z)}w(z)] - t_0[r(z,w) - r(z,0)] \\
\quad = (t - t_0)[r(z,w) - r(z,0)] + R(z) + h(z), \ z \in \Gamma, \\
\mathrm{Im}[\overline{\lambda(a_j)}w(a_j)] - t_0[b_j(w) - b_j(0)] \\
\quad = (t - t_0)[b_j(w) - b_j(0)] + B_j, \ j \in J.
\end{cases}
\tag{1.27}
$$

Replacing $w(z)$ on the right–hand sides of (1.27) by a function $w_0(z) \in C_\alpha(\bar{D})$, especially, by $w_0(z) = 0$, it is obvious that the boundary value problem (1.27) then has a unique solution $w_1(z) \in C_\alpha(\bar{D})$. Using successive iteration, we obtain a sequence of solutions $w_n(z) \in C_\alpha(\bar{D}), n = 1, 2, \cdots$, which satisfy

$$
\begin{cases}
w_{n+1\bar{z}} - t_0[F(z, w_{n+1}) - F(z, 0)] \\
\quad = (t - t_0)[F(z, w_n) - F(z, 0)] + A(z), \ z \in D, \\
\mathrm{Re}[\overline{\lambda(z)}w_{n+1}] - t_0[r(z, w_{n+1}) - r(z, 0)] \\
\quad = (t - t_0)[r(z, w_n) - r(z, 0)] + R(z) + h(z), \ z \in \Gamma, \\
\mathrm{Im}[\overline{\lambda(a_j)}w_{n+1}(a_j)] - t_0[b_j(w_{n+1}) - b_j(0)] \\
\quad = (t - t_0)[b_j(w_n) - b_j(0)] + B_j, \ j \in J.
\end{cases}
\tag{1.28}
$$

From the above formulae, it follows that

$$
\begin{cases}
[w_{n+1} - w_n]_{\bar{z}} - t_0[F(z, w_{n+1}) - F(z, w_n)] \\
\quad = (t - t_0)[F(z, w_n) - F(z, w_{n-1})], \ z \in D, \\
\mathrm{Re}[\overline{\lambda(z)}(w_{n+1} - w_n)] - t_0[r(z, w_{n+1}) - r(z, w_n)] \\
\quad = (t - t_0)[r(z, w_n) - r(z, w_{n-1})] + h(z), \ z \in \Gamma, \\
\mathrm{Im}[\overline{\lambda(a_j)}(w_{n+1}(a_j) - w_n(a_j))] - t_0[b_j(w_{n+1}) - b_j(w_n)] \\
\quad = (t - t_0)[b_j(w_n) - b_j(w_{n-1})], \ j \in J.
\end{cases}
\tag{1.29}
$$

Noting that

$$\begin{cases}
L_p[(t - t_0)(F(z, w_n) - F(z, w_{n-1})), \bar{D}] \leq |t - t_0| k_0 C_\alpha[w_n - w_{n-1}, \bar{D}], \\
C_\alpha[t_0(r(z, w_{n+1}) - r(z, w_n)) + (t - t_0)(r(z, w_n) - r(z, w_{n-1})), \Gamma] \\
\leq \varepsilon C_\alpha[w_{n+1} - w_n, \bar{D}] + |t - t_0|\varepsilon C_\alpha[w_n - w_{n-1}, \bar{D}], \\
C_\alpha[t_0(b_j(w_{n+1}) - b_j(w_n)) + (t - t_0)(b_j(w_n) - b_j(w_{n-1})), \Gamma] \\
\leq \varepsilon C_\alpha[w_{n+1} - w_n, \bar{D}] + |t - t_0|\varepsilon C_\alpha[w_n - w_{n-1}, \bar{D}],
\end{cases} \tag{1.30}$$

and applying Lemma 1.1, we have

$$\begin{aligned}
C_\alpha[w_{n+1} - w_n, \bar{D}] &\leq 2\varepsilon M_1 C_\alpha[w_{n+1} - w_n, \bar{D}] \\
&+ |t - t_0| M_1[k_0 + 2\varepsilon] C_\alpha[w_n - w_{n-1}, \bar{D}].
\end{aligned} \tag{1.31}$$

Assuming the constant ε small enough namely $2\varepsilon M_1 < 1$ and choosing $|t - t_0| \leq \delta = (1 - 2\varepsilon M_1)/[2M_1(k_0 + 2\varepsilon)]$, it follows that

$$\begin{aligned}
C_\alpha[w_{n+1} - w_n, \bar{D}] &\leq M_1(k_0 + 2\varepsilon)/(1 - 2\varepsilon M_1) \\
&\times |t - t_0| C_\alpha[w_n - w_{n-1}, \bar{D}] \leq C_\alpha[w_n - w_{n-1}, \bar{D}]/2,
\end{aligned} \tag{1.32}$$

and when $n, m \geq N_0 + 1$ (N_0 is a positive integer),

$$\begin{aligned}
C_\alpha[w_{n+1} - w_n, \bar{D}] &\leq 2^{-N_0} \sum_{j=0}^{\infty} 2^{-j} C_\alpha[w_1 - w_0, \bar{D}] \\
&\leq 2^{-N_0+1} C_\alpha[w_1 - w_0, \bar{D}].
\end{aligned} \tag{1.33}$$

Hence $\{w_n(z)\}$ is a Cauchy sequence. According to the completeness of the Banach space $C_\alpha(\bar{D})$, there exists a function $w_*(z) \in C_\alpha(\bar{D})$, so that $C_\alpha[w_n - w_*, \bar{D}] \to 0$ for $n \to \infty$. From (1.28), we can see that $w_*(z)$ is a solution of Problem B_1 for every $t \in T_\delta = \{|t - t_0| \leq \delta\}$. Because the constant δ is independent of $t_0 \, (0 \leq t_0 < 1)$, therefore from the solvability of Problem B_1 when $t_0 = 0$, we can derive the solvability of Problem B_1 when $t = \delta, 2\delta, \cdots, [1/\delta]\, \delta, 1$. In particular, when $t = 1$ and $A(z) = F(z, 0)$, $R(z) = r(z, 0)$, $B_j = b_j(0)$, $j \in J$, Problem B_1 is solvable. This completes the proof.

By using the method from the proof of Theorem 2.5 below, we can weaken condition (1.3) in Theorem 1.4 to: $F(z, w)$ is continuous in $w \in \mathbb{C}$ for almost every point $z \in D$.

1.4   Nonlinear Haseman problem for quasilinear elliptic complex equations

We discuss the quasilinear elliptic complex equation (1.1) in the z-plane $\mathbb{C}$ and suppose that (1.1) satisfies Condition C_*, i.e. $A_j(z, w) \in L_{p,2}(\mathbb{C})$ for any sectionally continuous function

$$w(z) = \begin{cases} w^+(z), \ z \in \overline{D^+}, \\ w^-(z), \ z \in \overline{D^-}, \end{cases} \tag{1.34}$$

where D^+ is the bounded domain with the given boundary $\Gamma \in C_\nu \, (0 < \nu < 1)$, $z = 1 \in \Gamma$, $z = 0 \in D^+$, and D^- is the complement of $\overline{D^+}$ with respect to $\mathbb{C}$, and

$$L_{p,2}[A_j(z, w^+, w^-), \mathbb{C}] \leq k_4, \; j = 1, 2, \; L_{p,2}[z^{-K} A_3(z, w^+, w^-), \mathbb{C}] \leq k_5, \qquad (1.35)$$

$$|F(z, w_1^+, w_1^-) - F(z, w_2^+, w_2^-)| \leq k_4[|w_1^+ - w_2^+| + |w_1^- - w_2^-|], \qquad (1.36)$$

where $p(> 2), k_4$ and k_5 are nonnegative constants, and the integer K is as stated in (1.39) below.

Problem H   The so–called Haseman boundary value problem for (1.1) is to find a sectionally continuous solution $w(z)$ of (1.1) in $\overline{D^\pm} \backslash \{\infty\}$ satisfying the nonlinear boundary condition

$$w^+[\alpha(z)] = G(z)w^-(z) + g(z, w^+, w^-), \; z \in \Gamma, \qquad (1.37)$$

where $\alpha(z)$ is a positive shift on Γ as stated in Section 3, Chapter 2 in [140] and $\alpha(z), G(z), g(z, w^+, w^-)$ satisfy

$$\begin{cases} C_\nu[G, \Gamma] \leq k_6, \; |G(z)| \geq k_6^{-1} > 0, \; C_\nu[g(z,0,0), \Gamma] \leq k_6, \\[2mm] C_\nu[g(z, w_1^+, w_1^-) - g(z, w_2^+, w_2^-), \Gamma] \leq \varepsilon[C_\nu(w_1^+ - w_2^-, \Gamma) + C_\nu(w_1^- - w_2^-, \Gamma)], \\[2mm] C_\mu^1[\alpha, \Gamma] \leq k_6, \; |\alpha'(z)| \geq k_6^{-1} > 0, \; 0 < \mu, \nu < 1, \end{cases}$$

$$(1.38)$$

for all $w_1^\pm(z), w_2^\pm(z) \in C_\nu(\Gamma), \mu, \nu, k_6, \varepsilon$ are positive constants.

$$K = \frac{1}{2\pi} \Delta_\Gamma \arg G(z) \qquad (1.39)$$

is called the index of Problem H. When $K \geq 0$, we require that the solution $w(z)$ is bounded in D^-, and when $K < 0$, we allow that $w(z)$ has a pole of order $|K| - 1$ at $z = \infty$. If $\alpha(z) = z, z \in \Gamma$, then Problem H is the Riemann boundary value problem.

Similarly to Theorem 3.6 and Theorem 3.7, Chapter 2 in [140], we can prove the following lemma.

Lemma 1.5   *Let the complex equation (1.1) satisfy Condition C_*.*

(1) *The function $w(z)$ in (1.34) is a solution of Problem H for (1.1) if and only if $w^*(\zeta) = w[z(\zeta)]$ is a solution of the following problem (Problem R)*

$$w_{\bar\zeta}^* = \overline{z'(\zeta)} F[z(\zeta), w^*] = F^*(\zeta, w^*) \text{ in } \Omega^\pm = \zeta(D^\pm), \qquad (1.40)$$

$$w^{*+}(\zeta) = G^*(\zeta)w^{*-}(\zeta) + g^*(\zeta, w^{*+}, w^{*-}) \text{ on } L = \zeta(\Gamma), \qquad (1.41)$$

in which

$$\zeta(z) = \begin{cases} \zeta^+(z), \; z \in D^+, \\ \zeta^-(z), \; z \in D^- \end{cases}$$

is the conformal glue function satisfying the boundary condition $\zeta^+[\alpha(z)] = \zeta^-(z)$, $z \in \Gamma$ and $\zeta(0) = 0$, $\zeta(1) = 1$, $\zeta(\infty) = \infty$, $z(\zeta)$ is the inverse function of $\zeta(z)$, $G^*(\zeta)$ and $g^*(\zeta)$ satisfy conditions on L similar to those for $G(z)$ and $g(z)$ on Γ.

(2) The function $w^*(z)$ is a solution of Problem R for (1.40) if and only if

$$W(\zeta) = w(\zeta)/X(\zeta) \tag{1.42}$$

is a solution of the following problem(Problem S)

$$W_{\bar{\zeta}} = F^*[\zeta, X(\zeta)W(\zeta)]/X(\zeta) = f(\zeta, W), \tag{1.43}$$

$$W^+(\zeta) - W^-(\zeta) = r[\zeta, W^+(\zeta), W^-(\zeta)], \ \zeta \in L, \ W(\infty) = 0, \tag{1.44}$$

where $X(\zeta)$ is an analytic function in $\Omega^\pm \backslash \{\infty\}$ satisfying the boundary condition

$$X^+(\zeta) = G^*(\zeta)X^-(\zeta), \ \zeta \in L \tag{1.45}$$

as stated in (3.32), Chapter 2 in [140], and

$$r(\zeta, W^+, W^-) = g^*(\zeta, X^+W^+, X^-W^-)/X^+(\zeta) \tag{1.46}$$

satisfies the condition

$$\left\{ \begin{array}{l} C_\nu[r(\zeta,0,0), L] \leq k_7, \ C_\nu[r(\zeta, W_1^+, W_1^-) - r(\zeta, W_2^+, W_2^-), L] \\[2mm] \leq \varepsilon k_7[C_\nu(W_1^+ - W_2^+, L) + C_\nu(W_1^- - W_2^-, L)], \end{array} \right. \tag{1.47}$$

where $\Omega^\pm = \zeta(D^\pm)$, $k_7 = k_7(\nu, k, K, D)$, $k = (k_4, k_5, k_6)$.

Next, we give the estimate of solutions of Problem S for the complex equation (1.43).

Theorem 1.6   *Under the same condition as in Lemma* 1.5 *and for a small enough constant* ε, *any solution* $W(\zeta)$ *of Problem* S *for* (1.43) *satisfies the estimate*

$$C_\nu[W, \overline{D^\pm}] \leq M_3 = M_3(p, \nu, k, K, D). \tag{1.48}$$

There is no harm in assuming $\nu \leq 1 - 2/p$.

Proof   Let the solution $W(\zeta)$ of Problem S be substituted into the complex equation (1.43) and the boundary condition (1.44). It is not difficult to see that the function $A_3^*(\zeta) = \overline{z'(\zeta)}A_3[z(\zeta), X(\zeta)W(\zeta)]/X(\zeta)$ satisfies

$$L_{p,2}[A_3^*, \overline{\Omega^\pm}] \leq k_8 = k_8(p, \nu, k, K, D).$$

According to the representation of the solution $W(\zeta)$ for (1.43)

$$W(\zeta) = \Phi(\zeta)e^{\phi(\zeta)} + \psi(\zeta), \tag{1.49}$$

in which $\phi(\zeta), \psi(\zeta)$ satisfy the estimates

$$C_\nu[\phi, \mathcal{C}] \le k_9, \ C_\nu[\psi, \mathcal{C}] \le k_9 = k_9(p, \nu, k, D),$$

the boundary condition (1.44) can be reduced to the boundary condition for the sectionally analytic function $\Phi(\zeta)$

$$\Phi^+(\zeta) - \Phi^-(\zeta) = R(\zeta), \ R(\zeta) = r(\zeta, \Phi^+ e^\phi + \psi, \Phi^- e^\phi + \psi)e^{-\phi}, \ \zeta \in L. \tag{1.50}$$

Because the function $\Phi(\zeta)$ can be expressed as the integral of Cauchy type

$$\Phi(\zeta) = \frac{1}{2\pi i} \int_L \frac{R(t)}{t - z} dt, \tag{1.51}$$

it can be derived that

$$C_\alpha[\Phi, \overline{\Omega^\pm}] \le k_{10} + \varepsilon k_{10} C_\alpha[\Phi, \overline{\Omega^\pm}]. \tag{1.52}$$

where $k_{10} = k_{10}(p, \nu, k, K, D)$. Choosing ε sufficiently small, we can obtain

$$C_\alpha[\Phi, \overline{\Omega^\pm}] \le M_4 = M_4(p, \nu, k, K, D). \tag{1.53}$$

From (1.49) and (1.53), it follows that (1.48) holds.

Theorem 1.7 *Under the same conditions as in Theorem 1.6, there exists a solution $W(\zeta)$ of Problem S for (1.43), and*

$$w(z) = X[\zeta(z)]W[\zeta(z)] \tag{1.54}$$

is a solution of Problem H for (1.1).

Proof We consider the boundary value problem (Problem S^*) with a parameter $t(0 \le t \le 1)$:

$$\begin{cases} W_{\bar{\zeta}} = tf(\zeta, W), \ \zeta \in \mathcal{C}, \ 0 \le t \le 1, \\ W^+(\zeta) - W^-(\zeta) = t\, r(\zeta, W^+, W^-), \ \zeta \in L, \ W^-(\infty) = 0. \end{cases} \tag{1.55}$$

It is clear that when $t = 0$, Problem S^* possesses a unique solution $W(\zeta) = 0$. If Problem S^* for $t = t_0 (0 \le t_0 < 1)$ has a unique solution $W_0(\zeta) \in C_\nu(\overline{\Omega^\pm})$, then we know Problem S^* in the form

$$\begin{cases} W_{\bar{\zeta}} - t_0 f(\zeta, W) = (t - t_0)f(\zeta, W_0), \ \zeta \in \mathcal{C}, \\ W^+(\zeta) - W^-(\zeta) - t_0 r(\zeta, W^+, W^-) = (t - t_0)r(\zeta, W_0^+, W_0^-), \ \zeta \in L, \ W^-(\infty) = 0 \end{cases} \tag{1.56}$$

has a unique solution $W_1(\zeta) \in C_\alpha(\overline{\Omega^\pm})$. Similarly to Theorem 1.4, by using successive iteration, we can obtain a sequence of solutions $W_n(\zeta) \in C_\nu(\overline{\Omega^\pm}), n = 1, 2, 3, \cdots$, and prove that Problem S^* for $t = 1$, i.e. Problem S has a solution $W(\zeta)$. Obviously, $w(z)$ in (1.54) is a solution of Problem H for (1.1). The contents in this section are continuations and generalizations of results in [11]1),2),3) and [10]1),3).

2　Nonlinear Riemann–Hilbert Problem for Nonlinear Elliptic Complex Equations in a Multiply Connected Domain

2.1　Formulation of nonlinear Riemann–Hilbert problem in a multiply connected domain

Let D be an $(N+1)$–connected circular domain with the boundary $\Gamma = \cup_{j=0}^{N}\Gamma_j$ as stated in Section 3, Chapter 1. We discuss the nonlinear elliptic complex equation

$$w_{\bar{z}} = F(z, w, w_z),\ F = Qw_z + A_1 w + A_2 \bar{w} + A_3, \tag{2.1}$$

where

$$Q = Q_1 + Q_2 \overline{w_z}/w_z,\ Q_j = Q_j(z, w, w_z),\ j = 1, 2,\ A_j = A_j(z, w),\ j = 1, 2, 3,$$

which satisfy Condition C similar to Condition C in Section 1, Chapter 1, in which the main conditions are as follows:

$$|F(z, w, U_1) - F(z, w, U_2)| \le q_0 |U_1 - U_2|, \tag{2.2}$$

for almost every point $z \in D$ and $w, U_1, U_2 \in \mathbb{C}$, and

$$L_p[A_j(z, w(z)), \bar{D}] \le k_0,\ j = 1, 2,\ L_p[A_3(z, w(z)), \bar{D}] \le k_1, \tag{2.3}$$

for all continuous functions $w(z)$ on $\bar{D}$, where $q_0(< 1), p(> 2), k_0, k_1$ are nonnegative constants.

Problem A　Find a solution $w(z) \in C_\alpha(\bar{D})$ of the complex equation (2.1), which satisfies the nonlinear Riemann–Hilbert boundary condition

$$\mathrm{Re}[\overline{\lambda(z)}w(z)] = r(z, w),\ z \in \Gamma, \tag{2.4}$$

in which $|\lambda(z)| = 1$, $\lambda(z)$, $r_0(z) = r(z, 0)$ and $r(z, w)$ satisfy

$$\begin{cases} C_\alpha[\overline{\lambda(z(\zeta))}, L] \le k_0,\ C_\alpha[r_0(z(\zeta)), L] \le k_2,\ L = \zeta(\Gamma), \\ C_\alpha[r(z(\zeta), w_1) - r(z(\zeta), w_2), L] \le \varepsilon C_\alpha[w_1 - w_2, L], \end{cases} \tag{2.5}$$

for all $w_1(z(\zeta)), w_2(z(\zeta)) \in C_\alpha(L)$, herein $\zeta(z)$ is the homeomorphic solution of the Beltrami equation

$$\zeta_{\bar{z}} = Q(z)\zeta_z,\ |Q(z)| \le q_0 < 1, \tag{2.6}$$

which maps D onto an $(N+1)$–connected circular domain G such that $\zeta(0) = 0$, $\zeta(1) = 1$, $z(\zeta)$ is the inverse function of $\zeta(z)$, $\alpha\,(0 < \alpha < 1)$, $k_2, \varepsilon\,(> 0)$ are nonnegative constants. Problem A with $r(z, w) = 0$ and $A_3(z, w) = 0$ will be called Problem A_0.

Problem B The corresponding modified Riemann–Hilbert problem for (2.1) is to find a solution $w(z) \in C_\alpha(\bar{D})$ satisfying

$$\mathrm{Re}[\overline{\lambda(z)}w(z)] = r(z,w) + h(z), \quad z \in \Gamma, \tag{2.7}$$

$$L_j w = \begin{cases} \mathrm{Im} \int_{\Gamma_j} \overline{\lambda(z)}w(z)ds = b_j(w), j = \begin{cases} 1, \cdots, N+1, K \geq N, \\ N-K+1, \cdots, N+1, 0 \leq K < N, \end{cases} \\ \mathrm{Re} \int_{\Gamma_{N+1}} z^{j-N-1}\overline{\lambda(z)}w(z)ds = b_j(w), j = N+2, \cdots, K+1, \\ \mathrm{Im} \int_{\Gamma_{N+1}} z^{j-K-1}\overline{\lambda(z)}w(z)ds = b_j(w), j = K+2, \cdots, 2K-N+1, \\ \qquad\qquad\qquad\qquad\qquad K \geq N, \end{cases} \tag{2.8}$$

where $K = \Delta_\Gamma \arg \lambda(z)/2\pi$ and

$$h(z) = \begin{cases} 0, z \in \Gamma, \ K \geq N, \\ \left.\begin{array}{l} h_j, z \in \Gamma_j, j = 1, \cdots, N-K \\ 0, z \in \Gamma_j, j = N-K+1, \cdots, N+1 \end{array}\right\} 0 \leq K < N, \\ \left.\begin{array}{l} h_j, z \in \Gamma_j, j = 1, \cdots, N, \\ h_0 + \mathrm{Re} \sum_{m=1}^{-K-1}(h_m^+ + ih_m^-)z^m, z \in \Gamma_0 \end{array}\right\} K < 0, \end{cases} \tag{2.9}$$

in which $h_j \ (j = 0, 1, \cdots, N)$, $h_m^{\pm} \ (m = 1, \cdots, -K-1, K < 0)$ are undetermined real constants, K is the index of Problem A and Problem B, $a_j \in \Gamma_j \ (j = 1, \cdots, N)$, $a_j \in \Gamma_0 \ (j = N+1, \cdots, 2K-N+1)$ are distinct points, $b_j(w)$ satisfies

$$|b_j(0)| \leq k_3, \ |b_j(w_1) - b_j(w_2)| \leq \varepsilon|w_1 - w_2|,$$

$$j \in J = \begin{cases} 1, \cdots, 2K-N+1, \ K \geq N, \\ N-K+1, \cdots, N+1, \ 0 \leq K < N \end{cases} \tag{2.10}$$

for $w_1, w_2 \in \mathbb{C}$, where k_3 is a nonnegative constant. Problem B with $r(z,w) = 0$ and $A_3(z,w) = 0$ will be called Problem B_0.

In order to obtain the uniqueness for solutions of Problem B for (2.1), we add a condition: for any function $w_1(z), w_2(z) \in C_\alpha(\bar{D})$ and any measurable function $U(z) \in L_{p_0}(\tilde{D})$, there is

$$F(z, w_1, U) - F(z, w_2, U) = A(z, w_1, w_2, U)(w_1 - w_2), \ A \in L_p(\bar{D}), p > 2, \tag{2.11}$$

where $p_0 \ (0 < p_0 < p)$ is a constant and $\tilde{D}$ is any closed subset in D.

By using a priori estimate of solutions for Problem B, the continuity method and the Schauder fixed–point theorem, we shall prove the existence and uniqueness of solutions for Problem B and then derive the result of solvability for Problem A.

2.2 A priori estimate of solutions for Problem B

Lemma 2.1 *Denote by Problem B^* the linear boundary value problem B for the nonlinear complex equation (2.1) with Condition C, i.e. $r(z,w) = r(z,0)$, $b_j(w) = b_j(0)(j \in J)$ in (2.7) and (2.8). Then any solution $w(z)$ of Problem B^* satisfies the estimate*

$$C_{\alpha_0}[w, \bar{G}] \leq M_1(k_1 + k_2 + k_3), \tag{2.12}$$

where $M_1 = M_1(q_0, p_0, \alpha, k_0, K, D)$ is a nonnegative constant, and $p_0\,(2 < p_0 < p)$, $\alpha_0 = \alpha\beta\,(\beta = 1 - 2/p_0)$ are nonnegative constants.

Proof Let $w(z)$ be a solution of Problem B^*. In a similar way as in the proof of Lemma 1.1, when $k = k_1 + k_2 + k_3 > 0$, we denote $W(z) = w(z)/k$. It is clear that $W(z)$ is a solution of the boundary value problem

$$\begin{cases} W_{\bar{z}} = F(z, w, w_z)/k, \ F/k = QW_z + A_1 W + A_2 \overline{W} + A_3/k, \\[2mm] \mathrm{Re}[\overline{\lambda(z)}W(z)] = [r(z,0) + h(z)]/k, \\[2mm] L_j W = b_j(0)/k, \ j \in J. \end{cases} \tag{2.13}$$

Noting that

$$L_p[A_3/k, \bar{D}] \leq 1, \ C_\alpha[r(z,0)/k, \Gamma] \leq 1, \ |b_j(0)/k| \leq 1, \ j \in J, \tag{2.14}$$

and following the proof of Theorem 4.3, Chapter 2 in [140], we can prove

$$C_{\alpha_0}[W, \bar{D}] \leq M_1. \tag{2.15}$$

Consequently, (2.12) is true. When $k = k_1 + k_2 + k_3 = 0$, on the basis of Theorem 4.1, Chapter 2 in [140], it is seen that $w(z) = 0$ in D. Thus (2.12) holds.

Theorem 2.2 *Let the complex equation (2.1) satisfy Condition C and the constant ε in (2.5) and (2.10) be small enough. Then any solution $w(z) \in C_{\alpha_0}(\bar{D})$ of Problem B for (2.1) satisfies the estimate*

$$C_{\alpha_0}[w, \bar{D}] \leq M_2 = M_2(q_0, p_0, \alpha, k, K, D), \tag{2.16}$$

where $k = (k_0, k_1, k_2, k_3)$, $\alpha_0 = \alpha\beta$, $\beta = 1 - 2/p_0$.

Proof According to Theorem 2.4, Chapter 2 in [140], the solution $w(z)$ can be expressed as

$$w(z) = \Phi[\zeta(z)]e^{\phi(z)} + \psi(z), \tag{2.17}$$

where $\phi(z), \psi(z), \zeta(z)$ and its inverse function $z(\zeta)$ satisfy

$$C_\beta[\phi, \bar{D}] \leq k_4, \ C_\beta[\psi, \bar{D}] \leq k_4, \ C_\beta[\zeta, \bar{D}] \leq k_4, \ C_\beta[z(\zeta), \bar{G}] \leq k_4, \tag{2.18}$$

in which $G = \zeta(D)$, $k_4 = k_4(q_0, p_0, D)$, and the analytic function $\Phi(\zeta)$ satisfies the boundary and integral conditions

$$\mathrm{Re}[\overline{\lambda(z(\zeta))}e^{\phi(z(\zeta))}\Phi(\zeta)] = R(\zeta, \Phi), \ \zeta \in L = \zeta(\Gamma),$$

$$R(\zeta, \Phi) = r[z(\zeta), \Phi(\zeta)e^{\phi(z(\zeta))} + \psi(z(\zeta))] - \mathrm{Re}[\overline{\lambda(z(\zeta))}\Psi(z(\zeta))], \qquad (2.19)$$

$$L_j\{\Phi[\zeta(z)]e^{\phi(z)}\} = b_j\{\Phi[\zeta(z)]e^{\phi(z)} + \psi(z)\} - L_j[\psi(z)] = B_j(\Phi), \ j \in J.$$

By (2.5) and (2.10), $R(\zeta, \Phi)$ and B_j satisfy

$$C_\alpha[\overline{\lambda(z(\zeta))}e^{\phi(z(\zeta))}, L] \leq k_5, \ C_\alpha[R(\zeta, 0), L] \leq k_5,$$

$$C_\alpha[R(\zeta, \Phi_1) - R(\zeta, \Phi_2), L] \leq \varepsilon k_5 C_\alpha[\Phi_1 - \Phi_2, L], \qquad (2.20)$$

$$|B_j(0)| \leq k_5, \ |B_j(\Phi_1) - B_j(\Phi_2)| \leq \varepsilon k_5 |\Phi_1 - \Phi_2|, j \in J,$$

where $k_5 = k_5(q_0, p_0, \alpha, k, D)$. In accordance with Lemma 2.1, the estimate

$$C_\alpha[\Phi, \bar{D}] \leq M_3 = M_3(q_0, p_0, \alpha, k, K, G) \qquad (2.21)$$

can be obtained. From (2.17) and (2.21), the estimate (2.16) is derived.

2.3 Uniqueness and existence of solutions of Problem B

Theorem 2.3 *Suppose that the same hypotheses as in Theorem 2.2 are satisfied and (2.11) holds. Then the solution of Problem B is unique.*

Proof Let $w_1(z), w_2(z)$ be two solutions of Problem B for (2.1). It is evident that $w(z) = w_1(z) - w_2(z)$ is a solution of the boundary value problem

$$\begin{cases} w_{\bar{z}} = Qw_z + Aw, \ z \in D, \\[2mm] \mathrm{Re}[\overline{\lambda(z)}w(z)] = r(z, w_1) - r(z, w_2) + h(z), \ z \in \Gamma, \\[2mm] L_j w = b_j(w_1) - b_j(w_2), \ j \in J, \end{cases} \qquad (2.22)$$

where

$$Q = \begin{cases} [F(z, w_1, w_{1z}) - F(z, w_1, w_{2z})]/w_z, \ \text{for } w_z \neq 0, \\[2mm] 0, \ \text{for } w_z = 0, \ z \in D, \end{cases}$$

$$A = \begin{cases} [F(z, w_1, w_{2z}) - F(z, w_2, w_{2z})]/w, \ \text{for } w \neq 0, \\[2mm] 0, \ \text{for } w(z) = 0, \ z \in D. \end{cases}$$

Afterwards, similarly to the proof of Theorem 1.3, from (2.16) we can get $w(z) \equiv 0$, namely $w_1(z) \equiv w_2(z)$ on $\bar{D}$.

To prove the solvability of Problem B, we first prove that Problem B for analytic functions has a unique solution.

Theorem 2.4 *Let the constant ε in (2.5) and (2.10) be sufficiently small. Then Problem B for analytic functions possesses a solution and its solution is unique.*

Proof The uniqueness of solutions of Problem B for analytic functions is known by Theorem 2.3. In order to apply the continuity method to prove the solvability of Problem B, we discuss the following boundary condition and integral conditions with a parameter $t \in [0,1]$:

$$\begin{cases} \operatorname{Re}[\overline{\lambda(z)}\Phi(z)] = t\,r(z,\Phi) + R(z) + h(z),\ z \in \Gamma, \\ L_j\Phi = t\,b_j(\Phi) + B_j,\ j \in J, \end{cases} \qquad (2.23)$$

where $R(z) \in C_{\alpha_0}(\Gamma)$, $B_j(j \in J)$ are real constants. The boundary value problem for analytic functions is denoted by Problem B'. Let T be a point set in $0 \le t \le 1$ such that for every $t \in T$, Problem B' is solvable, i.e. Problem B' for analytic functions has a unique solution. It is obvious that when $t = 0$, Problem B' has a solution $\Phi(z)$. Hence T is not empty. If we can prove that T is both open and closed in $0 \le t \le 1$, then we can derive that when $t = 1$, Problem B' is solvable. In the special case $R(z) = 0$, $B_j = 0\,(j \in J)$, Problem B' has also a solution. Consequently this theorem holds.

From Theorem 2.2, we can verify that T is closed. To prove that T is an open set in $0 \le t \le 1$, we choose an arbitrary $t_0 \in T$. Let t_0 replace t in (2.23), then Problem B' with any function $R(z) \in C_{\alpha_0}(\Gamma)$ and any real constants $B_j\ (j \in J)$ is solvable. Next, we shall find a neighborhood $T_\delta = \{|t - t_0| < \delta, 0 \le t \le 1, \delta > 0\}$, so that Problem B' for every $t \in T_\delta$ is solvable. (2.23) can be rewritten in the form

$$\begin{cases} \operatorname{Re}[\overline{\lambda(z)}\Phi(z)] - t_0 r(z,\Phi) = (t - t_0)r(z,\Phi) + R(z) + h(z),\ z \in \Gamma, \\ L_j\Phi - t_0 b_j(\Phi) = (t - t_0)b_j(\Phi) + B_j,\ j \in J. \end{cases} \qquad (2.24)$$

Since $t_0 \in T$, Problem B' with any function $\Phi(z)$, $R(z) \in C_{\alpha_0}(\Gamma)$ and any constants $B_j\,(j \in J)$ on the right–hand side of (2.24) is solvable. In fact, choosing a function $\Phi_0(z) \in C_{\alpha_0}(\Gamma)$, especially $\Phi_0(z) = 0$ and substituting $\Phi_0(z) \equiv 0$ into the position of Φ of the right–hand side of (2.24), it is evident that

$$(t - t_0)r(z,\Phi_0) + R(z) \in C_{\alpha_0}(\Gamma),\ (t - t_0)b_j(\Phi_0) + B_j \in \mathbb{R}\,(j \in J).$$

Hence Problem B' has a unique solution $\Phi_1(z)$. Thus by using successive iteration, we obtain a sequence of solution $\Phi_n(z), n = 1, 2, \cdots$, which satisfy

$$\begin{cases} \operatorname{Re}[\overline{\lambda(z)}\Phi_{n+1}] - t_0 r(z,\Phi_{n+1}) = (t - t_0)r(z,\Phi_n) + R(z) + h(z),\ z \in \Gamma, \\ L_j\Phi_{n+1} - t_0 b_j(\Phi_{n+1}) = (t - t_0)b_j(\Phi_n) + B_j,\ j \in J. \end{cases} \qquad (2.25)$$

It follows from the above formulae

$$\begin{cases} \mathrm{Re}[\overline{\lambda(z)}(\Phi_{n+1} - \Phi_n)] - t_0[r(z, \Phi_{n+1}) - r(z, \Phi_n)] \\ = (t - t_0)[r(z, \Phi_n) - r(z, \Phi_{n-1})] + h(z),\ z \in \Gamma, \\ L_j(\Phi_{n+1} - \Phi_n) - t_0[b_j(\Phi_{n+1}) - b_j(\Phi_n)] \\ = (t - t_0)[b_j(\Phi_n) - b_j(\Phi_{n-1})],\ j \in J. \end{cases} \quad (2.26)$$

On the basis of Lemma 2.1, it follows that $\Phi_{n+1}(z) - \Phi_n(z)$ satisfies the estimate

$$C_{\alpha_0}[\Phi_{n+1} - \Phi_n, \bar{D}] \le 2M_1[\varepsilon C_{\alpha_0}(\Phi_{n+1} - \Phi_n, \bar{D}) + |t - t_0|\varepsilon C_{\alpha_0}(\Phi_n - \Phi_{n-1}, \Gamma)]. \quad (2.27)$$

Selecting ε appropriately small such that $2M_1\varepsilon < 1$, and $\delta = (1 - 2\varepsilon M_1)/4(M_1 + 1)\varepsilon$, we get

$$C_{\alpha_0}[\Phi_{n+1} - \Phi_n, \bar{D}] \le \frac{2M_1|t - t_0|\varepsilon}{1 - 2\varepsilon M_1}C_{\alpha_0}[\Phi_n - \Phi_{n-1}, \bar{D}] \le \frac{1}{2}C_{\alpha_0}[\Phi_n - \Phi_{n-1}, \bar{D}], \quad (2.28)$$

for $t \in T_\delta = \{|t - t_0| < \delta, 0 \le t \le 1\}$. Consequently, $C_{\alpha_0}[\Phi_n - \Phi_m, \bar{D}] \to 0$ when $n, m \to \infty$, thus there exists a function $\Phi^*(z) \in C_{\alpha_0}(\bar{D})$ such that $C_{\alpha_0}[\Phi_n - \Phi^*, \bar{D}] \to 0$ when $n \to \infty$. Therefore $\Phi^*(z)$ is a solution of Problem B' for any $t \in T_\delta$. It shows that T is an open set in $0 \le t \le 1$.

Theorem 2.5 *Under the same assumptions as in Theorem 2.2, Problem B for the complex equation (2.1) has a solution $w(z) \in C_{\alpha_0}(\bar{D})$, where $\alpha_0 = \alpha\beta, \beta = 1 - 2/p_0$.*

Proof We introduce a bounded and closed convex set B in the Banach space $L_{p_0}(\bar{D}) \times L_{p_0}(\bar{D}) \times L_{p_0}(\bar{D}), 2 < p_0 < p$, the elements of which are all systems of measurable functions $\omega = [q(z), f(z), g(z)]$ satisfying the following condition

$$|q(z)| \le q_0 < 1,\ L_{p_0}[f, \bar{D}] \le q_1,\ L_{p_0}[g, \bar{D}] \le q_1, \quad (2.29)$$

where q_0 is the constant in (2.2) and q_1 is a constant similar to the constant k_4 stated in (1.15) and (1.16), Chapter 1. We select an arbitrary $\omega = [q(z), f(z), g(z)] \in B$ and discuss five integral equations as stated in (3.40)–(3.44) of Chapter 1, in which $\Phi(\zeta)$ is an analytic function in $G = \zeta(D)$ satisfying the following boundary conditions

$$\begin{cases} \mathrm{Re}\{\overline{\lambda(z(\zeta))}[\Phi(\zeta)e^{\phi(z(\zeta))} + \psi(z(\zeta))]\} \\ = r[z(\zeta), \Phi(\zeta)e^{\phi(z(\zeta))} + \psi(z(\zeta))],\ \zeta \in L = \zeta(\Gamma), \\ L_j[\Phi(\zeta(z))e^{\phi(z)} + \psi(z)] = b_j[\Phi(\zeta)e^{\phi(z(\zeta))} + \psi(z)],\ j \in J, \end{cases} \quad (2.30)$$

where $z(\zeta)$ is the inverse function of $\zeta(z)$. Noting condition (2.5), the functions $\lambda(z(\zeta)), R_0(\zeta) = r[z(\zeta), \psi(z(\zeta))] - \mathrm{Re}[\overline{\lambda(z(\zeta))}\psi(z(\zeta))]$ satisfy

$$C_{\alpha_0}[\lambda(z(\zeta)), L] \le k_6,\ C_{\alpha_0}[R_0(\zeta), L] \le k_6. \quad (2.31)$$

Moreover, we may require that the function $R(\zeta, \Phi) = r[z(\zeta), \Phi(\zeta)e^{\phi(z(\zeta))} + \psi(z(\zeta))] - \mathrm{Re}[\overline{\lambda(z(\zeta))}\psi(z(\zeta))]$ satisfies

$$C_\alpha[R(\zeta, \Phi_1) - R(\zeta, \Phi_2), L] \leq \varepsilon_1 C_{\alpha_0}[\Phi_1 - \Phi_2, L], \qquad (2.32)$$

in which $k_6 = k_6(\alpha, k, q_0, p_0, K, G)$, $\varepsilon_1 = \varepsilon_1(\alpha, k, q_0, p_0, K, G)$, $\beta = 1 - 2/p_0$, $W(z) = \Phi[\Psi(\chi)]e^{\phi(z)}$, $S'(\chi) = [\Phi(\Psi(\chi))]'_\chi$.

By the principle of contracting mappings or the continuity method, we can seek a unique system of solutions $[h(z), f^*(z), g^*(z), h^*(z), q^*(z)]$ of the integral equations (3.40)-(3.44) of Chapter 1. Denote by $\omega^* = S_B(\omega)$ the mapping from $\omega = [q(z), f(z), g(z)]$ onto $\omega^* = [q^*(z), f^*(z), g^*(z)]$. Afterwards, by using a similar method as in the proof of Theorem 3.5 in Chapter 1, we see that there exists a system of functions $\omega = [q(z), f(z), g(z)] \in B$, so that $\omega = S_B(\omega)$, and can construct a function $w(z) = \Phi[\zeta(z)]e^{\phi(z)} + \psi(z)$, which is just a solution of Problem B for (2.1).

We mention that the condition (2.11) is not used in the proof of Theorem 2.5.

By using the same method as in the proof of Theorem 1.6 in Chapter 1, we can obtain the next result.

Theorem 2.6   *Under the same conditions as in Theorem 2.2, a solvability result for Problem A is as follows:*

(1) If the index $K \geq N$, Problem A is solvable.

(2) If $0 \leq K < N$, the total number of solvability conditions for Problem A does not exceed $N - K$.

(3) If $K < 0$, Problem A has $N - 2K - 1$ solvability conditions, see [147]3).

In the following, we sometimes will denote $M_j = M_j(q_0, p_0, \alpha, k, K, G)$ by $M_j = M_j(q_0, p_0, \alpha, k, G)$ for simplicity.

3   Nonlinear Discontinuous Boundary Value Problem for Nonlinear Elliptic Complex Equations of First Order

First of all, we propose the nonlinear discontinuous boundary value problem.

We discuss still the nonlinear elliptic complex equation (2.1) in the $(N{+}1)$–connected circular domain D. The boundary condition of nonlinear discontinuous boundary value problem is

$$\mathrm{Re}[\overline{\lambda(z)}w(z)] = r(z, w) + s(z), \ z \in \Gamma^*, \qquad (3.1)$$

where $|\lambda(z)| = 1, r(z, w) = r_0(z)r_1(z, w), s(z) = s_0(z)\Pi_{j=1}^n|z - c_j|^{-\beta_j}, 0 \leq \beta_j/\beta^2 < 1, \beta = 1 - 2/p_0$, and $\lambda(z), r_0(z), s_0(z)$ satisfy the conditions

$$C_\alpha[\lambda, \Gamma] \leq k_0, \ C_\alpha[r_0, \overline{\Gamma^j}] \leq k_0, \ C_\beta[s_0, \Gamma] \leq k_0, \qquad (3.2)$$

in which $\alpha\,(0 < \alpha < 1), k_0, \beta_j$ are nonnegative constants and $\Gamma^j (j = 1, \cdots, m)$ are arcs on Γ such that $\cup_{j=1}^m \Gamma^j = \Gamma\backslash\{c_1, \cdots, c_n\} = \Gamma^*$. Moreover, we assume that $\tilde{r}_1[z, \tilde{w}(z)] = \tilde{\Pi}(z)r_1(z, w), \tilde{w}(z) = \tilde{\Pi}(z)w(z), \tilde{\Pi}(z) = \Pi(z)^{\beta^2}$ satisfy

$$
\begin{cases}
C_{\tau_0}[\tilde{r}_1(z(\zeta), \tilde{w}(z(\zeta))), L] \le k_0 + \varepsilon C_{\tau_0}[\tilde{w}(z(\zeta)), L], \\[2mm]
C_{\tau_0}[\tilde{r}_1(z(\zeta), \tilde{w}_1(z(\zeta))) - \tilde{r}_1(z(\zeta), \tilde{w}_2(z(\zeta))), L] \\[2mm]
\quad \le \varepsilon C_{\tau_0}[\tilde{w}_1(z(\zeta)) - \tilde{w}_2(z(\zeta)), L], \, 0 < \tau_0 < \alpha\beta,
\end{cases}
\tag{3.3}
$$

where $\zeta(z)$ is a homeomorphic solution of the Beltrami equation (2.6), and $z(\zeta)$ is the inverse function of $\zeta(z)$, $L = \zeta(\Gamma)$, ε is a positive constant. We denote by Problem $\tilde{A}$ the boundary value problem (2.1), (3.1), in which the solution $w(z)$ is required to satisfies $w(z)\Pi(z)^{-\beta^2} \in C_{\beta\tau_0}(\bar{G})$, where $\Pi(z)$ is as stated in (3.12) below. The constant

$$
K = \frac{1}{2}(K_1 + \cdots + K_m)
\tag{3.4}
$$

is called the index of Problem $\tilde{A}$, where K_j is stated in (3.5) of Chapter 1. There is no harm in assuming that $2K_j\,(j = 1, \cdots, N_0, N_0 \le N)$ are even numbers and $2K_j\,(j = N_0 + 1, \cdots, N)$ are odd numbers.

We need the corresponding modified boundary value problem $\tilde{B}$ for analytic functions. Taking

$$
W(z) = \frac{w(z)}{X(z)}, \text{ i.e. } w(z) = X(z)W(z),
\tag{3.5}
$$

where

$$
X(z) = \prod_{j=0}^{n_0}(z - c_j)^{\phi_j} \prod_{j=n_0+1}^{n_1}\left(\frac{z - c_j}{z - z_j}\right)^{\phi_j} \cdots \prod_{j=n_{N-1}+1}^{m}\left(\frac{z - c_j}{z - z_N}\right)^{\phi_j},
$$

herein $c_j \in \Gamma_0, j = 1, \cdots, n_0$, $c_{n_k+1} \in \Gamma_{k+1}, k = 0, 1, \cdots, N-1$, the boundary condition (3.1) can be reduced to

$$
\text{Re}[\overline{\lambda_1(z)}W(z)] = \{r[z, X(z)W(z)] + s(z)\}/|X(z)|, \, z \in \Gamma,
\tag{3.6}
$$

in which $\lambda_1(z) = \overline{\lambda(z)}X(z)/|X(z)|$, and $\lambda_1(c_j - 0)/\lambda_1(c_j + 0) = e^{K_j\pi i} = \pm 1$, $j = 1, \cdots, m$. It is easy to see that $C_\beta[\lambda_1, \Gamma^j] \le k_1 = k_1(\alpha, k_0, \Gamma^j)$, and the index of $\lambda_1(z)$ is K. The boundary value problem for analytic functions with the boundary condition (3.6) is called Problem $\hat{A}$. Because Problem $\hat{A}$ and Problem $\tilde{A}$ may not be solvable, we introduce the corresponding modified boundary value problem $\tilde{B}$ and problem $\hat{B}$ for analytic functions with the boundary conditions

$$
\text{Re}[\overline{\lambda(z)}w(z)] = r[z, w(z)] + s(z) + h(z), \, z \in \Gamma,
\tag{3.7}
$$

$$
\text{Re}[\overline{\lambda_1(z)}W(z)] = \{r[z, X(z)W(z)] + s(z)\}/|X(z)| + h(z), \, z \in \Gamma,
\tag{3.8}
$$

respectively, where

$$h(z) = \begin{cases} 0, \ z \in \Gamma, \ \text{if } K \geq N - 1/2, \\ \left. \begin{array}{l} h_j, \ z \in \Gamma_j, 1 \leq j \leq N - K', \\ 0, \ z \in \Gamma_j \ N - K' < j \leq N + 1 \end{array} \right\} \ \text{if } 0 \leq K \leq N - 1, \\ h_j, \ z \in \Gamma_j, \ 0 \leq j \leq N, \ \text{if } K < 0, \end{cases} \qquad (3.9)$$

in which $K' = [K + 1/2] - 1$, $h_j (j = 0, 1, \cdots, N)$ are undetermined real constants, and when $2K$ is an odd integer, $h_{N+1} = h_0 = 0$. If $K < 0$, we allow the solution $W(z)$ to have a pole of order $\leq K' - 1 = [-K + 1/2]$ at $z = 0 \in G$. If $K \geq 0$, we may require that the solution $W(z)$ satisfies the following point condition

$$\mathrm{Im}[\overline{\lambda(a_j)}w(a_j)] = |X(a_j)| \, b_j, \ j \in J, \qquad (3.10)$$

$$\mathrm{Im}[\overline{\lambda_1(a_j)}W(a_j)] = b_j, \ j \in J, \qquad (3.11)$$

where

$$J = \begin{cases} 1, \cdots, 2K - N + 1, \ \text{if } K > N - 1, \\ N_0 + 1 \cdots, N_0 + [K] + 1, \ 0 \leq K' \leq N - N_0, \\ N - K' + 1, \cdots, N - K' + [K] + 1, N - N_0 < K' \leq N - 1, \\ \qquad \qquad \text{if } 0 \leq K \leq N - 1, \end{cases}$$

in which $a_j (\neq c_k) \in \Gamma_j, j = 1, \cdots, N_0$, $a_j (\neq c_k) \in \Gamma_0, j = N_0 + 1, \cdots, 2K - N + 1$, and $b_j \ (j \in J)$ are real constants with the conditions $|b_j| \leq k_0, j \in J$. We can require that the solution $w(z)$ satisfies the condition

$$w(z) = O(|z - c_j|^{-\beta_j - \tau}), \ \Pi(z)w_*(z) \in C_{\tau_0}(\bar{G}), \qquad (3.12)$$

in which

$$w_*(z) = \begin{cases} w(z), & \text{for } K \leq 0, \\ w(z)z^{K'-1}, & \text{for } K < 0, \end{cases}$$

herein $0 < \tau_0 \leq \beta \min(\alpha, \tau)$, $\sigma_j = \max(\beta_j, |\phi_j|)$, $j = 1, \cdots, m$, $\Pi(z) = \Pi_{j=1}^n |z - c_j|^{\sigma_j + \tau}$, τ is an arbitrary small positive constant.

Theorem 3.1 *Let $w(z)$ be a solution of Problem $\tilde{B}$ for analytic functions and the constant ε in (3.3) be sufficiently small. Then $w(z)$ satisfies the estimates*

$$C_*(w_*) = C[w_*, \bar{D}_m] + C[w_*'(z), \bar{D}_m] \leq M_1, \qquad (3.13)$$

$$C^*(w_*) = C_{\tau_0}[\Pi w_*, \bar{D}] \leq M_2, \ 0 < \tau_0 \leq \min(\alpha, \tau), \qquad (3.14)$$

where $w_*(z)$ is as stated in (3.12), $\bar{D}_m$ is a set of points z in D such that $|z| < 1/m \, (m > 1)$ and $|z - c_j| > 1/m \, (j = 1, \cdots, n)$, $\tau_0, \ \tau$ are positive constants, $M_1 = M_1(\lambda, r, s, D, D_m)$, $M_2 = M_2(\lambda, r, s, D, \Pi, \tau_0)$.

Proof Let the solution $w(z)$ of Problem $\tilde{B}$ for analytic functions be substituted into the boundary condition (3.7) and denote $Y(z) = \{r(z, w(z)) + s(z)\}/|X(z)|$, $z \in \Gamma$. From (3.2),(3.3) and (3.12), we know that $Y(z)$ satisfies

$$\Pi(z)|X(z)|Y(z) \in C_{\tau_0}(\Gamma), \ 0 < \tau_0 < \tau. \tag{3.15}$$

Following the method of the proof of Theorem 3.3, Chapter 1 in [140], $W(z) = w(z)/X(z)$ can be expressed as

$$W(z) = w(z)/X(z) = \frac{1}{2\pi} \int_{\Gamma} T(z, t) Y(t) d\theta + W_0(z), \tag{3.16}$$

where $T(z, t)$ is the Schwarz kernel for Problem $\hat{B}$. If the constant ε in (3.3) is small enough, then the function

$$W_*(z) = \begin{cases} W(z), & \text{for } K \geq 0, \\ W(z) z^{K'-1}, & \text{for } K < 0 \end{cases}$$

satisfies the estimate

$$C_*(W_*) = C[W_*, D_m] + C[W_*', D_m] \leq M_3 = M_3(\lambda, r, s, D, D_m). \tag{3.17}$$

Using the results and the method in [139]17) and [140], we can obtain

$$C_{\tau_0}[\Pi X W_*, \bar{D}] \leq M_4 = M_4(\lambda, r, s, D, \Pi, \tau_0). \tag{3.18}$$

From (3.17) and (3.18), it follows that (3.13) and (3.14) hold.

Theorem 3.2 *If the constant ε in (3.3) is small enough, then Problem $\tilde{B}$ for analytic functions has a solution.*

Proof To apply the method of parameter extension, we consider the boundary condition and point conditions

$$\text{Re}[\overline{\lambda(z)} w(z)] - t[r(z, w(z)) + s(z)] = g(z) + |X(z)| h(z), \ z \in \Gamma \tag{3.19}$$

$$\text{Im}[\overline{\lambda(a_j)} w(a_j)] = |X(a_j)| \, b_j, \ j \in J, \tag{3.20}$$

where $g(z)$ satisfies the condition $\Pi(z) g(z) \in C_{\tau_0}(\Gamma)$, $0 < \tau_0 < \tau$, for any positive number τ. The above boundary value problem for analytic functions is still called Problem $\tilde{B}$.

We first discuss the case of $K > -1$. It is evident that Problem $\tilde{B}$ for $t = 0$ is solvable. Moreover, Problem $\tilde{B}$ has a unique solution

$$w(z) = X(z)[\frac{1}{2\pi} \int_{\Gamma} T(z, t) \frac{g(t)}{|X(z)|} d\theta + W_0(z)].$$

If Problem $\tilde{B}$ for $t = t_0 \, (0 \le t_0 < 1)$ is solvable, we can prove that there exists $T_\delta = \{|t - t_0| \le \delta, 0 \le t \le 1, \delta > 0\}$ independent of t_0 such that Problem $\tilde{B}$ for every $t \in T_\delta$ is solvable. In fact, the boundary condition (3.19) can be rewritten in the form

$$\mathrm{Re}[\overline{\lambda(z)}w(z)] = t_0[r(z, w(z)) + s(z)]$$
$$= (t - t_0)[r(z, w(z)) + s(z)] + g(z) + |X(z)|h(z), \ z \in \Gamma. \tag{3.21}$$

Replacing $w(z)$ by $w_0(z) = 0$ on the right–hand side of (3.21) and observing that $\Pi(z)\{(t - t_0)[r(z, 0) + s(z)] + g(z)\} \in C_{\tau_0}(\Gamma), \ 0 < \tau_0 < \tau$, we see that there exists an analytic function $W_1(z)$, which satisfies the boundary and point conditions

$$\begin{cases} \mathrm{Re}[\overline{\lambda(z)}w_1(z)] - t_0[r(z, w_1(z)) + s(z)] \\[2mm] = (t - t_0)[r(z, 0) + s(z)] + g(z) + |X(z)|h(z), \ z \in \Gamma, \\[2mm] \mathrm{Im}[\overline{\lambda(a_j)}w_1(a_j)] = |X(a_j)| \, b_j, \ j \in J. \end{cases} \tag{3.22}$$

According to Theorem 3.1, $w_1(z)$ satisfies an estimate similar to (3.13) and (3.14), where the constants M_1, M_2 are replaced by other ones. By using successive iteration, we obtain a sequence of solutions $w_n(z), n = 1, 2, \cdots$, which satisfy

$$\begin{cases} \mathrm{Re}[\overline{\lambda(z)}w_{n+1}(z)] - t_0[r(z, w_{n+1}(z)) + s(z)] \\[2mm] = (t - t_0)[r(z, w_n(z)) + s(z)] + g(z) + |X(z)|h(z), \ z \in \Gamma, \\[2mm] \mathrm{Im}[\overline{\lambda(a_j)}w_{n+1}(a_j)] = |X(a_j| \, b_j, \ j \in J. \end{cases} \tag{3.23}$$

From the above formulae

$$\mathrm{Re}[\overline{\lambda(z)}(w_{n+1} - w_n)] - t_0[r(z, w_{n+1}) - r(z, w_n)]$$
$$= (t - t_0)[r(z, w_n) - r(z, w_{n-1})] + |X(z)|h(z), \ z \in \Gamma, \tag{3.24}$$

$$\mathrm{Im}[\overline{\lambda(a_j)}(w_{n+1}(a_j) - w_n(a_j))] = 0, \ j \in J \tag{3.25}$$

follow. By condition (3.3),

$$C_{\tau_0}\{\Pi(z)[r_1(z, w_n(z)) - r_1(z, w_{n-1}(z))], \Gamma\} \le \varepsilon C_{\tau_0}\{\Pi(z)[w_n(z) - w_{n-1}(z)], \Gamma\} \tag{3.26}$$

can be derived. In accordance with the proof of Theorem 3.1, from (3.24) and (3.25), the estimate

$$s_{n+1} = C[w_{n+1} - w_n, D_m] + C[w'_{n+1} - w'_n, D_m]$$
$$+ C_{\tau_0}\{\Pi(z)[w_{n+1} - w_n], \bar{D}\} \le \varepsilon M_5|t - t_0|s_n \tag{3.27}$$

can be obtained, where $M_5 = M_5(M_1, M_2)$. Choosing $\delta = 1/2\varepsilon M_5$, we have $s_{n+1} \le s_n/2$, when $|t - t_0| \le \delta, 0 \le t \le 1$. Thus when $n \ge l \ge N > 1, s_{n+1} \le 2^{-N}s_1$ and

$$C[w_n - w_l, D_m] + C[w'_n - w'_l, D_m] + C_{\tau_0}\{\Pi(z)[w_n - w_l], D\}$$
$$\le s_n + \cdots + s_{l+1} \le 2^{-N+1}s_1. \tag{3.28}$$

Hence there exists a function $w_*(z)$ similar to $w_n(z)$, such that

$$C[w_n - w_*, D_m] + C[w_n' - w_*', D_m] + C_{T_0}[\Pi(z)(w_n - w_*), \bar{D}] \to 0 \text{ for } n \to \infty, \quad (3.29)$$

and then $w_*(z)$ is a solution of Problem $\tilde{B}$ for analytic functions corresponding to $t \in T_\delta$.

Setting $N = [1/\delta] + 1$, we obtain that (3.19) is solvable for analytic function when $t_1 = \delta, t_2 = 2\delta, \cdots, t_{N-1} = (N-1)\delta, t = 1$. In particular, when $t = 1$ and $g(z) = 0$, Problem $\tilde{B}$ is solvable.

As for the case of $K \le -1$, we may take $w_*(z) = w(z)z^{K'-1}$. It is not difficult to see that $w_*(z)$ satisfies the boundary condition and point conditions with the index $K = 0$ or -1. Therefore we can derive that Problem $\tilde{B}$ with $K < 0$ is solvable.

Now, we prove the uniqueness of solutions for Problem $\tilde{B}$. Let $w_1(z), w_2(z)$ be two solution of Problem $\tilde{B}$. Then $w(z) = w_1(z) - w_2(z)$ satisfies the boundary condition and point conditions

$$\mathrm{Re}[\overline{\lambda(z)}w(z)] = r(z, w_1(z)) - r(z, w_2(z)) + |X(z)|h(z), \ z \in \Gamma, \quad (3.30)$$

$$\mathrm{Im}[\overline{\lambda(a_j)}w(a_j)] = 0, \ j \in J. \quad (3.31)$$

The estimate

$$s_* := C[w_*, D_m] + C[w_*', D_m] + C_{T_0}[\Pi w_*, \bar{D}] \le \varepsilon M_5 s_* \quad (3.32)$$

can be concluded. Choosing the constant ε so small that $\varepsilon M_5 < 1$, it is clear that $s_* = 0, w_*(z) = w(z) = 0$, i.e. $w_1(z) = w_2(z), \ z \in D$.

Next, we consider the nonlinear elliptic complex equation (2.1). On the basis of Theorem 2.4, Chapter 2 in [140], the solution $w(z)$ for (2.1) in D can be expressed as (2.17), where $\Phi(\zeta)$ is an analytic function in $\zeta(D)\backslash\{0\}$ continuous on $\zeta(\bar{D})\backslash\{0, c_1', \cdots, c_n'\}, c_j' = \zeta(c_j), j = 1, \cdots, n$. Similarly to Section 2, it can be derived that $\Phi(\zeta)$ satisfies the boundary and point conditions

$$\mathrm{Re}[\overline{\Lambda(\zeta)}\Phi(\zeta)] = R[\zeta, \Phi(\zeta)] + S(\zeta), \ \zeta \in L = \zeta(\Gamma), \quad (3.33)$$

$$\mathrm{Im}[\overline{\Lambda(a_j')}\Phi(a_j')] = b_j', \ j \in J, \quad (3.34)$$

in which

$$\overline{\Lambda(\zeta)} = \overline{\lambda[z(\zeta)]}e^{\phi[z(\zeta)]}, \ R[\zeta, \Phi(\zeta)] = r[z(\zeta), \Phi(\zeta)e^{\phi[z(\zeta)]} + \psi[z(\zeta)]],$$

$$S(\zeta) = s[z(\zeta)] - \mathrm{Re}[\overline{\lambda[z(\zeta)]}\psi[z(\zeta)]], \ c_j' = \zeta(c_j), j = 1, \cdots, n,$$

$$X_1(\zeta) = \prod_{j=0}^{n_0}(\zeta - c_j')^{\phi_j} \prod_{j=n_0+1}^{n_1}\left(\frac{\zeta - c_j'}{\zeta - \zeta_1}\right)^{\phi_j} \cdots \prod_{j=n_{N-1}+1}^{n}\left(\frac{\zeta - c_j'}{\zeta - \zeta_N}\right)^{\phi_j},$$

ζ is the center of $L_j = \zeta(\Gamma_j)(j = 1, \cdots, N), b'_j = b_j - \text{Im}[\overline{\lambda(a_j)}\psi(a_j)], j \in J$, and

$$
h(\zeta) = \left\{
\begin{array}{l}
0, \ \zeta \in L, \ \text{for } K > N - 1, \\
\left.
\begin{array}{l}
h_j, \ \zeta \in L_j, \ 1 \leq j \leq N - K', \\
0, \ \zeta \in L_j, \ N - K' < j \leq N + 1,
\end{array}
\right\} \text{for } 0 \leq K \leq N - 1, \\
h_j, \ \zeta \in L_j, \ 0 \leq j \leq N, \ \text{for } K < 0,
\end{array}
\right.
\tag{3.35}
$$

where $h_j \, (j = 0, 1, \cdots, N)$ are unknown constants as before. If the analytic function $\Phi(\zeta)$ is found, then the $h_j \, (j = 0, 1, \cdots, N)$ are determined. When $K \geq 0$, $\Phi(\zeta)$ is analytic at $\zeta = 0$ and when $K \leq -1$, $\Phi(\zeta)$ has a pole of order $\leq K' - 1$.

In the following, with a similar method as used in the proof of Theorem 2.5, we shall prove the existence of solutions of Problem $\tilde{B}$ for the nonlinear complex equation (2.1) and afterwards derive a result of solvability of Problem $\tilde{A}$ for (2.1).

Theorem 3.3   *Suppose that the complex equation (2.1) satisfies Condition C and the constant ε in (3.3) is small enough. Then Problem $\tilde{B}$ for (2.1) is solvable.*

Proof   Here we only indicate the difference with the proof of Theorem 2.5. In the formulae (2.30)–(2.32), the function $R(\zeta, \Phi)$ shall be replaced by $R(\zeta, \Phi) = R_0(\zeta)R_1(\zeta, \Phi)$, and $S(\zeta) = S_0(\zeta)\Pi_{j=0}^{n}|\zeta - c'_j|^{-\beta_j/\beta}$, where $\Lambda(\zeta), R_0(\zeta), S_0(\zeta)$ satisfy

$$
C_{\alpha\beta}[\Lambda, L_j] \leq k_2, \ C_{\alpha\beta}[R_0, L_j] \leq k_2, \ C_{\alpha\beta}[S_0, L^j] \leq k_2,
\tag{3.36}
$$

and $|b'_j| \leq k_2, \ L^j = \zeta(\Gamma^j), \ j = 1, \cdots, m, \ k_2 = k_2(q_0, p_0, \alpha, k_0, G)$, and $\tilde{R}_1(\zeta, \tilde{\Phi}) = \tilde{\Pi} R_1(\zeta, \Phi)$ satisfies

$$
\left\{
\begin{array}{l}
C_{\tau_0}[\tilde{R}_1(\zeta, \tilde{\Phi}(\zeta)), L] \leq k_3 + \varepsilon k_3 C_{\tau_0}[\tilde{\Phi}, L], \\
C_{\tau_0}[\tilde{R}_1(\zeta, \tilde{\Phi}_1(\zeta)) - \tilde{R}_1(\zeta, \tilde{\Phi}_2(\zeta)), L] \leq \varepsilon k_3 C_{\tau_0}[\tilde{\Phi}_1 - \tilde{\Phi}_2, L],
\end{array}
\right.
\tag{3.37}
$$

where $\tilde{\Pi}(\zeta) = \Pi_{j=1}^{n}|\zeta - c'_j|^{\beta_j/\beta+\tau}, \ \tilde{\Phi} = \tilde{\Pi}\Phi, \ k_3 = k_3(q_0, p_0, \alpha, k_0, G)$. On the basis of Theorem 3.2, it can be seen that there exists a unique analytic function $\Phi(\zeta)$ in $\zeta(G)\backslash\{0\}$, which satisfies the boundary conditions (2.30). Thus, we can find a solution $w(z) = \Phi[\zeta(z)]e^{\phi(z)} + \psi(z)$ of Problem $\tilde{B}$ for (2.1).

According to the proof of Theorem 2.6, from Theorem 3.3, we can derive the following result.

Theorem 3.4   *Under the same hypothesis as in Theorem 3.3, we have*

(1) *When $K > N - 1$, Problem $\tilde{A}$ for the complex equation (2.1) is solvable.*

(2) *When $0 \leq K \leq N - 1$, the total number of solvability conditions for Problem $\tilde{A}$ does not exceed $N - [K + 1/2]$.*

(3) *When $K < 0$, Problem $\tilde{A}$ has $N - 2K - 1$ solvability conditions.* (See [139]10))

4 Nonlinear Irregular Oblique Derivative Problems for Nonlinear Elliptic Equations of Second Order

4.1 Formulation of nonlinear irregular oblique derivative problems

This section deals with the nonlinear irregular oblique derivative boundary value problem for nonlinear elliptic equations of the general form

$$F[u] = F(z, u, u_z, u_{zz}, u_{z\bar{z}}) = 0 \text{ in } D, \tag{4.1}$$

$$G[u] = G(z, u, u_z) = 0 \text{ on } \Gamma, \tag{4.2}$$

where D is an $(N+1)$-connected bounded domain in the z-plane with the boundary $\Gamma = \cup_{j=0}^{N}\Gamma_j \in C_\alpha^4 \, (0 < \alpha < 1)$. We may assume that D is a circular domain as stated in Section 3, Chapter 1. F and G are real valued functions on $S = \bar{D} \times \mathbb{R} \times \mathbb{C} \times \mathbb{C} \times \mathbb{R}$ and $S' = \Gamma \times \mathbb{R} \times \mathbb{C}$ respectively. Suppose that the equation (4.1) satisfies

Condition C $F(z, u, p, q, s)$ is of class C_α^2 with respect to z, u, p, q, s in any bounded subset of S. Then (4.1) can be written as

$$A_0 u_{z\bar{z}} - \text{Re}[Q u_{zz} + A_1 u_z] - A_2 u - A_3 = 0, \tag{4.3}$$

where

$$A_0 = \int_0^1 F_{\tau s}(z, u, p, \tau q, \tau s)d\tau, \; Q = -2\int_0^1 F_{\tau q}(z, u, p, \tau q, \tau s)d\tau,$$

$$A_1 = -2\int_0^1 F_{\tau p}(z, u, \tau p, 0, 0)d\tau, \; A_2 = -\int_0^1 F_{\tau u}(z, \tau u, 0, 0, 0)d\tau,$$

$$A_3 = -F(z, 0, 0, 0, 0),$$

and for $z \in \bar{D}, u, s \in \mathbb{R}, p, q \in \mathbb{C}$, (4.1) and (4.3) satisfy

$$F_u(z, u, 0, 0, 0) < 0, \tag{4.4}$$

$$|F(z, u, p, 0, 0)| \le k_0(1 + |p|^2), \tag{4.5}$$

$$0 < \delta \le A_0 \le \delta^{-1}, \; |Q/A_0| \le q_0 < 1, \tag{4.6}$$

$$|F| + |F_z|/(1 + |p|) + |F_u| + |F_p|(1 + |p|) \le k_1(u)(1 + |p|^2 + |r|), \tag{4.7}$$

$$F_{(\eta)(\eta)} \le k_2(u, p)\{(1 + |r|^3)(|\tilde{u}|^2 + |\tilde{z}|^2) + (1 + |r|)|\tilde{p}|^2 + [(1 + |r|)(|\tilde{u}| + |\tilde{z}|) + |\tilde{p}|]|\tilde{r}|\}, \tag{4.8}$$

in which $\delta, q_0(\ge 0), k_0(> 0)$ are real constants, $k_1(u), k_2(u, p)$ are bounded functions of u and u, p respectively, and $|r| = (|q|^2 + |s|^2)^{1/2}$, $\eta = (\tilde{z}, \tilde{u}, \tilde{p}, \tilde{q}, \tilde{s})$, $F_{(\eta)(\eta)}$ denotes the second directional derivative along the direction η.

Problem M The so-called nonlinear irregular oblique derivative boundary value problem for (4.1) is to find a classical solution $u(z) \in C_\beta^2(\bar{D}) \, (0 < \beta \le \alpha)$ of (4.1) satisfying the nonlinear boundary condition (4.2), supposing that (4.2) satisfies the following conditions.

Condition C' $G(z, u, p)$ can be extended into a small neighborhood $\tilde{S}$ of S', which is of class C_α^3 with respect to z, u, p in any subset of $\tilde{S}$, and then

$$\mathrm{Re}[a_0 u_z] + a_1 u + a_2 = 0, \text{ i.e. } \mathrm{Re}[a_0 u_z] + G(z, u, 0) = 0 \text{ on } \Gamma, \tag{4.9}$$

where

$$a_0 = 2 \int_0^1 G_{\tau p}(z, u, \tau p) d\tau, \ a_1 = \int_0^1 G_{\tau u}(z, \tau u, 0) d\tau, \ a_2 = G(z, 0, 0),$$

and

$$\pm u G(z, 0, 0) > 0 \text{ on } E^\pm, \text{ when } |u| \geq k_3 > 0, \tag{4.10}$$

$$|a_0| \geq \delta > 0, \ a_0/|a_0| = \cos(\nu, x) + i\cos(\nu, y), \ \nu = (\mathrm{Re}\, a_0, \mathrm{Im}\, a_0)/|a_0|, \tag{4.11}$$

$$\begin{cases} |G| + |G_u| + |G_z|/(1 + |p|) + |G_{uu}| + |G_{xp}| + (|G_p| + |G_{up}|)(1 + |p|) \\ +(|G_{zz}| + |G_{z\bar{z}}| + |G_{pp}| + |G_{p\bar{p}}|)(1 + |p|^2) \leq k_4(u)(1 + |p|), \end{cases} \tag{4.12}$$

where α, δ, k_3 are positive constants, and $k_4(u)$ is a bounded function. The boundary Γ can be divided into two parts, namely $E^+ \subset \{z \in \Gamma, \tilde{\nu} \cdot n \geq 0, a_1 \geq 0\}$ and $E^- \subset \{z \in \Gamma, \tilde{\nu} \cdot n \leq 0, a_1 \leq 0\}$, where $\tilde{\nu} = (\mathrm{Re}\, \tilde{a}_0, \mathrm{Im}\, \tilde{a}_0)/|\tilde{a}_0|$, $\tilde{a}_0, \tilde{a}_1$ are as stated in (4.16) below, and $E^+ \cap E^- = \phi$, $E^+ \cup E^- = \Gamma$, $\overline{E^+} \cap \overline{E^-} = E^0$. For any component L of Γ, there are three possible cases: 1) $L \subset E^+$. 2) $L \subset E^-$. 3) There exists at least one point on each of $E^+ \cap L$ and $E^- \cap L$, such that $\tilde{\nu} \cdot n \neq 0$, and $E^0 = \{a_1, \cdots, a_m\}$. When the direction of $\tilde{\nu}$ at a_j is equal to the direction of L, a curve $L_j \subset L$ with the initial point a_j belongs to $E^+ \cap L(1 \leq j \leq m)$. When the direction of $\tilde{\nu}$ is opposite to the direction of L, a curve $L_{j-1} \subset L$ with the terminal point a_j belongs to $E^+ \cap L(1 \leq j \leq m)$. In this case, we may assume that

$$u(z) = g(z) \text{ on } E^0, \tag{4.13}$$

in which $g(z)$ is a known function satisfying the condition $C_\alpha^4[g(z), E^0] \leq k_0$. We suppose that E^0 is a set of fixed points. In addition, if $\tilde{\nu} \cdot n \equiv 0$, $G_u \equiv 0$ on $\Gamma_j, 0 \leq j \leq N$, then (4.2) can be replaced by

$$u(z) = g(z) \text{ on } \Gamma_j \tag{4.14}$$

with the condition $C_\alpha^4[g(z), \Gamma_j] \leq k_0, 0 \leq j \leq N$, in which $\alpha, 0 < \alpha < 1$ and k_0 are positive constants. There is no harm in assuming that $\tilde{\nu} \cdot n \equiv 0$ and $G_u \equiv 0$ on $\Gamma_* = \Gamma_1 \cup \cdots \cup \Gamma_{N_0}, N_0 \leq N + 1$, and that this condition does not hold on $\Gamma_j, N_0 < j \leq N + 1$.

In the book[140], the linear irregular oblique derivative problem for nonlinear elliptic complex equations of second order with measurable coefficients is considered, but there condition (4.5) is replaced by $|F(z, u, p, 0, 0)|/(1 + |p|) \leq A$, where

$L_p[A, \bar{D}] \leq k_0 < \infty, 2 < p < \infty$. In [128], N.S. Trudinger discussed the nonlinear regular oblique derivative problem for fully nonlinear elliptic equations of second order in an n–dimensional domain D. Here, we extend the nonlinear regular oblique derivative boundary condition to the case of nonlinear irregular oblique derivative boundary condition as stated in (4.2) and (4.9), where the direction ν in the boundary condition may be tangent to the boundary Γ, and may direct to the outside or inside of the domain D. Hence the boundary value problem for (4.1) is an irregular oblique derivative problem, which includes the Dirichlet problem and the regular oblique derivative problem as special cases.

Now we prove the uniqueness of solutions of Problem M for equation (4.1).

Theorem 4.1 *Under Condition C and Condition C', the solution of Problem M for (4.1) is unique.*

Proof Let $u_1(z), u_2(z)$ be two solutions of Problem M for (4.1). It is clear that $u(z) = u_1(z) - u_2(z)$ is a solution of the following boundary value problem

$$\tilde{A}_0 u_{z\bar{z}} - \text{Re}[\tilde{Q} u_{zz} + \tilde{A}_1 u_z] - \tilde{A}_2 u = 0 \text{ in } D, \tag{4.15}$$

$$\text{Re}[\tilde{a}_0 u_z] + \tilde{a}_1 u = 0 \text{ on } \Gamma \backslash \{E_0 \cup \Gamma_*\}, \ u(z) = 0 \text{ on } E_0 \cup \Gamma_*, \tag{4.16}$$

where

$$\tilde{A}_0 = \int_0^1 F_s(z, u, p, q, s) d\tau, \quad \tilde{Q} = -2 \int_0^1 F_q(z, u, p, q, s) d\tau,$$

$$\tilde{A}_1 = -2 \int_0^1 F_p(z, u, p, q, s) d\tau, \quad \tilde{A}_2 = -\int_0^1 F_u(z, u, p, q, s) d\tau,$$

$$a_0 = 2 \int_0^1 G_p(z, u, p) d\tau, \quad \tilde{a}_1 = \int_0^1 G_u(z, u, p) d\tau,$$

$$s = u_{2z\bar{z}} + \tau(u_1 - u_2)_{z\bar{z}}, \quad q = u_{2zz} + \tau(u_1 - u_2)_{zz},$$

$$p = u_{2z} + \tau(u_1 - u_2)_z, \quad u = u_2 + \tau(u_1 - u_2),$$

Suppose that $u(z) \not\equiv 0$ on $\bar{D}$. From the maximum principle for solutions of (4.15), there exists a point $t_* \in \Gamma$ or $t_{**} \in \Gamma$, such that

$$M = u(t_*) = \max_{z \in \bar{D}} u(z) > 0 \text{ or } m = u(t_{**}) = \min_{z \in \bar{D}} u(z) < 0. \tag{4.17}$$

If $M = u(t_*) > 0$, it is clear that $t_* \in \Gamma_{**} = \Gamma \backslash \{\Gamma_* \cup E_0\}$, and

$$\text{Re}[\tilde{a}_0 \, u_z] \geq 0 \text{ as } z = t_* \in E^+; \ \text{Re}[\tilde{a}_0 \, u_z] \leq 0 \text{ as } z = t_* \in E^-.$$

So $\tilde{a}_1(t_*) = 0$ and $\cos(\tilde{\nu}, n) = 0$ at $z = t_*$. Let $\tilde{\Gamma}$ be the longest circular arc of Γ including the point t_* so that $\cos(\tilde{\nu}, n) = 0$, $\tilde{a}_1(z) = 0$ on $\tilde{\Gamma}$. Therefore $\text{Re}[\tilde{a}_0 u_z] = 0$, $u(z) = M$ on $\tilde{\Gamma}$. Obviously, a_j is not an end point of $\tilde{\Gamma}$, thus there exists a point $t'_* \in \Gamma \backslash \tilde{\Gamma}$, such that $\cos(\tilde{\nu}, n) > 0$, $\cos(\tilde{\nu}, s) > 0 (< 0)$, $\partial u / \partial n > 0$ and $\partial u / \partial s \geq 0 (\leq 0)$ at $z = t'_*$. It follows that

$$\text{Re}[\tilde{a}_0 \, u_z] = [\frac{\partial u}{\partial n} \cos(\tilde{\nu}, n) + \frac{\partial u}{\partial s} \cos(\tilde{\nu}, s)]|_{z=t'_*} > 0.$$

If $\cos(\tilde{\nu}, n) < 0$ at $z = t'_*$, in a similar way we can derive that $\mathrm{Re}[\tilde{a}_0 u_z]|_{z=t'_*} < 0$. These estimates are contradictory to (4.16). If $m = u(t_{**}) < 0$, we can also derive a contradiction. Hence $u(z) \equiv 0$, i.e. $u_1(z) \equiv u_2(z)$ in $\bar{D}$.

In the following, we first give a priori estimates of solutions for Problem M for (4.1), and then prove the existence of solutions for Problem M by using the method of parameter extension.

4.2   A priori estimates of solutions of Problem M for (4.1)

Theorem 4.2   *If Condition C and Condition C' hold, then any solution $u(z)$ of Problem M for (4.1) satisfies the estimate*

$$C[u, \bar{G}] \leq M_0 = \max\{\max_{\bar{D}} |F(z,0,0,0,0)| / \min_{\bar{D}} A_2, k_0, k_3\}, \tag{4.18}$$

where k_0, k_3 are positive constants as stated in (4.10), and in connection with (4.13) and (4.14).

Proof   If $u^2 = [u(z)]^2$ takes its maximum at a point $z_* \in D$, then multiplying (4.3) by u, we get

$$\frac{1}{2}\{A_0(u^2)_{z\bar{z}} - \mathrm{Re}[Q(u^2)_{zz} + A_1(u^2)_z - 2Q(u_z)^2] - 2A_0|u_z|^2\} = A_2 u^2 + A_3 u. \tag{4.19}$$

Taking into account $u_z = 0$, $(u^2)_z = 0$, $A_0(u^2)_{z\bar{z}} - \mathrm{Re}Q(u^2)_{zz} \leq 0$ at z_*, from (4.19) it follows that at z_*,

$$A_2 u^2 + A_3 u \leq 0, \text{ i.e. } \max_{\bar{D}} |u| \leq \max_{\bar{D}} |A_3| / \min_{\bar{D}} A_2. \tag{4.20}$$

If u^2 takes the maximum at a point $z_* \in E_0 \cup \Gamma_*$, then it is evident that

$$\max_{\bar{D}} |u| \leq \max_{E_0 \cup \Gamma_*} |u(z)| = \max_{E_0 \cup \Gamma_*} |g(z)| \leq k_0. \tag{4.21}$$

If u^2 takes the maximum at a point $z_* \in \Gamma \backslash \{E_0 \cup \Gamma_*\}$, from (4.9) it follows that

$$\pm\frac{1}{2}\mathrm{Re}[a_0(u^2)_z] = \mp uG(z, u, 0) \geq 0 \text{ at } z_* \in E^{\pm}.$$

Hence,

$$\max_{\bar{D}} |u| \leq k_3. \tag{4.22}$$

Combining (4.20)–(4.22), the estimate (4.18) is derived.

Theorem 4.3   *Under the same conditions as in Theorem 4.2, any solution $u(z)$ of Problem M for (4.1) satisfies the estimate*

$$C_\beta[u, \bar{D}] = C[u, \bar{D}] + H_\beta[u, \bar{D}] \leq M_\beta, \tag{4.23}$$

where $\beta = \beta(\delta, q_0, \alpha, k, D), 0 < \beta \leq \alpha$, $M_\beta = M_\beta(\delta, q_0, \alpha, k, D)$ *are nonnegative constants,* $k = (k_0, k_1, \cdots, k_4)$.

Proof Similarly to [42], we can estimate the Hölder coefficient H_β of the solution $u(z)$ of Problem M for (4.1) in D and near the boundary part $\Gamma' = \{z \in \Gamma, |\cos(\nu, n)| \geq \varepsilon\}$. We have

$$\sup_{z_1, z_2 \in D_\varepsilon} \frac{|u(z_1) - u(z_2)|}{|z_1 - z_2|^\beta} \leq M_\beta^1, \tag{4.24}$$

$$\sup_{z_1 \in D, z_2 \in \Gamma'} \frac{|u(z_1) - u(z_2)|}{|z_1 - z_2|^\beta} \leq M_\beta^2, \tag{4.25}$$

in which ε is a sufficiently small positive number, $D_\varepsilon = \{z | z \in D, \text{dist}(z, \Gamma) \geq \varepsilon\}$, $M_\beta^1 = M_\beta^1(\delta, q_0, \alpha, k, D_\varepsilon)$, $M_\beta^2 = M_\beta^2(\delta, q_0, \alpha, k, D, \Gamma')$. Now, we prove the estimate near the boundary $\Gamma'' = \Gamma \backslash \Gamma'$, i.e.

$$\sup_{z_1 \in D, z_2 \in \Gamma''} \frac{|u(z_1) - uz_2)|}{|z_1 - z_2|^\beta} \leq M_\beta^3, \tag{4.26}$$

where $M_\beta^3 = M_\beta^3(\delta, q_0, \alpha, k, D, \Gamma'')$. For this, we need a lemma.

Lemma 4.4 *Let the coefficients of the linear operator*

$$Lu = A_0 u_{z\bar{z}} - \text{Re}[Q u_{zz} + A_1 u_z] \tag{4.27}$$

in the domain $K_R = \{0 < \text{Re}\, z < R \leq 1, |\text{Im}\, z| < R\}$ *satisfy the condition*

$$0 < \delta \leq A_0 \leq \delta^{-1}, |Q/A_0| \leq q_0 < 1, |A_1| \leq k_0, \tag{4.28}$$

and the function $u(z) \in C^2(K_R) \cap C^1(\overline{K_R})$ *satisfy the condition*

$$u \geq 0,\ Lu \leq 0 \text{ in } K_R,\ Nu = \frac{\partial u}{\partial \nu} \geq 0 \text{ on } \partial K_R \cap \{\text{Re}\, z = 0\}, \tag{4.29}$$

where $0 \leq \cos(\nu, n) \leq \sigma, z \in \partial K_R \cap \{\text{Re}\, z = 0\}$, δ, q_0 *and* k_0 *are real constants as stated before,* $\sigma(< 1)$ *and* R *are positive constants. If*

$$\text{meas}\,[K_R \cap \{u \geq 1\}] \geq \eta\, \text{meas}\, K_R, \tag{4.30}$$

where $\eta(\leq 1/2)$ *is a positive constant, then there exist two constants* $\rho, \varepsilon \in (0, 1)$, *such that*

$$u(z) \geq \gamma > 0,\ z \in \hat{K} = \{0 \leq x \leq \rho R, |y| \leq (1 - \varepsilon)R\}, \tag{4.31}$$

in which $\gamma = \gamma(\delta, q_0, k_0, \eta, \sigma, \varepsilon)$.

Proof In order to prove the lemma, we introduce a domain $K' = \{\rho R/2 < x < R, |y| < R\}$, where ρ is a small positive constant. It is easy to see that

$$\text{meas}\,[K' \cap \{u \geq 1\}] \geq (\eta - \rho/2)\, \text{meas}\, K_R \geq 3\eta\, \text{meas}\, K'/4,$$

in which $\rho < \eta/2$. According to the ordinary density theorem [42], we can obtain

$$u(z) \geq \gamma' > 0, \text{ on } K'' = \{\rho R \leq x \leq (1-\varepsilon)R, \ |y| \leq (1-\varepsilon/2)R\}. \tag{4.32}$$

Introduce an auxiliary function

$$w(z) = u(z) + \gamma'[8x^2/(\rho R)^2 - x/(\rho R) \pm 2(y-y_0)/(\varepsilon R)]/18$$

$$\text{on } \tilde{K} = \{0 < x < 2\rho R, \ |y - y_0| < \varepsilon R/2\}, \tag{4.33}$$

where $|y_0| \leq (1-\varepsilon)R$ and $\pm 1 = \text{sign} \cos(\nu, y)$. Through direct calculation, we have

$$Lw \leq Lu + \gamma'[-\lambda/2(\rho R)^2 + 5k_0/(2\rho R) + k_0/(\varepsilon R)]/18$$

$$\leq \gamma'[-1/4(\rho R)^2 + k_0/(\varepsilon R)]/18 < 0 \text{ in } \tilde{K},$$

$$Nw = Nu + \gamma'[(2x/(\rho R)^2 + 1/(\rho R))\cos(\nu, x)$$

$$\pm 2\cos(\nu, y)/(\varepsilon R)]/18 > 0 \text{ on } \bar{\tilde{K}} \cap \{x = 0\},$$

where $\lambda = (1 - q_0)\delta$, and ρ is chosen according to

$$\rho < \min\left[\frac{\lambda}{10k_0 R}, \left(\frac{\lambda\varepsilon}{8k_0 R}\right)^{1/2}\right], \text{ i.e. } \frac{5k_0}{2\rho R} < \frac{\lambda}{4(\rho R)^2}, \ \frac{k_0}{\varepsilon R} < \frac{\lambda}{8(\rho R)^2}.$$

By the above inequelities, we see that $w(z)$ cannot take its minimum on $\tilde{K} \cup [\tilde{K} \cap \{x = 0\}]$. From (4.32), it follows that

$$w(z) \geq \gamma' \text{ on } \{x = 2\rho R, \ |y - y_0| = \varepsilon R/2\}.$$

Hence $w(z) \geq \gamma'$ in $\tilde{K}$. Next we choose x, y, such that $0 \leq x \leq 2\rho R$, $y = y_0$. By (4.33),

$$u(z) + \gamma'/2 \geq \gamma', \text{ i.e. } u(z) \geq \gamma'/2 = \gamma \text{ in } \hat{K}$$

can be obtained. Thus Lemma 4.4 is proved.

Now we continue to prove the proof of Theorem 4.3. According to the method in [42], we have to find a conformal mapping, which maps D into $\text{Re}\,\zeta > 0$ and a boundary component onto $\text{Re}\,\zeta = 0$. Choosing $w(z) = u(z) + c_1 y$, where c_1 is an undetermined constant, (4.26) can be concluded. Combining (4.18),(4.24)–(4.26), the estimate (4.23) is derived.

Secondly we show the boundness and the Hölder continuity of u_z.

Theorem 4.5   *Under the same conditions as in Theorem 4.2, any solution $u(z)$ of Problem M for (4.1) satisfies the estimate*

$$C[u_z, \bar{D}] \leq M_1 = M_1(\delta, q_0, \alpha, k, D). \tag{4.34}$$

Proof The estimates of u_z in the inner part of D and near the boundary $\Gamma' = \{z \in \Gamma, |\cos(\nu, n)| \geq \varepsilon\}$ are similar as in [42], namely

$$C[u_z, D_\varepsilon \cup D'_\varepsilon] \leq M'_1 = M'_1(\delta, q_0, \alpha, k_0, D_\varepsilon, D'_\varepsilon), \tag{4.35}$$

where $D_\varepsilon = \{z \in D, \operatorname{dist}(z, \Gamma) \geq \varepsilon\}$, $D'_\varepsilon = \{z \in \bar{D}, \operatorname{dist}(z, \Gamma'') \geq \varepsilon\}$, in which ε is sufficiently small. It remains to estimate u_z near $\Gamma'' = \Gamma \backslash \Gamma'$.

Denote by M_1 the maximum of $|u_z|$ on $\bar{D}$, where $u(z)$ is any solution of Problem M for (4.1). There is no harm in assuming that there exists a point $z_0 \in \Gamma''$ such that $M_1 = |u_z(z_0)|$. We may assume that the domain D lies entirely in $\operatorname{Re} z = x > 0$ and Γ has a part ω lying on $x = 0$, herein $z_0 \in \omega$. The conformal mapping can be chosen such that $|u_z(z_0)| \geq M_1/5$, $\max_{\bar{D}} |u_z| \leq 2M_1$, and $|\cos(\nu, y)| \geq 1/2$ on $K = \{|z - z_0| \leq R_0, x \geq 0\}$, where R_0 is a small positive number so that $\{|z - z_0| \leq R_0, x = 0\} \subset \omega$ and $\{|z - z_0| \leq R_0, x > 0\} \subset D$. Let $\eta(z) = 1 - |z - z_0|^2/R^2$, in which $0 < R < \min(R_0, 1/M_1)$. Let $u_y = h(z, u, u_x) = au_x + bu + c$ be a solution of $G(z, u, u_z) = 0$ and denote $f = u_y - h(z, u, u_x)$. We introduce an auxiliary function

$$w(z) = \eta^2(z)(u_x^2 + c_2 f^2) + c_3 M_1^2 [u(z) - u(z_0)]^2$$

$$\mp \begin{cases} c_4 M_1^3 (y - y_0)(1 - c_4 M_1^2 x^2)^2, & \text{for } x < 1/(\sqrt{c_4} M_1), \\ 0, & \text{for } x \geq 1/(\sqrt{c_4} M_1), \end{cases} \tag{4.36}$$

where $c_2, c_3 (\geq 1), c_4$ are undetermined positive constants, and $\pm 1 = -\operatorname{sign} \cos(\nu, y)$. By the inequality $(A + B)^2 \geq A^2/2 - 2B^2$, it can be seen that

$$f^2 = (u_y - au_x - bu - c)^2 \geq u_y^2/2 - c_5 u_x^2 - c_6,$$

where c_5 and c_6 are positive constants. Choosing $c_2 \leq \min(1, 1/(2c_5))$, if

$$M_1 < 10\sqrt{c_6}, \tag{4.37}$$

then M_1 is bounded. Otherwise $M_1^2 \geq 100 c_6$. Then we have

$$w(z) \geq u_x^2 + c_2 f^2 \geq \frac{c_2}{2}\left(\frac{M_1}{5}\right)^2 - c_2 c_6 \geq \frac{c_2}{100} M_1^2 = c_7 M_1^2 \text{ at } z = z_0. \tag{4.38}$$

When $|y - y_0| \leq c_7 R/(2c_4)$, then

$$\pm c_4 M_1^3 (y - y_0)(1 - c_4 M_1^2 x^2)^2 \leq c_7 M_1^2/2. \tag{4.39}$$

Denote by $\tilde{z}$ a maximum point of $w(z)$ on K. From (4.38) and (4.39), we have

$$c_7 M_1^2 \leq w(z_0) \leq w(\tilde{z}) \leq \eta^2(\tilde{z}) c_8 M_1^2 + c_3 M_\beta^2 M_1^{2-2\beta} + c_7 M_1^2/2,$$

where c_8 is a positive constant. When $c_3 M_\beta^2 M_1^{2-2\beta} \geq c_7 M_1^2/4$, i.e.

$$M_1 \leq [4c_3 M_\beta^2/c_7]^{1/2\beta}. \tag{4.40}$$

68   *Heinrich Begehr and Guo Chun Wen*

This shows that M_1 is bounded. Otherwise $c_3 M_\beta^2 M_1^{2-2\beta} < c_7 M_1^2/4$. Then we have

$$\eta^2(\tilde{z}) \geq c_7/(4c_8), \tag{4.41}$$

$$c_7 M_1^2/4 \leq \eta^2(z)(u_x^2 + f^2) \leq c_9|u_z(\tilde{z})|^2, \tag{4.42}$$

$$|u_z(\tilde{z})|^2 \leq c_{10}(u_x^2 + f^2) \leq c_{11}M_1^2. \tag{4.43}$$

Here and in the following, $c_j(j = 9, \cdots, 21)$ are positive constants. When $\tilde{x} = \mathrm{Re}\,\tilde{z} = 0$, from (4.43) it can be seen that

$$0 \leq G_{u_x}w_x + G_{u_y}w_y \leq 2\eta(G_{u_x}\eta_x + G_{u_y}\eta_y)u_x^2 - 2\eta^2 u_x[G_u u_x$$

$$+G(z,0,0)_x] + 2c_3 M_1^2[u(z) - u(z_0)](G_{u_x}u_x + G_{u_y}u_y) - c_4 M_1^3$$

$$\leq c_{12}M_1^3 + c_3 c_{13}M_1^{3-\beta} - \delta c_4 M_1^3/2.$$

Setting $c_4 = 6c_{12}/\delta$, from the above inequality

$$M_1 \leq [c_3 c_{13}/c_{12}]^{1/\beta} \tag{4.44}$$

follows. When $\tilde{x} > 0$, denote $L = A_0(\ )_{z\bar{z}} - \mathrm{Re}[Q(\ )_{zz} + A_1(\ )_z]$. Then at $\tilde{z}$, we have

$$0 \geq Lw = (u_x^2 + c_2 f^2)L\eta^2 + 2\mathrm{Re}(\eta^2)_z[A_0(u_x^2 + c_2 f^2)_z$$

$$-Q(u_x^2 + c_2 f^2)_z] + 2\eta^2[u_x L u_x + \mathrm{Re}(A_0|u_{xz}|^2 - Q(u_{xz})^2)]$$

$$+2c_2\eta^2[fLf + \mathrm{Re}(A_0|f_z|^2 - Q(f_z)^2] + 2c_3 M_1^2\{(u(z) - u(z_0))Lu$$

$$+\mathrm{Re}[A_0|u_z|^2 - Q(u_z)^2]\} \mp c_4 M_1^3(y - y_0)L(1 - c_4 M_1^2 x^2)^2$$

$$\mp(1 - c_4 M_1^2 x^2)^2\mathrm{Im}A_1 c_4 M_1^3/2 \mp \mathrm{Im}\,c_4 M_1[A_0(1 - c_4 M_1^2 x^2)_{\bar{z}}^2$$

$$-Q(1 - c_4 M_1^2 x^2)_z^2]/2 \text{ for } x \leq 1/(\sqrt{c_4}M_1). \tag{4.45}$$

If $x > 1/(\sqrt{c_4}M_1)$, the last three terms in the above inequality are replaced by zero. Taking into account

$$Lu_y = -F_u u_y - F_y = O[1 + |u_z|^3 + |r|(1 + |u_z|)],$$

$$\mathrm{Re}[A_0|u_{xz}|^2 - Q(u_{xz})^2] \geq \delta(1 - q_0)|u_{xz}|^2 = \lambda_0|u_{xz}|^2,$$

and

$$fLf + \mathrm{Re}[A_0|f_z|^2 - Q(f_z)^2] = f\{Lu_y - aLu_x - bLu - Lc - u_x La$$

$$-uLb - 2\mathrm{Re}[a_z(A_0 u_{x\bar{z}} - Qu_{xz} - A_1 u_x) + b_z(A_0 u_{\bar{z}} - Qu_z - A_1 u)]\}$$

$$+2\mathrm{Re}\{A_0|u_{yz} - (au_x)_z - (bu)_z - c_z|^2 - Q[u_{yz} - (au_x)_z - (bu)_z - c_z]^2\}$$

$$\geq \lambda_0|u_{yz}|^2 - c_{14}|r|^2 - c_{15}(1 + |u_z|^4),$$

where we use Condition C, (4.3) and the Cauchy inequality, and appropriately choose the positive constants. Moreover, selecting $c_2 = \min[1/(2c_5), \lambda_0/(2c_{14})]$, from (4.45) it follows that

$$0 \geq Lw \geq c_{16}|r|^2 + c_{17}c_3|u_z|^4 - (c_{18} + c_{19}c_3|u_z|^{-\beta})(|u_z|^2|r| + |u_z|^4)$$

for $|u_z| \geq c_{20}$. Finally, choosing c_3 such that $c_3c_{17} > 2c_{18}(1 + 2c_{18}/c_{16})$, we obtain

$$|u_z| \leq [c_3/c_{21}]^{1/2\beta}. \tag{4.46}$$

When $|u_z| \geq c_{20}$, combining (4.35),(4.37),(4.40),(4.44) and (4.46), the estimate (4.34) is derived.

Moreover, according to the method in [128], we can prove the following result.

Theorem 4.6 *Under the same conditions as in Theorem 4.2, any solution $u(z)$ of Problem M for (4.1) satisfies the estimate*

$$C_\beta^2[u, \bar{D}] = C_\beta[u, \bar{D}] + C_\beta[u_z, \bar{D}] + C[u_{z\bar{z}}, \bar{D}] + C_\beta[u_{zz}, \bar{D}] \leq M_2, \tag{4.47}$$

where $\beta = \beta(\delta, q_0, \alpha, k, D), 0 < \beta \leq \alpha, M_2 = M_2(\delta, q_0, \alpha, k, D)$ are nonnegative constants.

4.3   Existence of solutions of Problem M for (4.1)

Now we use the foregoing estimates of solutions and the method of parameter extension to prove the solvability of Problem M for (4.1).

Theorem 4.7 *Suppose that Condition C and Condition C' hold. Then Problem M for (4.1) has a solution $u(z) \in C_\beta^2(\bar{D})$.*

Proof   Introduce the boundary value problem M_τ with the parameter $\tau(0 \leq \tau \leq 1)$

$$(1 - \tau)u_{z\bar{z}} + \tau F(z, u, u_z, u_{zz}, u_{z\bar{z}}) = A(z) \text{ in } \bar{D}, \tag{4.48}$$

$$(1 - \tau)\frac{\partial u}{\partial \mu} + \tau G(z, u, u_z) = B(z) \text{ on } \Gamma, \tag{4.49}$$

where $A(z) \in C_\beta^2(\bar{D})$, $B(z) \in C_\beta^3(\Gamma)$, $0 < \beta \leq \alpha$, and μ is a special case of ν corresponding to the linear case of the boundary condition (4.2). When $\tau = 0$, it is clear that Problem M_0 has a unique solution $u(z) \in C_\beta^4(\bar{D})$(see [50],[81]). Supposing that when $\tau = \tau_0, 0 \leq \tau_0 < 1$, Problem M_{τ_0} is solvable, we shall prove that there exists a positive constant ε such that for any $\tau \in E = \{|\tau - \tau_0| \leq \varepsilon, 0 \leq \tau \leq 1\}$, Problem M_τ has a solution. For this, we rewrite (4.48), (4.49) as

$$u_{z\bar{z}} - \tau_0[u_{z\bar{z}} - F(z, u, u_z, u_{zz}, u_{z\bar{z}})]$$
$$= (\tau - \tau_0)[u_{z\bar{z}} - F(z, u, u_z, u_{zz}, u_{z\bar{z}})] + A(z) \text{ in } \bar{D}, \tag{4.50}$$

$$\frac{\partial u}{\partial \mu} - \tau_0[\frac{\partial u}{\partial \mu} - G(z, u, u_z)]$$

$$= (\tau - \tau_0)[\frac{\partial u}{\partial \mu} - G(z, u, u_z)] + B(z) \text{ on } \Gamma. \tag{4.51}$$

We choose an arbitrary function $u_0(z) \in C_\beta^4(\bar{D})$ and substitute it into the position of u in the right–hand sides of (4.50),(4.51). There is no harm in assuming that $u_0(z) \equiv 0$. It is easy to see that

$$(\tau - \tau_0)[u_{0z\bar{z}} - F(z, u_0, u_{0z}, u_{0zz}, u_{0z\bar{z}})] + A(z) \in C_\beta^2(\bar{D}),$$

$$(\tau - \tau_0)[\frac{\partial u}{\partial \mu} - G(z, u_0, u_{0z})] + B(z) \in C_\beta^3(\Gamma).$$

On the basis of the hypothesis, the boundary value problem (4.50),(4.51) has a solution $u_1(z) \in C_\beta^4(\bar{D})$. Applying successive iteration, we obtain a sequence of solutions $\{u_n(z)\}$, which satisfy

$$u_{n+1z\bar{z}} - \tau_0[u_{n+1z\bar{z}} - F(z, u_{n+1}, u_{n+1z}, u_{n+1zz}, u_{n+1z\bar{z}})]$$

$$= (\tau - \tau_0)[u_{nz\bar{z}} - F(z, u_n, u_{nz}, u_{nzz}, u_{nz\bar{z}})] + A(z) \text{ in } \bar{D}, \tag{4.52}$$

$$\frac{\partial u_{n+1}}{\partial \mu} - \tau_0[\frac{\partial u_{n+1}}{\partial \mu} - G(z, u_{n+1}, u_{n+1z})]$$

$$= (\tau - \tau_0)[\frac{\partial u_n}{\partial \mu} - G(z, u_n, u_{nz})] + B(z) \text{ on } \Gamma. \tag{4.53}$$

From (4.52),(4.53), it can be derived that $U_{n+1} = u_{n+1}(z) - u_n(z)$ satisfies

$$U_{n+1z\bar{z}} - \tau_0[U_{n+1z\bar{z}} - \tilde{F}(z, u_{n+1}, u_n)]$$

$$= (\tau - \tau_0)[U_{nz\bar{z}} - \tilde{F}(z, u_n, u_{n-1})] \text{ in } \bar{D}, \tag{4.54}$$

$$\frac{\partial U_{n+1}}{\partial \mu} - \tau_0[\frac{\partial U_{n+1}}{\partial \mu} - \tilde{G}(z, u_{n+1}, u_n)]$$

$$= (\tau - \tau_0)[\frac{\partial U_n}{\partial \mu} - \tilde{G}(z, u_n, u_{n-1})] \text{ on } \Gamma, \tag{4.55}$$

where

$$\tilde{F}(z, u_{n+1}, u_n) = F(z, u_{n+1}, u_{n+1z}, u_{n+1zz}, u_{n+1z\bar{z}}) - F(z, u_n, u_{nz}, u_{nzz}, u_{nz\bar{z}}),$$

$$\tilde{G}(z, u_{n+1}, u_n) = G(z, u_{n+1}, u_{n+1z}) - G(z, u_n, u_{nz}).$$

By using the method of proofs of Theorem 4.2 to Theorem 4.6, we can prove that

$$C_\beta^2[U_{n+1}, \bar{D}] \leq |\tau - \tau_0| M_3 C_\beta^2[U_n, \bar{D}], \tag{4.56}$$

where $M_3 = M_3(\delta, q_0, \alpha, k, D)$. Choosing $\varepsilon = 1/2(M_3 + 1)$, we obtain when $|\tau - \tau_0| \le \varepsilon$, and $n \ge m > N + 1 > 2$, that

$$C_\beta^2[U_{n+1}, \bar{D}] \le 2^{-1} C_\beta^2[U_n, \bar{D}], \ C_\beta^2[u_n - u_m, \bar{D}] \le 2^{-N+1} C_\beta^2[u_1 - u_0, \bar{D}]. \qquad (4.57)$$

This shows that $\{u_n(z)\}$ is a Cauchy sequence. Due to the compactness of the Banach space $C_\beta^2(\bar{D})$, there exists a function $u_*(z) \in C_\beta^2(\bar{D})$, such that $C_\beta^2[u_n - u_*, \bar{D}] \to 0$ for $n \to \infty$. Hence, $u_*(z)$ is a solution of (4.50),(4.51), i.e. of Problem M_τ for $\tau \in E$. Thus from the solvability of Problem M_0, we can derive the solvability of Problem M_1. In particular, when $\tau = 1$ and $A(z) = 0, B(z) = 0$, (4.50),(4.51) i.e. Problem M for (1.1),(1.2) has a solution $u(z) \in C_\beta^2(\bar{D}), 0 < \beta \le \alpha$, see [139]20).

Finally, we mention that some solvability results of nonlinear oblique derivative problems for nonlinear elliptic systems of second order equations in a multiply connected domain can be found in [139]11) and [149].

III   Boundary Value Problems for Degenerate Elliptic Equations and Systems

In this chapter, we first consider the Riemann–Hilbert problem for nonlinear degenerate elliptic complex equations of first order. Afterwards we introduce the mixed boundary value problem, the oblique derivative problem and the Poincaré boundary value problem for degenerate elliptic equations of second order. By using a priori estimates of solutions for the foregoing boundary value problems, the method of auxiliary functions and the principle of compactness, we can prove the solvability of these problems. Under some additional conditions, the uniqueness of solutions can be derived.

1   Riemann–Hilbert Problem for Degenerate Elliptic Complex Equations of First Order

1.1   Riemann–Hilbert problem for degenerate elliptic complex equations and its solvability

We consider the complex equation of first order

$$w_{\bar{z}} = F(z, w, w_z), \ F = Q(z, w, w_z)w_z + A(z, w), \tag{1.1}$$

and suppose that (1.1) satisfies the following conditions.

Condition C^*   The equation (1.1) satisfies the degenerate ellipticity conditions which are

$$F(z, w, U_1) - F(z, w, U_2) = Q(z, w, U_1 - U_2)(U_1 - U_2), \tag{1.2}$$

$$0 \leq |Q(z, w, U)| \leq 1, \ \varlimsup_{U \to \infty} |Q(z, w, U))| = q_0 < 1, \ z \in D, \ w \in \mathbb{C}, \tag{1.3}$$

where D is an $(N+1)$–connected circular domain as stated in Section 3 of Chapter 1, q_0 is a nonnegative constant, $Q(z, w, U)$ and $A(z, w)$ are continuous in $w, U \in \mathbb{C}$ for almost every point $z \in D$, and measurable in D for all continuous functions $w(z)$ and measurable functions $U(z)$ in D, and satisfies the condition

$$L_p[A(z, w), \bar{D}] \leq k_0 < \infty, \ p > 2, \tag{1.4}$$

in which p, k_0 are real constants.

It is not difficult to see that

$$F(z, w, U) = \begin{cases} U^2/2, & \text{for } |U| \leq 1, \\ \bar{U}^{-2}/2, & \text{for } |U| > 1 \end{cases}$$

and

$$F(z, w, U) = \begin{cases} |U| \text{ for } |U| \leq 1, \\ (|z| + |w| + 2)|U|/(|z| + |w| + 1 + |U|) \text{ for } |U| > 1 \end{cases}$$

satisfy Condition C^*.

The so–called Riemann–Hilbert boundary value problem for the complex equation (1.1) may be formulated as follows.

Problem A^*   Find a solution $w(z) \in C(\bar{D}) \cap W_2^1(D)$ of (1.1) satisfying the boundary condition

$$\text{Re}[\overline{\lambda(z)}w(z)] = r(z) + h(z), \ z \in \Gamma, \tag{1.5}$$

where $\lambda(z)$ is of standard form

$$\lambda(z) = \begin{cases} z^K, \ z \in \Gamma_0, \\ e^{i\theta_j}, \ t \in \Gamma_j, \ j = 1, \cdots, N, \end{cases} \tag{1.6}$$

herein K is an integer, $\theta_j (j = 1, \cdots, N)$ are real constants,

$$C_\alpha[r, \Gamma] \leq k_0, \tag{1.7}$$

in which $\alpha (1/2 < \alpha < 1)$, k_0 are nonnegative constants, $h(z)$ is as stated in (1.8), Chapter 1.

Problem B^*   If the solution $w(z)$ of Problem A^* satisfies the point conditions

$$\text{Im}[\overline{\lambda(a_j)}w(a_j)] = b_j, \ j \in J = \begin{cases} 1, \cdots, 2K - N + 1, \text{ for } K \geq N, \\ N - K + 1, \cdots, N + 1, \text{ for } 0 \leq K < N, \end{cases} \tag{1.8}$$

where $a_j \in \Gamma_j (j = 1, \cdots, N)$, $a_j(j = N + 1, \cdots, 2K - N + 1)$ are distinct points on Γ_0, $b_j(j = 1, \cdots, 2K - n + 1)$ are real constants and satisfy $|b_j| \leq k_0, \ j \in J$, then $w(z)$ is said to solve Problem B^*.

In order to discuss the solvability of Problem A^* and Problem B^* for the complex equation (1.1), we introduce a function

$$F^{(m)}(z, w, U) = \begin{cases} F(z, w, U), \ |Q(z, w, U)| \leq 1 - 1/m, \\ Q_m(z, w, U)U + A(z, w), \ |Q(z, w, U)| > 1 - 1/m, \end{cases} \tag{1.9}$$

where $Q_m(z, w, U) = (1 - 1/m)e^{i \arg Q(z,w,U)}$, m is a positive integer, and consider the uniformly elliptic complex equation

$$w_{\bar{z}} = F^{(m)}(z, w, w_z). \tag{1.10}$$

By Theorem 4.7, Chapter 2 in [140], we know that Problem A^* and Problem B^* for (1.10) are solvable, and if the complex equation (1.1) satisfies Condition C^*, then any solution $w(z)$ of Problem B^* for (1.1) satisfies some estimates formulated in the next theorem.

Theorem 1.1 *Let the complex equation (1.1) satisfy Condition C^*. Then Problem A^* and Problem B^* for the corresponding complex equation (1.10) are solvable, and any solution $w(z)$ of Problem $B^*(K \leq 0)$ for the complex equation (1.1) satisfies the estimate*

$$C_\beta[w, \bar{D}] \leq M_1, \quad L_{p_0}[|w_{\bar{z}}| + |w_z|, \bar{D}] \leq M_2, \tag{1.11}$$

where $M_j = M_j(q_0, p_0, \alpha, k_0, M, K, D)(j = 1, 2)$ are constants, $\beta = 1 - 2/p_0$, M, p_0 $(2 < p_0 < \min(p, 1/(1 - \alpha))$ are constants satisfying (1.12) and (1.16) below respectively. If in the case of $K > 0$, we assume that the constant q_0 in (1.3) is small enough, the estimate (1.11) remains true.

Proof By using condition (1.3), there exists a positive constant M, such that

$$|Q(z, w, U)| \leq \delta = (1 + q_0)/2 < 1, \quad \text{when } |U| > M. \tag{1.12}$$

We first discuss the case of $K \leq 0$. The solution $w(z)$ of Problem B^* for the complex equation

$$w_{\bar{z}} = \omega(z), \quad \omega(z) \in L_{p_0}(\bar{D}) \tag{1.13}$$

can be expressed as

$$w(z) = \Phi(z) + \hat{T}\omega, \tag{1.14}$$

where $\Phi(z)$ is a unique solution of Problem B^* for analytic functions, $\hat{T}\omega$ is a double integral defined by

$$\begin{cases} \hat{T}\omega = -\dfrac{1}{\pi} \int\int_D [P(z, \zeta)\omega(\zeta) + Q(z, \zeta)\overline{\omega(\zeta)}]d\sigma_\zeta + \Psi(z), \\[2ex] P(z, \zeta) = \dfrac{1}{2}[G_1(z, \zeta) + G_2(z, \zeta) + H_1(z, \zeta) - H_2(z, \zeta)], \; z \in \bar{D}, \\[2ex] Q(z, \zeta) = \dfrac{1}{2}[G_1(z, \zeta) - G_2(z, \zeta) + H_1(z, \zeta) + H_2(z, \zeta)], \; \zeta \in D, \end{cases} \tag{1.15}$$

in which $\Psi(z)$ is an analytic function in D and $G_j(z, \zeta)$, $H_j(z, \zeta)(j = 1, 2)$ are as stated in [140]. The integral $\hat{S}\omega = (\hat{T}\omega)_z$ possesses the following property

$$\delta\Lambda_{p_0} < 1, \quad \Lambda_{p_0} = \sup_{\omega(z)\in L_{p_0}(\bar{D})} \frac{L_{p_0}(\hat{S}\omega, \bar{D})}{L_{p_0}(\omega, \bar{D})}, \tag{1.16}$$

where p_0 is a proper constant satisfying $2 < p_0 < p$ (see Theorem 3.5, Chapter 1 in [140]) and $L_{p_0}[\omega, \bar{D}] \neq 0$. Substituting $w(z) = \Phi(z) + \hat{T}\omega$ into the complex equation (1.10), we obtain the integral equation

$$\begin{aligned} \omega(z) &= F^{(m)}(z, w, w_z) = F^{(m)}(z, w, w_z) - F^{(m)}(z, w, \Phi') \\ &+ F^{(m)}(z, w, \Phi') = Q^{(m)}(z, w, \hat{S}\omega)\hat{S}\omega + F^{(m)}(z, w, \Phi'). \end{aligned} \tag{1.17}$$

76   *Heinrich Begehr and Guo Chun Wen*

Taking

$$L_{p_0}[\Phi', \bar{D}] \leq k_1 = k_1(p_0, \alpha, k_0, K, D) < \infty \tag{1.18}$$

into account (see Theorem 3.7, Chapter 1 in [140]), we have

$$\begin{cases} |\omega(z)| \leq |Q^{(m)}||\hat{S}\omega| + |Q^{(m)}||\Phi'| + |A|, \\ \\ L_{p_0}[\omega, \bar{D}] \leq \delta L_{p_0}[\hat{S}\omega, \bar{D}] + \left(\iint_{|\hat{S}\omega| \leq M} |Q^{(m)}\hat{S}\omega|^{p_0} d\sigma_\zeta \right)^{1/p_0} \\ \\ + L_{p_0}[\Phi', \bar{D}] + L_{p_0}[A, \bar{D}] \leq \delta \Lambda_{p_0} L_{p_0}[\omega, \bar{D}] + k_0 + k_1 + \pi M, \end{cases} \tag{1.19}$$

consequently

$$L_{p_0}[\omega, \bar{D}] \leq (k_0 + k_1 + \pi M)/(1 - \delta \Lambda_{p_0}) = M_1. \tag{1.20}$$

From (1.20) and the result of Chapter 2 in [140], the estimate (1.11) can be derived.

Next, we discuss the case of $K > 0$ and suppose that constant q_0 in (1.3) is so small that

$$\delta \Lambda_{p_0} < 1, \quad \delta = q_0. \tag{1.21}$$

Thus, similarly to the case of $K \leq 0$, the estimate (1.11) can be obtained.

Now, we shall prove the result of solvability of Problem A^* and Problem B^* for the complex equation (1.1).

Theorem 1.2 *Under the hypothesis in Theorem 1.1, Problem A^* and Problem B^* for (1.1) have at least one solution.*

Proof Because the solution of Problem B^* is also a solution of Problem A^* for the complex equation (1.1), we only need to prove the solvability of Problem B^* for (1.1).

According to Theorem 1.1, Problem B^* for the complex equation (1.10) has a solution $w_m(z)$ and the solution $w_m(z)$ satisfies the estimate (1.11), i.e.

$$C_\beta[w_m, \bar{D}] \leq M_1, \ L_{p_0}[|w_{m\bar{z}}| + |w_{mz}|, \bar{D}] \leq M_2. \tag{1.22}$$

Hence we can select a subsequence of $\{w_m(z)\}$, which uniformly converges to $w_0(z)$ on $\bar{D}$, and $w_0(z)$ satisfies the boundary condition (1.5), the point condition (1.8) and the first estimate in (1.22). Moreover, from $\{w_{m\bar{z}}(z)\} = \{\omega_m(z)\}$ and $\{w_{mz}(z)\}$, we can choose subsequences $\{\omega_{m_k}(z)\}$ and $\{w_{m_k z}\}$, which weakly converge to $\omega_0(z)$ and $Y_0(z)$ on $\bar{D}$ respectively, and $w_m(z) = \Phi(z) + \hat{T}\omega_m$, where $\Phi(z)$, $\hat{\omega}_m$ are similar to (1.14),(1.15), and $\{\hat{T}\omega_{m_k}\}$ uniformly converges to $\hat{T}\omega_0$ on $\bar{D}$. For the sake of convenience, we denote still $\{w_{m_k}(z)\}$, $\{\omega_{m_k}(z)\}$ and $\{w_{m_k z}\}$ by $\{w_m(z)\}$, $\{\omega_m(z)\}$ and $\{w_{mz}(z)\}$ respectively. Because of the definition of generalized derivatives, for any function $\phi(z) \in D_1^0(D)$, the following formula holds:

$$\iint_D [w_m(z)\phi_z + w_{mz}\phi(z)]d\sigma_z = 0. \tag{1.23}$$

Then letting m tend to infinity, we obtain

$$\iint_D [w_0(z)\phi_z + Y_0(z)\phi(z)]d\sigma_z = 0. \tag{1.24}$$

So

$$[w_0(z)]_z = Y_0(z) = \Phi'(z) + \hat{S}\omega_0, \ z \in D.$$

In the following, we prove that $c_m(z) = F^{(m)}(z, w_m, w_{0z}) - F(z, w_0, w_{0z})$ possesses the property

$$L_{p_0}[c_m, \bar{D}] \to 0 \text{ as } m \to \infty. \tag{1.25}$$

In fact, $F^{(m)}(z, w_m, w_{0z}) - F^{(m)}(z, w_0, w_{0z})$ and $F^{(m)}(z, w_0, w_{0z}) - F(z, w_0, w_{0z})$ converge to 0 for almost every point $z \in D$, hence $c_m(z)$ possesses the same property. Thus for two arbitrary sufficiently small positive constants ε_1 and ε_2, there exists a subset D_* in D and a positive integer N, so that $\text{meas}D_* < \varepsilon_1$ and $|c_m| < \varepsilon_2$ on $\bar{D}\backslash D_*$ for $n > N$. By the Hölder inequality and the Minkowski inequality, we have

$$L_{p_0}[c_m, \bar{D}] \leq L_{p_0}[c_m, D_*] + L_{p_0}[c_m, \bar{D}\backslash D_*]$$

$$\leq L_{p_0}\{[Q_m(z, w_m, w_{0z}) - Q(z, w_0, w_{0z})]w_{0z} + A(z, w_m)$$

$$- A(z, w_0), D_*\} + \varepsilon_2(\text{meas}D)^{1/p_0} \tag{1.26}$$

$$\leq 2L_{p_0}[\Phi' + \hat{S}\omega_0, D_*] + k(\varepsilon_1) + \varepsilon_2\pi^{1/p_0}$$

$$\leq 2\varepsilon_1^{1/p_2}L_{p_1}[\Phi' + \hat{S}\omega_0, D_*] + k(\varepsilon_1) + \varepsilon_2\pi^{1/p_0} = \varepsilon,$$

where $k(\varepsilon_1) \to 0$ as $\varepsilon_1 \to 0, p_1$ is a constant such that $2 < p_0 < p_1 < \min(p, 1/(1-\alpha))$ and $p_2 = p_0 p_1/(p_1 - p_0)$. Therefore formula (1.25) is true.

It remains to prove that $F^{(m)}(z, w_m, w_{mz}) - F^{(m)}(z, w_m, w_{0z})$ weakly converges to 0 on $\bar{D}$. Due to

$$|F^{(m)}(z, w_m, w_{mz}) - F^{(m)}(z, w_m, w_{0z})| \leq |(w_m - w_0)_z|, \tag{1.27}$$

we only prove that $g_m(z) = |(w_m - w_0)_z|$ weakly converges to 0 on $\bar{D}$. Otherwise, we may choose a subsequence of $\{g_m(z)\}$ which weakly converges to $g_0(z)$ on $\bar{D}$ and $g_0(z) > 0$ on a positive measurable set $E \subset D$. However $(w_m - w_0)_z$ weakly converges to 0 on $\bar{D}$. This contradiction proves that $|F^{(m)}(z, w_m, w_{mz}) - F^{(m)}(z, w_m, w_{0z})|$ and $F^{(m)}(z, w_m, w_{mz}) - F^{(m)}(z, w_m, w_{0z})$ weakly converge to 0 on $\bar{D}$, i.e. $F^{(m)}(z, w_m, w_{mz})$ weakly converges to $F(z, w_0, w_{0z})$ on $\bar{D}$. On the basis of the definition of the generalized derivative $w_{m\bar{z}}$, when $m \to \infty$, for any function $\phi(z) \in D_1^0(D)$, we have

$$\iint_D [w_0(z)\phi_{\bar{z}} + F(z, w_0, w_{0z})\phi(z)]d\sigma_z = 0. \tag{1.28}$$

Hence $w_{0\bar{z}} = F(z, w_0, w_{0z})$. This shows that $w_0(z)$ is a solution of the complex equation (1.1) in D. This completes the proof of Theorem 1.2.

If $A(z,w)$ in the complex equation (1.1) is replaced by $A_1(z,w)w + A_2(z,w)\bar{w} + A_3(z,w)$, in this case, the complex equation (1.1) is represented by

$$w_{\bar{z}} = F(z,w,w_z), \ F = Qw_z + A_1 w + A_2 \bar{w} + A_3, \tag{1.29}$$

where $Q = Q(z,w,w_z)$, $A_j = A_j(z,w)(j=1,2,3)$ satisfy

$$L_{p_0}[A_j(z,w), \bar{D}] \leq \varepsilon k_0, \ j = 1,2, \ L_{p_0}[A_3(z,w), \bar{D}] \leq k_0, \tag{1.30}$$

in which $\varepsilon \, (0 < \varepsilon \leq 1)$ is a constant. Then using a method similarly as before, we can obtain the result of solvability of Problem A^* and Problem B^* for (1.29), as a corollary.

Corollary 1.3 *Suppose that the complex equation* (1.29) *satisfies the condition* (1.2), (1.3) *and* (1.30), *that the constant* ε *in* (1.30) *and the constant* $q_0(K \geq 0)$ *in* (1.3) *are small enough. Then Problem A^* and Problem B^* for* (1.29) *are solvable* (see [139]14))

1.2 Uniqueness of solutions of Riemann–Hilbert problem for degenerate elliptic complex equations

Finally, we discuss the uniqueness of solutions of Problem A^* and Problem B^* for (1.29). We assume that $F(z,w,U)$, $Q(z,w,U)$ satisfy

$$F(z,w_1,U) - F(z,w_2,U) = B(z,w_1,w_2,U)(w_1 - w_2),$$
$$|B(z,w_1,w_2,U)| \leq \varepsilon, \ |Q(z,w,U)| < 1 \text{ for } 0 < |U| \leq \infty, \tag{1.31}$$

and $|Q(z,w,U)|$ is a continuous function in the third variable U on $0 \leq |U| \leq \infty$, $|Q(z,w,|U|^{1/2})|^2|U|$ is increasing and concave in U, for any $w, w_1, w_2 \in C(\bar{D})$, $U(z) \in L_2(\bar{D})$ and almost every point $z \in D$.

Theorem 1.4 *If the complex equation* (1.29) *satisfies the conditions stated in Corollary* 1.3 *and* (1.31) *holds, then the solution of Problem A^* (or Problem B^*) for* (1.29) *is unique.*

Proof As in (1.14), we denote by $\Phi(z)$ the unique solution of Problem B^* for analytic functions. Let $w_1(z)$, $w_2(z)$ be two solutions of Problem B^* for (1.29), and substitute $w_1(z)$ and $w_2(z)$ into (1.29). Since $w_j(z)(j=1,2)$ can be expressed as

$$w_j(z) = \Phi(z) + \hat{T}\omega_j, \ \omega_j(z) = w_{j\bar{z}} \in L_2(\bar{D}), \ j = 1,2, \tag{1.32}$$

by condition (1.31), we have

$$\omega_1 - \omega_2 = F(z, \Phi + \hat{T}\omega_1, \Phi' + \hat{S}\omega_1) - F(z, \Phi + \hat{T}\omega_2, \Phi' + \hat{S}\omega_2) =$$
$$= Q[z, w_1, \hat{S}(\omega_1 - \omega_2)]\hat{S}(\omega_1 - \omega_2) + B(z, w_1, w_2, \Phi' + \hat{S}\omega_2)\hat{T}(\omega_1 - \omega_2). \tag{1.33}$$

According to (1.31) and using the Jensen inequality, from

$$|\omega_1 - \omega_2|^2 \leq |Q[z, w_1, \hat{S}(\omega_1 - \omega_2)]|^2 |\hat{S}(\omega_1 - \omega_2)|^2$$

$$+\varepsilon[2|\hat{S}(\omega_1 - \omega_2)| + \varepsilon|\hat{T}(\omega_1 - \omega_2)|]|\hat{T}(\omega_1 - \omega_2)|$$

it can be derived that

$$\| \omega_1 - \omega_2 \|_{L_2}^2 \leq Q^2(\| \hat{S}(\omega_1 - \omega_2) \|_{L_2}) \| \hat{S}(\omega_1 - \omega_2) \|_{L_2}^2 + \varepsilon M_3 \| \omega_1 - \omega_2 \|_{L_2}^2$$

$$\leq [Q^2(\| \hat{S}(\omega_1 - \omega_2) \|_{L_2})\Lambda_2 + \varepsilon M_3] \| \omega_1 - \omega_2 \|_{L_2}^2, \tag{1.34}$$

where M_3 is a real constant. Choosing ε sufficiently small, there exists a constant $q_1 < 1$, so that

$$\| \omega_1 - \omega_2 \|_{L_2} \leq q_1 \| \omega_1 - \omega_2 \|_{L_2} < \| \omega_1 - \omega_2 \|_{L_2} . \tag{1.35}$$

Hence $\omega_1(z) = \omega_2(z)$ for almost every point $z \in D$. Thus $w_1(z) = w_2(z)$ in D.

Obviously, the solution of Problem A^* for (1.29) is also unique (see [73]).

2 Mixed Boundary Value Problem for a Class of Nonlinear Degenerate Elliptic Equations of Second Order

2.1 Formulation of the mixed boundary value problem

Boundary value problems for degenerate elliptic equations of second order were investigated by many authors (see [109],[132],[65],[90],[125]), which include some linear and nonlinear degenerate elliptic equations. In this section, we mainly discuss the following nonlinear degenerate elliptic equations of second order:

$$L_1(u) = a_1 y^m u_{xx} + 2a_2 u_{xy} + a_3 u_{yy} + d_1 u_x + d_2 u_y + d_3 u = 0, \tag{2.1}$$

$$L_2(u) = a_1 u_{xx} + 2a_2 u_{xy} + a_3 y^m + d_1 u_x + d_2 u_y + d_3 u = 0, \tag{2.2}$$

where the coefficients

$$a_j = a_j(x, y, u, u_x, u_y, u_{xx} - u_{yy}, u_{xy}),$$

$$d_j = d_j(x, y, u, u_x, u_y), \ j = 1, 2, 3 \tag{2.3}$$

with the condition

$$a_1 a_3 - a_2^2 > \Delta_0 > 0, \ a_1 > 0, \ |a_1| + |a_2| + |a_3| \leq k \text{ in } D, \tag{2.4}$$

in which Δ_0 and k are positive constants, and D is a simply connected domain bounded by $\partial D \in C_\mu^2 (0 < \mu < 1)$ consisting of a curve Γ in $\text{Im } z > 0$ and a segment $\gamma = [0, 1]$ on the x-axis. Using the same method as in Section 1 of Chapter 3, [140], the equations (2.1),(2.2) can be transformed into the complex forms

$$\begin{cases} u_{z\bar{z}} = F(z, u, u_z u_{zz}), \ F = \mathrm{Re}[Qu_{zz} + A_1 u_z] + A_2 u, \\ Q = Q(z, u, u_z, u_{zz}) = -(a_1 y^m - a_3 + 2ia_2)/(a_1 y^m + a_3), \\ A_1 = A_1(z, u, u_z) = -(d_1 + id_2)/(a_1 y^m + a_3), \\ A_2 = A_2(z, u, u_z) = d_3/2(a_1 y^m + a_3), \end{cases} \tag{2.5}$$

$$\begin{cases} u_{z\bar{z}} = F(z, u, u_z, u_{zz}), \ F = \mathrm{Re}[Qu_{zz} + A_1 u_z] + A_2 u, \\ Q = Q(z, u, u_z, u_{zz}) = -(a_1 - a_3 y^m + 2ia_2)/(a_1 + a_3 y^m), \\ A_1 = A_1(z, u, u_z) = -(d_1 + id_2)/(a_1 + a_3 y^m), \\ A_2 = A_2(z, u, u_z) = -d_3/2(a_1 + a_3 y^m), \end{cases} \tag{2.6}$$

respectively. We assume that (2.5) and (2.6) satisfy

Condition C_* (1) The functions $Q(z, u, u_z, V)$ and $A_j(z, u, u_z)(j = 1, 2)$ are continuous in $u \in I\!\!R$ and $u_z \in \mathbb{C}$ for $V \in \mathbb{C}$ and almost every point $z \in D$.

(2) The above functions are measurable in $z \in D$ for any continuously differentiable function $u(z)$ and measurable function $V(z)$ in D and satisfy

$$|A_j(z, u, u_z)| \le k_0 < \infty, \ j = 1, 2, \ A_2(z, u, u_z) \ge 0, \tag{2.7}$$

and $A_j = 0$ for $z \notin D$, $j = 1, 2$, where k_0 is a real constant.

(3) Equations (2.5) and (2.6) satisfy the ellipticity condition

$$|F(z, u, u_z, V_1) - F(z, u, u_z, V_2)| \le q_0(D_*)|V_1 - V_2| \tag{2.8}$$

for almost every point $z \in D$, $u \in I\!\!R$, $u_z, V_1, V_2 \in \mathbb{C}$, where D_* is any closed subset on $\bar{D} \cap \{y > 0\}$ and $q_0(D_*)(0 \le q_0(D_*) \le 1)$ is a constant which depends only to D_*.

It is easy to see that $|Q| < 1$ for $z \in D$ and $|Q| = 1$ for $z \in \gamma$, this shows that the equations (2.5) and (2.6) are degenerate on γ.

Problem M The mixed boundary value problem M for (2.5) (or (2.6)) is to find a continuously differentiable solution $u(z)$ in D, which is continuous in $\bar{D}$ and satisfies the boundary condition

$$l(u) = \alpha_1(t)\frac{\partial u}{\partial \nu} + \alpha_2(t)u(t) = r(t), \ t \in \partial D, \tag{2.9}$$

where $\alpha_1 \ge 0$, $\alpha_2 > 0$, $\alpha_1^2 + \alpha_2^2 = 1$, $\alpha_1(x) = 0$, $0 \le x \le 1$, $\nu = \nu_1 + i\nu_2$, and the angle between $\nu(t)$ and the outer normal direction of Γ is less than $\pi/2$, and we suppose that

$$C_\mu^1[\alpha_j(t), \partial D] \le l_0, \ C_\mu^1[r(t), \partial D] \le l_0, \ C_\alpha^1[\nu_j(t), \Gamma] \le l_0, \ j = 1, 2, \tag{2.10}$$

in which $\mu(1/2 < \mu < 1)$, $l_0\,(0 \leq l_0 < \infty)$ are real constants.

Problem M^*   The mixed boundary value problem M^* for (2.6) is to find a bounded solution $u(z)$, which is continuous on $D \cap \Gamma$ and satisfies the boundary condition

$$l(u) = \alpha_1(t)\frac{\partial u}{\partial \mu} + \alpha_2(t)u(t) = r(t), \ t \in \Gamma, \tag{2.11}$$

where $\nu(t)$, $\alpha_j(t)(j = 1, 2)$, $r(t)$ coincide on Γ with those in (2.9) and (2.10).

2.2   Existence of solutions for the mixed boundary value problems

Theorem 2.1   *Let equation (2.6) satisfy Condition C_*. Then Problem M^* for (2.6) is solvable.*

Proof   Let $\{\varepsilon_n\}$ be a decreasing sequence of positive numbers with the condition $\lim_{n\to\infty} \varepsilon_n = 0$. For each ε_n, we find a domain D_n as follows:

(1) $\overline{D_n} \subset D_{n+1} \subset D$ and $D_n \subset \{y \geq \varepsilon_n\}$, $\partial D_n \in C_\mu^2\,(0 < \mu < 1, n = 1, 2, \cdots)$.

(2) In $\{y \geq 2\varepsilon_n\}$, $D_n \cap \{y \geq 2\varepsilon_n\} = D \cap \{y \geq 2\varepsilon\}$; in $\{\varepsilon_n < y < 2\varepsilon_n\}$, the boundary of D_n consists of smooth curves S_n^1 in $\varepsilon_n < y < 3\varepsilon_n/2$ and S_n^2 in $3\varepsilon_n/2 < y < 2\varepsilon_n$. It is clear that (2.6) on $\overline{D_n}$ is a uniformly elliptic equation.

We continuously extend $r(t)$ on the closed domain $\bar{D}$ so that

$$C_\mu^1[r, \bar{D}] \leq M_1 = C_\mu^1[r, \partial D] + 1. \tag{2.12}$$

Now we discuss the following boundary value problem

$$\begin{cases} u_{z\bar{z}} = \mathrm{Re}[Qu_{zz} + A_1 u_z] + A_2 u \ \text{in} \ D_n, \\[2mm] l_n(u) = \alpha_{1n}(t)\dfrac{\partial u}{\partial \nu_n} + \alpha_{2n}(t)u(t) = r(t) \ \text{on} \ \partial D_n, \end{cases} \tag{2.13}$$

where $\nu_n = \nu$, $\alpha_{1n} = \alpha_1$, $\alpha_{2n} = \alpha_2$ for $t \in \{\partial D_n\} \cap \{y \geq 2\varepsilon_n\}$, $\nu_n \in C_\mu$; $\alpha_{1n} = 1$, $\alpha_{2n} = 0$ for $t \in \{\partial D_n\} \cap \{\varepsilon_n \leq y < 3\varepsilon_n/2\}$, and $\nu_n \in C_\mu$; $\alpha_{1n} \geq 0$, $0 < \alpha_{2n} \leq 1$, $\alpha_{1n}, \alpha_{2n} \in C_\mu^1$ for $t \in \{\partial D_n\} \cap \{3\varepsilon/2 \leq y < 2\varepsilon_n\}$, and the angle between ν_n and the outer normal direction of ∂D_n is less than $\pi/2$. According to Theorem 4.6, Theorem 4.4 in Chapter 1, there exists a solution $u_n(z) \in C(\overline{D_n})$ of the above boundary value problem (2.13), and $u_n(z)$ satisfies the estimate

$$C_\beta^1[u_n, D_n^*] \leq M_2, \ L_{p_0}[|u_{nz\bar{z}}| + |u_{nzz}|, D_n^*] \leq M_3, \tag{2.14}$$

where $\beta = 1 - 2/p_0$, $D_n^* = \overline{D_n} \cap \{\cap_{t^* \in \partial \Gamma_*}(|z - t^*| \geq \varepsilon > 0\}$, $2 < p_0 < 1/(1 - \mu)$, $M_j = M_j(q_0, p_0, k_0, l_0, \mu, D_n, \Gamma_*, \varepsilon)$, $j = 2, 3$, and $\Gamma_* = \{t \in \partial D_n \,|\, \alpha_1(t) = 0\}$, $\Gamma_{**} = \{t \in \Gamma \,|\, \alpha_1(t) = 0\}$. By Theorem 4.2 of Chapter 1, it can be derived that

$$|u_n(z)| \leq M_4 \ \text{in} \ D_n^* \tag{2.15}$$

in which M_4 is a constant independent of n. Hence, from $\{u_n(z)\}$ we can select a subsequence $\{u_{n_k}(z)\}$, which uniformly converges to $u_0(z)$ on any closed domain D_n^{**} in $D\backslash\Gamma_*$, and $\{u_{n_k z}\}$ uniformly converges to u_{0z} on D_n^{**}. It can be seen that $u_0(z)$ is a bounded solution of equation (2.6) and satisfies the boundary condition $l(u_0) = r(t)$ on $\Gamma\backslash\Gamma_{**}$. It remains to prove that $u_0(z)$ is continuous on Γ and satisfies $l(u_0) = r(t)$ on Γ_{**}.

Choosing an arbitrary point $z^* \in \Gamma_{**}$ and denoting $\Gamma_\delta = \bar{D}\cap\{|z - z^*| < \delta\}\cap\Gamma(0 < \delta < y^*/2 = \mathrm{Im}\, z^*/2)$, we define a real-valued function $g(z) \in C_\mu^1(\Gamma)$ as follows:

$$g(z) = \begin{cases} M_5 + 1, & \text{for } z \in \Gamma\backslash\Gamma_{\delta/2}, \\ \eta > 0, & \text{for } z \in \Gamma_{\delta/4}, \\ \text{a monotone function}, & \text{for } z \in \Gamma_{\delta/2}\backslash\Gamma_{\delta/4}, \end{cases} \tag{2.16}$$

where $\eta \le |g(z)| \le M_5 + 1$, M_5 is an unknown constant. It is clear that $g(z)$ satisfies the estimate

$$C_{\mu-\varepsilon}^1[g, \Gamma] \le M_6/\delta^{1+\mu-\varepsilon}, \tag{2.17}$$

in which $1/2 < \mu-\varepsilon < 1$, $M_6 = M_6(M_5, \Gamma)$. Let $u_n(z)$ be substituted in the coefficients of (2.6), and $\hat{u}_n(z)$ be a solution of equation (2.6) on $\overline{D_n}$ satisfying the boundary condition

$$\hat{u}_n(t) = f(t),\ t \in \partial D_n,\ f(t) = \begin{cases} g(t), & \text{for } z \in \Gamma, \\ M_5 + 1, & \text{for } \partial D_n\backslash\Gamma. \end{cases} \tag{2.18}$$

We can prove that $\hat{u}_n(z)$ satisfies the estimate

$$C[\hat{u}_n(z), \overline{D_n}] \le M_7/\delta^{1+\mu-\varepsilon}, \tag{2.19}$$

where $M_7 = M_7(M_5, \Gamma)$. Next, let $u_n^*(z)$ be a solution of (2.6) on $\overline{D_n}$ satisfying the boundary condition

$$u_n^*(t) = r(t)/\alpha_{2n}(t),\ t \in \partial D_n, \tag{2.20}$$

and $\tilde{u}_n(z) = u_n(z) - u_n^*(z)$, where $u_n(z)$ is a solution of (2.13) i.e. the equation (2.6) satisfying the boundary condition

$$l_n(u_n) = \alpha_{1n}(t)\frac{\partial u_n}{\partial \nu_n} + \alpha_{2n}(t)u_n(t) = r(t),\ t \in \Gamma \cap \{y \ge \varepsilon_n\}. \tag{2.21}$$

Finally, we consider the solution $\tilde{u}(z) = \hat{u}_n(z) \pm \tilde{u}_n(z)$ of (2.6). It is evident that $\tilde{u}(z) = M_5 + 1 - M_5 > 0$ on $\partial D_n\backslash\Gamma_{\delta/2}$, where $M_5 = M_4 + \max_{t\in\partial D_n} r(t)/\alpha_{2n}(t)$. As in Theorem 4.6 of Chapter 1, we can prove $\tilde{u}(z) \ge 0$ on $\Gamma_{\delta/2}$. Hence $\hat{u}_n(z) \pm \tilde{u}_n(z) \ge 0$, i.e. $\hat{u}_n(z) \ge \mp\tilde{u}_n(z)$ on $\overline{D_n}$. In particular, we have $|\tilde{u}_n(z)| \le 2\eta$ on $\Gamma_{\delta/4}$. Noting that $u_n(z)(n = 1, 2, \cdots)$ are equicontinuous on $\overline{D_n}$, there exists a neighborhood of z^* on $\overline{D_n}$, so that $|\tilde{u}_n(z)| \le 2\eta$. Consequently the limit function $\tilde{u}_0(z)$ of $\tilde{u}_n(z)$ as $n \to \infty$ satisfies $|\tilde{u}_0(z)| \le 2\eta$. Taking into account the arbitrariness of η, it can be derived that $\tilde{u}_0(z)$ is continuous at $z^* \in \Gamma_{**}$ and $\tilde{u}_0(z^*) = 0$. Hence, $u_0(z) = \tilde{u}_0(z) + u_0^*(z)$ is continuous at $z^* \in \Gamma_{**}$, where $u_0^*(z)$ is the limit function of a subsequence of $\{u_n^*(z)\}$. So $u_0(z)$ satisfies the boundary condition $l(u_0) = r(t)$, $t \in \Gamma_{**}$.

Theorem 2.2 *Suppose that equation (2.5) satisfies Condition C^* and the angles between ν in (2.9) and the y–axis are less than $\pi/2$. Then Problem M for equation (2.5) is solvable.*

Proof As in the proof of Theorem 2.1, we choose the domain D_n and prove that there exists a solution $u_n(z)$ of the boundary value problem

$$
\begin{cases}
u_{z\bar{z}} - \mathrm{Re}[Qu_{zz} + A_1 u_z] - A_2 u = 0, \ z \in D_n, \\[2mm]
l_n(u) = \alpha_{1n}(t)\dfrac{\partial u}{\partial \nu} + \alpha_{2n}(t)u(t) = r(t), \ t \in \partial D_n,
\end{cases}
\tag{2.22}
$$

where we assume that the angles between ν and the y–axis on $3\varepsilon_n/2 \le y \le 2\varepsilon_n$ are less than $\pi/2$. The solution $u_n(z)$ satisfies the estimate (2.14). Hence, we can choose a subsequence of $\{u_n(z)\}$, which uniformly converges to $u_0(z)$, and $u_0(z)$ is a bounded solution of (2.5) in D satisfying the boundary condition $l(u_0) = r(t)$ on Γ. It remains to prove that $u_0(z)$ satisfies the boundary condition $l(u_0) = r(t)$ on γ, i.e.

$$
u_0(x) = r(x), \ 0 \le x \le 1.
\tag{2.23}
$$

We choose an arbitrary point $z = x_0\,(0 < x_0 < 1)$ and its one–sided neighborhood $U = \{|x - x_0| < \delta, 0 < y < 1/\eta\}$, in which δ and η are unknown positive constants to be determined appropriately. For arbitrary number $\beta > 0$, there exist positive constants $\delta(\beta)$ and η_0, so that $|r(z) - r(x_0)| < \beta$ on $|x - x_0| \le \delta(\beta), 0 < y \le 1/\eta_0$. Setting $\delta = \min\{x_0, 1 - x_0, \delta(\beta)\}$, it is easy to see that $\{z\,||x - x_0| \le \delta, y = \varepsilon_n\} \subset \partial D_n \cap \{y = \varepsilon_n\}$, where $0 < \varepsilon_n < 1/\eta$. We introduce the auxiliary real functions

$$
\begin{cases}
W^{\pm}(z) = CV(z) + \beta \pm r(x_0) \mp u_n(z), \\[2mm]
V(z) = (x - x_0)^2 + 3\eta y - \eta^2 y^2 \ \text{on} \ \overline{E_n} = \overline{U \cup D_n},
\end{cases}
\tag{2.24}
$$

in which $\eta(\eta > \eta_0, \varepsilon_n < 1/\eta)$ and $C(> 0)$ are unknown constants. We can verify that

$$
W_{z\bar{z}}^{\pm} - \mathrm{Re}[QW_{zz}^{\pm} + A_1 W_z^{\pm}] - A_2 W^{\pm} \le \{2k_2 C - 2\eta^2 k_1 C
$$
$$
+ k_3[Ck_0(\delta^2 + (3\eta/2)^2)^{1/2} + Ck_0(\delta^2 + 3\eta) + k_0(\beta + M_1)]\}/k_1 \le 0 \ \text{in} \ E_n,
\tag{2.25}
$$

where $\inf_{z \in \bar{D}}\{a_1 y^m + a_3\} \ge k_1 > 0$, $\sup_{z \in \bar{D}}\{a_1 y^m\} \le k_2 < \infty$, $\sup_{z \in \bar{D}}\{a_1 y^m + a_3\} \le k_3 < \infty$, k_1, k_2 and k_3 are real constants, and for the constant C given, we select that $\eta = \eta(\beta)$ is large enough, such that the second inequality of (2.25) holds. According to the extremum principle of solutions for the second equation as stated in Theorem 2.9, Chapter 3, [140], we know that $W^{\pm}(z)$ cannot attain a negative minimum in $E_n = U \cap D_n$. Now we prove that $W^{\pm}(z) \ge 0$ on ∂E_n. In fact, $u_n(z)$ satisfies the boundary condition $l_n(u_n) = r(t)$ on $\partial E_n \cap \{y = \varepsilon_n\}$, it implies that $u_n(z) = r(t)$, and then $W^\pm(z) \geq \beta - |r(x_0) - r(x)| > 0$. In addition, it is clear that

$$V(z) \geq \min\{\delta^2, 2\} \text{ on } \partial E_n \cap (\{|x - x_0| = \delta\} \cup \{y = 1/\eta\}). \tag{2.26}$$

Provided that the constant C is large enough, such that $C\delta^2 > M_1 + M_4$, where M_1, M_4 are as stated in (2.12), (2.15), we can obtain that $W^\pm(z) \geq C\delta^2 - |r(x_0) - u_n(z)| > 0$. In brief, first choosing the constant C, and then selecting a proper constant η, it can be obtained that $W^{\pm(z)} \geq 0$ on $\overline{E_n}$, namely

$$CV(z) + \beta \pm r(x_0) \mp u_n(z) \geq 0. \tag{2.27}$$

Let $\varepsilon_n \to 0$, we conclude that

$$-CV(z) - \beta \leq u_0(z) - r(x_0) \leq \beta + CV(z), \text{ and } \varlimsup_{z \to x_0} |u_0(z) - r(x_0)| \leq \beta.$$

Observing the arbitrariness of β, thus $u_0(x_0) = r(x_0)$.

Similarly, we consider the part $U = \{|x| < \delta, 0 < y < 1/\eta\}$ of a neighborhood of $z = 0$, where δ and $\eta\,(\varepsilon_n < 1/\eta)$ are undetermined constant. For arbitrary positive number β, there exists a positive constant $\delta(\beta)$, so that $|r(z) - r(0)| < \beta/6$, when $z \in \bar{U} \cap D$. We introduce the functions $W^\pm(z) = CV(z) + \beta \pm r(0) \mp u_n(z)$ on $\overline{E_n} = \overline{U \cap D_n}$, where $V(z) = x^2 + 3\eta y - \eta^2 y^2$, and C is an unknown positive constant. Similarly to the proof as stated before, it can be verified that $W^\pm(z)$ cannot attain a negative minimum in E_n, and $W^\pm(z) \geq 0$ on ∂E_n. In the following, we only prove that $W^\pm(z) \geq 0$ on $\partial E_n \cap (S_n^2 \cup \Gamma)$. When $z \in \partial E_n \cap (S_n^2 \cup \Gamma)$, the angle between ν and the y-axis is less than a positive number $\psi(< \pi/2)$ and $\cos\psi > 0$. Hence

$$\frac{\partial V}{\partial \nu_n} \geq -2(X_1 - X_2) + \eta\cos\psi,$$

where $X_1 = \max_{z \in \bar{D}} x$, $X_2 = \min_{z \in \bar{D}} x$. If we choose $\eta > 2[(X_1 - X_2) + 1]/\cos\psi$, then $\partial V/\partial \nu_n > 2$. We separate the boundary $\partial E_n \cap (S_n^2 \cup \Gamma)$ in two parts: $L_1 = \{t\,|\,\alpha_{1n} < \Delta = \beta/(3M_1)\}$ (we may assume that $\Delta < 1$) and $L_2 = \{t\,|\,\alpha_{1n} > \Delta\}$. Noting that $\alpha_{1n}^2 + \alpha_{2n}^2 = 1$, $\alpha_{2n} > 0$ on L_1, it is evident that $\alpha_{2n} > 1 - \alpha_{1n}^2/2 > 1/2$ and $\alpha_{2n} > 1 - \Delta$. Hence,

$$l_n(W^\pm) \geq \alpha_{2n}[\beta \pm r(0)] \mp r(z) > \beta/3 - M_1(1 - \alpha_{2n}) > 0 \text{ on } L_1.$$

Moreover, setting that

$$C > M_1/\Delta \text{ on } L_2, \tag{2.28}$$

it is clear that

$$l(W^\pm) \geq 2C\Delta - 2M_1 > 0,$$

where M_1 is a constant as stated in (2.12). This shows that $W^\pm(z)$ cannot attain a negative value. Thus, $W^\pm(z) \geq 0$ on $\overline{E_n}$, and then we can derive that $u_0(0) = r(0)$. Similarly, we can also prove that $u_0(1) = r(1)$. This completes the proof.

Theorem 2.3 *If equation (2.6) satisfies Condition C^* and one of the following conditions:*

(1) $0 < m < 1$;

(2) $m = 1$, $d_0 = \overline{\lim}_{y \to 0} d_2 < \inf_{z \to D}\{a_3\} = a_0$ and $d_2(x, y, \zeta, \xi, \eta) \in C(\bar{D} \times I\!\!R \times I\!\!R^2)$;

(3) $1 < m < 2$, $d_0 \leq 0$, and $d_2 \in C(\bar{D} \times I\!\!R \times I\!\!R^2)$;

(4) $m > 2$, $d_0 < 0$, and $d_2 \in C(\bar{D} \times I\!\!R \times I\!\!R^2)$.

Then Problem M for (2.6) is solvable, where the angles between ν and the $y-$axis is less than $\pi/2$ at $z = 0$ and $z = 1$.

Proof As in the proof of Theorem 2.2, we choose the domain D_n and prove that there exists a function $u_0(z)$ which satisfies the equation (2.6) and the boundary condition $l(u_0) = r_0(t)$.

Next, we prove that $u_0(z)$ satisfies $l(u_0) = r(t)$ on γ, i.e. $u_0(x) = r(x), 0 \leq x \leq 1$, if one of the conditions (1)–(4) holds. Similarly to the proof of Theorem 2.2, we select any point $z = x_0 \in [0, 1]$ and a part $U = \{|x - x_0| < \delta, 0 < y < 1/\eta\}$ of its neighborhood, and introduce a real function

$$\begin{cases} W^{\pm}(z) = CV(z) + \beta \pm r(x_0) \pm u_n(z), \\ V(z) = (x - x_0)^2 + Cy^{1/\ln \eta}, \end{cases} \tag{2.29}$$

where δ and η are undetermined positive constants so that $\varepsilon_n < 1/\eta$ and $\eta \geq \max[\eta_0, e^2]$. It can be concluded that $W^{\pm}(z) \geq 0$ on ∂E_n. We shall prove that $W^{\pm}(z) \geq 0$ in E_n. By computation, we obtain

$$W_{z\bar{z}}^{\pm} - \mathrm{Re}[QW_{zz}^{\pm} + A_1 W_z^{\pm}] - A_2 W^{\pm}$$

$$= \frac{a_1 C}{a_1 + a_3 y^m} + \frac{C}{2(a_1 + a_3 y^m) \ln \eta}(y^{\frac{1}{\ln \eta} - 1})^{2-m}$$

$$\times [C(\frac{1}{\ln \eta} - 1)y^{\frac{1}{\ln \eta}(m-1)} + \frac{d_2}{(y^{\frac{1}{\ln \eta} - 1})^{1-m}}] + \frac{d_1 C(x - x_0)}{a_1 + a_3 y^m} \tag{2.30}$$

$$- \frac{d_1}{2(a_1 + a_3 y^m)}[C(x - x_0)^2 + Cy^{\frac{1}{\ln \eta}} + \beta \pm r(x_0)].$$

If condition (1) holds, from (2.30) it follows that

$$W_{z\bar{z}}^{\pm} - \mathrm{Re}[QW_{zz}^{\pm} + A_1 W_z^{\pm}] - A_2 W^{\pm} \leq \frac{k_6 C}{k_4}$$

$$- \frac{C}{2k_6 \ln \eta}(y^{\frac{1}{\ln \eta} - 1})^{2-m}[k_5(1 - \frac{1}{\ln \eta})y^{\frac{1}{\ln \eta}(m-1)} \tag{2.31}$$

$$- \frac{k_0 k_6}{(y^{\frac{1}{\ln \eta} - 1})^{1-m}}] + k_0[C\delta + C\delta^2 + C + \beta + M_1],$$

where $\inf_{z \in D}\{a_1\} \geq k_4 > 0$, $\inf_{z \in D}\{a_3\} \geq k_5 > 0$, $\sup_{z \in D}\{a_1 + a_3 y^m\} \leq k_6 < \infty$. The first and third terms in (2.31) are bounded, and $y^{1/\ln \eta} < 1/e$, $y^{1/\ln \eta - 1} > \sqrt{\eta}$, when $0 < y < 1/\eta$, $\eta > e^2$. It can be seen that the second term tends to $-\infty$, as η tends to ∞. Hence $W^{\pm}(z)$ cannot attain a negative value. Similarly, we can also prove that if one of the conditions (2),(3),(4) holds, then $W^{\pm}(z) \geq 0$ on E_n. The remaining proof is the same as for Theorem 2.2. We have $u_0(x_0) = r(x_0)$, $0 \leq x_0 \leq 1$.

2.3   Uniqueness of solutions for the mixed boundary value problem

We assume that equation (2.5) satisfies

$$F(z, u_1, u_{1z}, V) - F(z, u_2, u_{2z}, V) = \mathrm{Re}[B_1(u_1 - u_2)_z] + B_2(u_1 - u_2), \qquad (2.32)$$

where $B_j = B_j(z, u_1, u_2, V)$, $|B_j| \leq k_0 < \infty$, $j = 1, 2$, $B_2 > 0$.

Theorem 2.4   *If equation (2.5) satisfies Condition C_* and (2.32), then the solution of Problem M for (2.5) is unique.*

Proof   Let $u_1(z)$ and $u_2(z)$ be two solutions of Problem M for (2.5). Denoting $u(z) = u_1(z) - u_2(z)$, it can be seen that $u(z)$ satisfies the equation

$$\begin{cases} u_{z\bar{z}} - \mathrm{Re}[qu_{zz} + B_1 u_z] = B_2 u, \ z \in D, \\[2mm] F(z, u_1, u_{1z}, u_{1zz}) - F(z, u_1, u_{1z}, u_{2zz}) = \mathrm{Re}[qu_{zz}] \end{cases} \qquad (2.33)$$

and

$$\begin{cases} l(u) = \alpha_1(t)\dfrac{\partial u}{\partial \nu} + \alpha_2(t)u(t) = 0, \ t \in \Gamma, \\[2mm] u(x) = 0, \ x \in \gamma. \end{cases} \qquad (2.34)$$

We choose an arbitrary positive number λ and let $U(z) = u(z)/\lambda$. On the basis of Theorem 3.2, the estimate

$$|u(z)| \leq \lambda M_8 \ \text{on} \ \bar{D} \qquad (2.35)$$

can be obtained, where M_8 is a constant. Letting λ tend to 0, it can be derived that $u(z) = 0$, i.e. $u_1(z) = u_2(z)$ in D.

Theorem 2.5   *Under the same hypotheses as in Theorem 2.3, the homogeneous boundary value problem M_0 for (2.6) has only the trivial solution, where the angles between ν and the y-axis at $z = 0, z = 1$ are less than $\pi/2$. In addition, if the boundary condition $u(t) = 0$ on γ is replaced by the boundedness of the solution on γ and $u(0) = u(1) = 0$, then equation (2.6) has a unique identically vanishing solution.*

Proof   We consider the auxitrary function

$$\begin{cases} W^{\pm}(z) = \beta V(z) \mp u(z), \ z \in D, \\[2mm] V(z) = y^{1/\ln \eta} - (x - x_*)^n + H, \end{cases} \qquad (2.36)$$

where η, n and H are unknown constants to be determined appropriately, $x_* = \min_{z \in D} x - 1$, and $\beta > 0$ is a constant.

We first prove that if one of the conditions (1)-(4) holds, then $W^{\pm}(z)$ cannot attain a negative minimum. In the following, we only prove that under the condition (2), the above conclusion is true. The remaining cases can be analoguously discussed.

Without loss of generality, we may assume that $\eta > e^2$ and $H \geq H_0$, where $H_0 = (X_1 - x_*)^n$, $X_1 = \max_{z \in \bar{D}} x$. Now, we distinguish between two steps:

(1) If $y < \varepsilon$, there is no harm in assuming that $\varepsilon > 1/\eta$. Since $d_0 < a_0$, there exist two constants $\mu_0 (0 < \mu_0 < a_0)$ and $\eta_2 > 0$, so that $d_2 \leq a_0 - \mu_0$, when $\eta > \eta_2$. Moreover,

$$
\begin{aligned}
W^{\pm}_{z\bar{z}} &- \mathrm{Re}[QW^{\pm}_{zz} + A_1 W^{\pm}_z] - A_2 W^{\pm} \\
&\leq -\frac{\beta n(X_1 - x_*)^{n-2}}{2(a_1 + a_3 y^m)}[(n-1)a_1 + d_2(X_1 - x_*)] \\
&\quad + \frac{\beta}{2(a_1 + a_3 y^m)}\left(\frac{1}{\ln\eta} y^{\frac{1}{\ln\eta}-1}\right)[C(\frac{1}{\ln\eta}-1) + d_2] \\
&\leq -\frac{\beta n}{2k_6}[(n-1)k_4 - k_6 k_0(X_1 - x_*)] - \frac{\beta}{2k_6}\left(\frac{1}{\ln\eta} y^{\frac{1}{\ln\eta}-1}\right) \\
&\quad \times [C(1 - \frac{1}{\ln\eta}) - d_2],
\end{aligned}
\tag{2.37}
$$

where $\sup_{z \in D}\{a_1 + a_3 y^m\} \leq k_6 < \infty$, $\inf_{z \in D}\{a_1 + a_3 y^m\} \geq k_4 > 0$. If we choose n according to

$$
n \geq 1 + \frac{k_6 k_0(X_1 - x_*)}{k_4},
\tag{2.38}
$$

then the first item on the right–hand side of (2.37) is not positive. Since $(1 - 1/\ln\eta) \to 1$ as $\eta \to \infty$, there exists a constant $\eta_3 > 0$, so that $C(1 - 1/\ln\eta) > a_0 - \mu_0/2$, for $\eta > \eta_3$. Thus the second item in the right–hand side of (2.37) is a negative number, for $\eta > \eta_4 = \max\{e^2, \eta_2, \eta_3\}$. Choosing $\eta = \eta_4 + 1$, it can be seen that the left–hand side in (2.37) is not positive.

(2) If $y \geq \varepsilon$, as before we select $\eta = \eta_4 + 1$, and then have

$$
\begin{aligned}
W^{\pm}_{z\bar{z}} &- \mathrm{Re}[QW^{\pm}_{zz} + A_1 W^{\pm}_z] - A_2 W^{\pm} \\
&\leq -\frac{\beta h}{2k_6}[(n-1)k_4 - k_6 k_0(X_1 - x_*)] + \frac{k_0 \beta}{\varepsilon \ln\eta} \leq 0,
\end{aligned}
\tag{2.39}
$$

for $n \geq \max\{1 + k_6 k_0(X_1 - x_*)/k_4, k_0 k_6/\varepsilon \ln\eta\}$. This shows that $W^{\pm}(z)$ cannot attain a negative minimum in D.

Next, we prove that $W^{\pm}(z) \geq 0$, $z \in \partial D$. Since $u(z)$ is the solution of the stated problem, it is obvious that $u(x) = 0$ on $\gamma = \{0 \leq x \leq 1, y = 0\}$, and $W^{\pm}(z) \geq 0$ on γ.

Now, we prove that $l(W^{\pm}) > 0$ on Γ. It can be assumed that $\alpha_2 \geq k_7 > 0$. When $y < \varepsilon$, the angle between ν and the y–axis is less than $\pi/2$, obviously $\cos(\nu, y) > 0$. Thus

$$l\left(W^{\pm}\right) \geq -\beta n (X_1 - x_*)^{n-1} \max_{z \in \Gamma} a_1 + k_7 \beta [H - (X_1 - x_*)^n]. \tag{2.40}$$

When $H > [n(X_1 - x_*)^{n-1}] \max_{z \in \Gamma} a_1 / k_7 + (X_1 - x_*)^n$, we have $l(W^{\pm}) > 0$. As for $y \geq \varepsilon$, we distinguish between two cases.

(1) If $y \geq 1$, then

$$l\left(W^m\right) \geq -[n(X_1 - x_*)^{n-1} + \frac{1}{y^{1-\frac{1}{\ln \eta}}}]\beta \max_{z \in \Gamma} a_1 + \beta k_7 [H - (X_1 - x_*)^n]. \tag{2.41}$$

Thus $l(W^{\pm}) > 0$, for $H > (X_1 - x_*)^n + [n(X_1 - x_*)^{n-1} + 1] \max_{z \in \Gamma} a_1 / k_7$.

(2) If $y < 1$, we choose $H > (X_1 - x_*)^n + [n(X_1 - x_*)^{n+1} + 1/\varepsilon] \max_{z \in \Gamma} a_1 / k_7$. Then

$$l\left(W^{\pm}\right) \geq -[n(X_1 - x_*)^{n-1} + \frac{1}{y}]\beta \max_{z \in \Gamma} a_1 + k_7 \beta [H - (X_1 - x_*)^n] > 0. \tag{2.42}$$

According to the extremum principle, we can obtain $W^{\pm}(z) \geq 0$ in D, namely $|u(z)| \leq \beta V(z)$ in D. Due to the arbitrariness of β, $u(z) = 0$ in D can be derived.

The statement of the second part can be similarly proved.

Finally, by a similar method as used in the proof of Theorem 2.5, we can prove

Theorem 2.6   *Let equation (2.6) satisfy Condition C^* and let one of the following conditions hold:*

(1) $m = 1$, $d_0 = \underline{\lim}_{y \to +\infty} d_2 \geq \sup_{z \in D} a_3 = a_0$, $d_2(x, y, \zeta, \xi, \eta) \in C(\bar{D} \times I\!\!R \times I\!\!R^2)$;

(2) $1 < m < 2$, $d_0 > 0$, $d_2 \in C(\bar{D} \times I\!\!R \times I\!\!R^2)$;

(3) $m \geq 2$, $d_0 \geq 0$, $d_2 \in C(\bar{D}) \times I\!\!R \times I\!\!R^2)$.

Then the homogeneous boundary value problem M_* (in which the boundary condition $u(t) = 0$ on γ is replaced by the boundedness of the solution on γ) has only the identically vanishing solution, where the angles between ν and the y–axis are not greater than $\pi/2$ at $z = 0$ and $z = 1$.(See [126]1))

3   Oblique Derivative Problem for a Class of Nonlinear Degenerate Elliptic Equations of Second Order

In this section, we discuss the oblique derivative boundary value problem(Problem P) for the nonlinear degenerate elliptic equations (2.5) and (2.6), the boundary condition of which is as follows:

$$l(u) = \frac{\partial u}{\partial \nu} + \sigma(t)u(t) = r(t), \ t \in \partial D = \Gamma \cup \gamma, \tag{3.1}$$

where $\nu \cdot n \geq 0$ on $\overline{\Gamma}$ and $\nu \cdot n > 0$ on γ, $\sigma(t) > 0$ and n is the outer normal direction of ∂D, and $\nu = \nu_1 + i\nu_2$, $\sigma(t)$, $r(t)$ satisfy

$$\begin{cases} C_\mu^1[\nu_j(t), \partial D] \leq l_0, \ j = 1, 2, \\ C_\mu^1[\sigma(t), \partial D] \leq l_0, \ C_\mu^1[r(t), \partial D] \leq l_0, \end{cases} \tag{3.2}$$

where $\mu(0 < \mu < 1)$ and $l_0(0 \leq l_0 < \infty)$ are real constants, and the angle between ν and the y–axis is greater than $\pi/2$ at $z = 0, z = 1$. The another boundary value problem(Problem Q) is to find a bounded solution $u(z)$ of equation (2.6) satisfying the boundary condition

$$l(u) = \frac{\partial u}{\partial \nu} + \sigma(t)u(t) = r(t), \ t \in \partial D, \tag{3.3}$$

in which ν, σ, r satisfy condition (3.2).

As in the proof of Theorem 2.1, we extend $\sigma(z), r(z)$ onto $\overline{D}$, such that

$$C_\mu^1[r, \overline{D}] \leq C_\mu^1[r, \partial D] + C_\mu^1[\sigma, \partial D] + 1 = M_1, \ C_\mu^1[\sigma, \overline{D}] \leq M_1, \ \sigma(t) > 0 \text{ on } \partial D \backslash \gamma, \tag{3.4}$$

and choose a decreasing sequence of numbers $\{\varepsilon_n\}$ so that ε_n tends to 0 as n tends to ∞. Similarly, we have the sequence $\{D_n\}$ of domains, $D_n \in C_\mu^2$, and in $\varepsilon_n < y < 2\varepsilon_n$, the boundary $S_n^1 \cup S_n^2$ of D_n consists of smooth curves. Moreover, $\gamma_n = \{y = \varepsilon_n\} \cap \partial D_n$ for $y = \varepsilon_n$, $\Gamma_n = \partial D_n \backslash \gamma_n$. Next corresponding to (2.13), we consider the following boundary value problem

$$\begin{cases} u_{z\bar{z}} = \text{Re}[Qu_{zz} + A_1 u_z] + A_2 u \text{ in } D_n, \\ l_n(u) = \dfrac{\partial u}{\partial \nu_n} + \sigma_n(t)u(t) = r(t), \ t \in \partial D_n, \end{cases} \tag{3.5}$$

where $\nu_n = \nu$, $\sigma_n = \sigma$, $r = r$ for $t \in (\partial D_n) \cap \{y \geq \varepsilon_n\}$ and the angles between ν_n and the y-axis are greater than $\pi/2$ for $t \in (\partial D_n) \cap \{y < 2\varepsilon_n\}$, the angles between ν_n and the outer normal direction of ∂D_n are less than $\pi/2$, and $\nu_n \in C_\mu$, $\sigma_n \in C_\mu^1$, $r \in C_\mu^1$ on ∂D_n. The equation in (3.5) is uniformly elliptic on D_n. On the basis of Theorem 4.4 of Chapter 1, there exists a solution $u_n(z)$ of problem (3.5) satisfying

$$C_\beta^1[u_n, \overline{D_n}] \leq M_2, \ L_{p_0}[|u_{nz\bar{z}}| + |u_{nzz}|, \overline{D_n}] \leq M_3, \tag{3.6}$$

where $\beta = 1 - 2/p_0$, $p_0(2 < p_0 < \infty)$, $M_j = M_j(q_0, p_0, k_0, \mu, l_0, D_n)$, $j = 2, 3$.

Lemma 3.1 *Let $u_n(z)$ be a solution of the boundary value problem (3.5). Then $u_n(z)$ satisfies the estimate*

$$|u_n(z)| \leq M_4 \text{ on } \overline{D_n}, \tag{3.7}$$

in which the constant M_4 is independent of n.

Proof   We consider the function

$$W^{\pm}(z) = V(z) \pm u_n(z), \quad V(z) = c_1 e^{\eta y} - M_4, \tag{3.8}$$

where M_4, η, c_1 are unknown positive constants with the condition $M_4 > c_1 e^{\eta Y}$, $Y = \max_{z \in \bar{D}} \mathrm{Im}\, z$. It can be derived that

$$W^{\pm}_{z\bar{z}} - \mathrm{Re}[QW^{\pm}_{zz} + A_1 W^{\pm}_z] - A_2 W^{\pm} = \frac{1}{4} c_1 \eta^2 e^{\eta y}$$

$$-\frac{1}{4}\mathrm{Re}[\frac{a_1 y^m + 2bi - a_3}{a_1 y^m + a_3} c_1 \eta^2 e^{\eta y}] - \mathrm{Re}[\frac{d_1 + id_2}{a_1 y^m + a_3} \frac{i}{2} c_1 \eta e^{\eta y}] - A_2 V \tag{3.9}$$

$$> \frac{1}{2} c_1 \eta e^{\eta y} \left[\frac{\eta a_3}{a_1 y^m + a_3} + \frac{d_2}{a_1 y^m + a_3} \right] \geq \frac{1}{2} c_1 \eta [\eta k_1 - k_0] > 0,$$

where $\inf_{z \in \bar{D}} a_3 / 2(a_1 y^m + a_3) = k_1 > 0$ and η is chosen so that

$$\eta > k_0 / k_1. \tag{3.10}$$

According to Theorem 2.5, Chapter 3 in [140], $W^{\pm}(z)$ cannot attain a positive maximum in D_n. On that boundary ∂D_n of D_n, there is

$$\begin{cases} l_n(W^{\pm}) &= \dfrac{\partial V}{\partial \nu_n} + \sigma_n(t) V(t) = c_1 \eta e^{\eta y} \cos(\nu_n, y) + \sigma_n(t)[c_1 e^{\eta y} - M_4] \\[2mm] &\leq c_1 \eta e^{\eta y} - \min_{t \in \Gamma_n} |\sigma_n|(M_4 - c_1 e^{\eta y}) > 0 \text{ on } \Gamma_n, \\[2mm] l_n(W^{\pm}) &\leq c_1 \eta e^{\eta y} \cos(\nu_n, y) \leq c_1 \eta \cos(\nu_n, y) < 0 \text{ on } \gamma_n, \end{cases} \tag{3.11}$$

in which we choose c_1, M_4 so that

$$M_4 > \frac{c_1 \eta e^{\eta Y} + \min_{z \in \Gamma_n} |\sigma_n| c_1 e^{\eta Y}}{\min_{z \in \Gamma_n} |\sigma_n|} \text{ on } \Gamma_n \text{ and } c_1 \eta \min_{t \in \gamma_n} |\cos(\nu_n, y)| > 0 \text{ on } \gamma_n.$$

This shows that $W^{\pm}(z)$ cannot attain a positive maximum on ∂D_n. Consequently $V(z) \pm u_n(z) \leq 0$, i.e. $|u_n(z)| \leq -V(z) \leq M_4$ on $\overline{D_n}$. This is just (3.7).

On the basis of Lemma 3.1 and (3.6) with a similar method as used in the proof of Theorem 2.1, we can select a subsequence $\{u_{n_k}(z)\}$ of $\{u_n(z)\}$, so that $\{u_{n_k}(z)\}$, $\{u_{n_k z}(z)\}$ uniformly converge to $u_0(z)$, u_{0z} respectively and $\{u_{n_k zz}\}$ weakly converges to $\{u_{0zz}\}$. Thus we have got the following statement.

The limit function $u_0(z)$ of $\{u_{n_k}(z)\}$ is a bounded solution of the equation (2.5) and satisfies the boundary condition (3.1) on Γ.

To prove the existence of solutions of Problem P, it is sufficient to verify that the solution $u_0(z)$ satisfies (3.1) on γ. We choose an arbitrary point $z = x_0 \, (0 < x_0 < 1)$

and the part $U = \{|x - x_0| < \delta, 0 < y < 1/\eta\}$ of its neighborhood, where δ $(\delta < \min(x_0, 1 - x_0))$ and η $(\varepsilon_n < 1/\eta)$ are positive constants and $U \subset D$.

Lemma 3.2 *The above solution $u_n(z)$ of the problem (3.5) satisfies the estimate*

$$|u_n(z)| \leq M_5 \text{ on } \overline{E_n} = \overline{U \cap D_n}, \tag{3.12}$$

where the constant M_5 is independent of n.

Proof The boundary condition (3.1) can be written in the complex form

$$\text{Re}\left[\overline{\lambda(t)}u_t\right] + \sigma(t)u(t)/2 = r(t)/2 \text{ on } \partial E_n, \tag{3.13}$$

in which $\lambda(t) = \cos(\nu, x) - i\cos(\nu, x)$. If the estimate (3.12) is not true, then there exists a sequence of second order equations

$$u_{mz\bar{z}} = \text{Re}\left[Q_m u_{mzz} + A_{1m}u_{mz}\right] + A_{2m}u_m, \; m = 1, 2, \cdots \tag{3.14}$$

satisfying Condition C^* and a sequence $\{u_m(z)\}$ of corresponding solutions satisfying the boundary condition (3.13), such that $\lim_{m \to \infty} H_m = \infty$, where $H_m = \max_{z \in \overline{E_n}} |u_{mz}(z)|$. We may suppose $H_m \geq 1, \; m = 1, 2, \cdots$. It is obvious that $U_m = u_m/H_m$ is a solution of the boundary value problem

$$\begin{cases} U_{mz\bar{z}} - \text{Re}\left[Q_m u_{mzz} + A_{1m}U_{mz}\right] = A_{2m}U_m/H_m \text{ in } E_n, \\ \text{Re}\left[\overline{\lambda(t)}U_{mt}\right] = [r(t) - \sigma(t)u_m(t)]/2H_m \text{ on } \partial E_n. \end{cases} \tag{3.15}$$

On the basis of Lemma 3.1, and Theorem 6.2 of Chapter 3 in [140], we have

$$C_\beta^1[U_m, \overline{E_n}] \leq M_6 = M_6(q_0, p_0, k_0, \mu, l_0, E), \tag{3.16}$$

where $\beta = 1 - 2/p_0$. Hence we can select a subsequence $\{u_{m_k}(z)\}$ of $\{u_m(z)\}$, so that $\{u_{m_k}(z)\}$ and $\{u_{m_k z}\}$ uniformly converge to 0, $v_0(z)$ respectively, and $(0)_z = v_0(z)$, $\max_{z \in \overline{E}} |v_0(z)| = 1$. This contradiction shows that (3.12) is true.

Theorem 3.3 *Suppose that equation (2.5) satisfies Condition C^*, and that the coefficients Q, A_1, A_2 and their first order partial derivatives belong to $C_\alpha(\bar{D})$, and satisfy*

$$\begin{cases} \max_{z \in D \cup \gamma} \{|A_{1z}|, |A_{1\bar{z}}|, |A_{2z}|, |A_{2\bar{z}}|, |A_{1u}|, |u_{2u}|, |A_{1u_z}|, |u_{1u_z}|, |A_{2u_z}|, |A_{2u_z}|\} \leq k_2, \\ \max_{z \in D \cup \gamma} \{|Q_z|, |Q_{\bar{z}}|, |Q_u|, |u_{u_z}|, Q_{u_z}|, |Q_{u_{zz}}|, |Q_{u_{\bar{z}z}}|\} \leq k_3 y^{m+2}, \end{cases} \tag{3.17}$$

where $\alpha(0 < \alpha < 1), k_2, k_3$ are nonnegative constants. Then Problem P for equation (2.5) is solvable.

Proof For any arbitrary given number $\beta > 0$, there exists a number $\delta(\beta) > 0$ such that $|\lambda(z) - \lambda(x_0)| < \beta/3M_5$, $|\sigma(z) - \sigma_0(x_0)| < \beta/3M_4$, $|r(z) - r(x_0)| < \beta/3$, if

$|x - x_0| < \delta(\beta)$. Obviously for $\delta \le \delta(\beta)$, $\{|x - x_0| \le \delta, y = \varepsilon_n\} \subset (\partial D_n \cap \{y = \varepsilon_n\}$, if n is large enough. We introduce a function

$$W^{\pm}(z) = CV(z) + \beta \pm r(x_0) \mp Y_n(z), \tag{3.18}$$

where $V(z) = (x - x_0)^2 + 3\eta y - \eta^2 y^2 > 0$, $Y_n(z) = 2\mathrm{Re}\,[\overline{\lambda(x_0)}u_{nz}] + \sigma(x_0)u_n(z) = 2\mathrm{Re}[\overline{\lambda}u_{uz}] + \sigma u_n(z)$ and C is an unknown positive constant. If it can be proved that $W^{\pm}(z) \ge 0$ on $\overline{E_n}$, then we have

$$|Y_n(z) - r(x_0)| < CV(z) + \beta,$$

in which $V(z)$ tends to 0 as n tends to ∞ and z tends to x_0, it follows that

$$Y_0(x_0) = 2\mathrm{Re}\,[\overline{\lambda(x_0)}u_{0z}(x_0)] + \sigma(x_0)u_0(x_0) = r(x_0),$$

i.e. $u_0(z)$ satisfies (3.1) on $\gamma \backslash \{0, 1\}$. Supposing that $W^{\pm}(z)$ attains a negative minimum at a point $z^* \in E$, it is clear that

$$W_z^{\pm}(z^*) = W_{\bar{z}}^{\pm}(z^*) = 0. \tag{3.19}$$

As in the proof of Lemma 2.1, Chapter 3 in [140],

$$W_{z\bar{z}}^{\pm} - \mathrm{Re}[QW_{zz}^{\pm} + A_1 W_z^{\pm}] - A_2 W^{\pm} \ge 0 \text{ at } z^* \tag{3.20}$$

can be derived. From (3.19), it follows that

$$\lambda^2 u_{n\bar{z}\bar{z}} + \mathrm{Re}[Qu_{nzz}] = \pm C\lambda V_{\bar{z}} - \mathrm{Re}A_1 u_{nz} - A_2 u_n - \lambda\sigma u_{n\bar{z}},$$

$$\bar{\lambda}^2 u_{nzz} + \mathrm{Re}[Qu_{nzz}] = \pm C\bar{\lambda} V_z - \mathrm{Re}A_1 u_{nz} - A_2 u_n - \bar{\lambda}\sigma u_{nz}$$

at $z^* \in E_n$. Thus we can show that

$$\begin{cases} \mathrm{Re}\,u_{nzz} = [\mathrm{Re}\lambda^2(\pm C\bar{\lambda} V_z - \bar{\lambda}\sigma u_{nz} - \mathrm{Re}A_1 u_{nz} - A_2 u_n) \\ \qquad + \mathrm{Im}Q\,\mathrm{Im}(\pm C\bar{\lambda} V_z - \bar{\lambda}\sigma u_{nz})]/(1 + \mathrm{Re}\lambda^2 Q), \\ \mathrm{Im}\,u_{nzz} = [\mathrm{Im}\lambda^2(\pm C\bar{\lambda} V_z - \bar{\lambda}\sigma u_{nz} - \mathrm{Re}A_1 u_{nz} - A_2 u_n) \\ \qquad + \mathrm{Re}Q\,\mathrm{Im}(\pm C\bar{\lambda} V_z - \bar{\lambda}\sigma u_{nz})]/(1 + \mathrm{Re}\lambda^2 Q) \end{cases} \tag{3.21}$$

at z^*. Consequently

$$W_{z\bar{z}}^{\pm} + \mathrm{Re}[QW_{zz}^{\pm} + A_1 W_z^{\pm}] - A_2 W^{\pm}$$

$$= C\{V_{z\bar{z}} - \mathrm{Re}[Q_1 V_{zz} + A_1 V_z] - A_2 V\} - A_2[\beta \pm r(x_0)]$$

$$\mp [Y_{nz\bar{z}} - \mathrm{Re}(QY_{nzz} + A_1 Y_{nz}) - A_2 Y_n] \tag{3.22}$$

$$\le C[1 - \eta^2 k_1 + k_0(\delta + \eta/2)] + k_0(\beta + M_1) + 8k_3 k_4 y^{m+2}[2C(2\delta + \eta)$$

$$+ 2(2M_1 + k_0)M_5 + k_0 M_4] + 8k_2(M_5 + M_4) < 0 \text{ at } z^*,$$

where $k_4 = 1/(1 - |Q(z^*)|)$, and after having fixed the positive constant C, η is chosen large enough, such that the last inequality holds. This contradiction proves that $W^{\pm}(z) \geq 0$ in E_n.

Next, we prove that $W^{\pm}(z)$ cannot attain a negative minimum on ∂E. Practically,

$$V(z) \geq \min(\delta^2, 2) = \delta^2 \text{ on } \partial E_n \cap (\{|x - x_0| = \delta\} \cup \{y = 1/\eta\}),$$

and

$$W^{\pm}(z) \geq C\delta^2 - \beta - M_1 - M_5 - M_1 M_4 > 0, \tag{3.23}$$

in which we select the constant C sufficiently large. In addition,

$$W^{\pm}(z) \geq \beta - |Y_n(z) - r(x_0)| > 0 \text{ on } \partial E \cap \{y = \varepsilon_n\}. \tag{3.24}$$

Combining the above results, we see that $W^{\pm}(z) \geq 0$ on $\overline{E_n}$.

Finally we discuss the cases: $z = 0$ and $z = 1$. Choosing the part $U = \{|x| < \delta, 0 < y < 1/\eta\}$ of a neighborhood of $z = 0$, we denote $E_0 = U \cup D_n$ and introduce a function

$$W^{\pm}(z) = CV(z) + \beta \pm r(0) \mp Y_n(z), \quad V(z) = x^2 + 3\eta y - \eta^2 y^2, \tag{3.25}$$

where β is a given positive constant and C is an unknown positive constant to be determined. Similarly as before, we can prove that $W^{\pm}(z) \geq 0$ in E_0. If we select the constant C large enough, then (3.23) remains true. If $z \in \partial E_0 \cap S_n^2$, we note that the angles between ν and the y–axis are greater than $\pi/2$ when $y \leq 2\varepsilon_n$. Thus there exists a constant $\theta(0 < \theta < \pi/2)$ such that the angles are greater than or equal to $\pi/2 + \theta$. This means that $\cos(\nu, y) \leq -\cos\theta < 0$. Due to $3\varepsilon_n/2 < y < 1/\eta$, it can be seen that

$$\frac{\partial V}{\partial \nu_n} \geq 3\varepsilon_n \eta^2 \cos\theta - 3\eta - 2X_1 < 1 + M_1(X_1^2 + 3), \tag{3.26}$$

where $X_1 = \max_{z \in \bar{D}} x$, and η is chosen so that

$$\eta \geq \{1 + [3 + 4\varepsilon_n \cos\theta\,(2X_1 + 1 + M_1(X_1^2 + 3))]^{1/2}\}/\{2\varepsilon_n \cos\theta\}.$$

Moreover, noting that

$$l_n(Y_n) = l_n[\text{Re}\lambda(0)u_{nx} - \text{Im}\lambda(0)u_{ny} + \sigma(0)u_n]$$

$$= \text{Re}\lambda(0)[r_x - \sigma_x u_n] - \text{Im}\lambda(0)[r_y - \sigma_y u_n] + \sigma(0)r(x),$$

we have

$$l_n(W^{\pm}) = C\frac{\partial V}{\partial \nu_n} \mp \frac{\partial Y_n}{\partial \nu_n} + \sigma(z)[CV(z) + \beta \pm r(0) \mp Y_n(z)]$$

$$> [1 + M_1(X_1^2 + 3)]C - M_1(2 + 2M_4 + l_0) - M_1[C(X_1^2 + 3) + \beta + l_0] > 0, \tag{3.27}$$

in which $C > M_1(2 + 2M_4 + \beta + 2l_0)$. Hence $W^{\pm}(z) \geq 0$ for $z \in \partial E_0 \cap S_n^2$. Using the same method, we can obtain $W^{\pm}(z) \geq 0$ on $\partial D_n \cap (S_n^1 \cup S_n^2 \cup \{y = \varepsilon_n\})$. It can be derived $l(u_0) = r(z)$ at $z = 0$.

Besides we can also prove that $l(u_0) = r(z)$ at $z = 1$. This completes the proof.

In order to discuss the solvability of Problem Q for equation (2.6), similarly to (3.5), we have the corresponding boundary value problem

$$\begin{cases} u_{z\bar{z}} = \mathrm{Re}[Qu_{zz} + A_1 u_z] + A_2 u \ \text{in} \ D_n, \\[2mm] l_n(u) = \dfrac{\partial u}{\partial \nu_n} + \sigma_n(t)u(t) = r(t) \ \text{on} \ \partial D_n \end{cases} \tag{3.28}$$

and its solution $u_n(z)$. The following lemma will be used.

Lemma 3.4   *The solution $u_n(z)$ of the boundary value problem (3.28) satisfies the estimate*

$$|u_n(z)| \leq M_7 \ \text{in} \ \overline{D_n}, \tag{3.29}$$

and

$$|u_{nz}(z)| \leq M_8 \ \text{in} \ \overline{E_n} = \overline{U \cap D_n}, \tag{3.30}$$

where D_n and E_n are as stated in Lemma 3.1 and Lemma 3.2, the constants M_7 and M_8 are independent of n.

Proof   We introduce an auxiliary function

$$W^{\pm}(z) = V(z) \pm u_n(z), \ \ V(z) = M_7 - (c_1 e^{nx} + c_2 y) \geq 0, \tag{3.31}$$

where c_1, c_2, n and M_7 are unknown positive constants to be determined appropriately. We shall prove that $W^{\pm}(z)$ cannot attain a negative minimum on $\overline{D_n}$. In fact,

$$\begin{aligned} W_{z\bar{z}}^{\pm} &- \mathrm{Re}\,[QW_{zz}^{\pm} + A_1 W_z^{\pm}] - A_2 W^{\pm} \\[1mm] &= V_{z\bar{z}} - \mathrm{Re}\,[QV_{zz} + A_1 V_z] - A_2 V \\[1mm] &< -\frac{a_1 c_1 n^2 e^{nx}}{2(a_1 + a_3 y^m)} - \frac{d_1 c_1 n e^{nx} - d_2 c_2}{2(a_1 + a_3 y^m)} \\[1mm] &< c_1 n e^{nX_2}(-k_5 n + k_0)/2 + c_2 k_0 < 0 \ \text{in} \ D_n, \end{aligned} \tag{3.32}$$

where $k_5 = \inf_{z \in D}\{a_1/(a_1 + a_3 y^m)\}$, $X_2 = \inf_{z \in D} x$, $n > k_0/k_5$, and $2c_2 k_0 < c_1 n e^{nX_2}$ $\times(k_5 n - k_0)$. By Lemma 2.6, Chapter 3 in [140], (3.32) shows that $W^{\pm}(z)$ cannot attain a negative minimum in D_n. Moreover,

$$l_n(W^{\pm}) = \frac{\partial V}{\partial \nu_n} + \sigma(z)V(z) \pm r(z)$$

$$\geq M_1(M_7 - c_1 e^{nX_1} - c_2 Y_1) - c_1 n e^{nX_1} - c_2 - M_1 > 0 \ \text{on} \ \partial D_n, \tag{3.33}$$

in which $Y_1 = \sup_{z \in D} y$, $M_7 > (1 + n/M_1)c_1 e^{nX_1} + c_2 Y_1 + c_2/M_1 + 1$. Hence we have

$$W^{\pm}(z) = V(z) \pm u_n(z) \geq 0 \text{ in } \overline{D_n}. \tag{3.34}$$

This implies that (3.29) holds.

Similarly to the proof of Lemma 3.2, we can also obtain the estimate (3.30).

Theorem 3.5 *Suppose that equation (2.6) satisfies Condition C_* and the coefficients Q, A_1, A_2 and their first order partial derivatives belong to $C_\alpha(\overline{D})$ and satisfy*

$$\begin{cases} \max_{z \in \gamma}\{|A_{1z}|, |A_{1\bar{z}}|, |A_{2z}|, |A_{2\bar{z}}|, |A_{1u}|, |A_{2u}|, |A_{1u_z}|, |A_{1u_{\bar{z}}}|, |A_{2u_z}|, |A_{2u_{\bar{z}}}|\} \leq k_6, \\[2mm] \max_{z \in D \cup \gamma}\{|Q_z|, |Q_{\bar{z}}|, |Q_u|, |Q_{u_z}|, |Q_{u_{\bar{z}}}|, |Q_{u_{zz}}|, |Q_{u_{z\bar{z}}}|\} \leq k_7 y^{m+3}, \end{cases}$$

$$\tag{3.35}$$

where k_6, k_7 are nonnegative constants. If one of the following conditions holds, then Problem Q for (2.6) is solvable.

(a) $0 < m < 1$;

(b) $m = 1$, $d_0 = \sup_{z \in D} d_2 < \inf_{z \in D} a_3 = a_0$, $d_2(x, y, \zeta, \xi, \eta) \in C^1(D \times \mathbb{R} \times \mathbb{R}^2)$;

(c) $1 < m < 2$, $d_0 \leq 0$;

(d) $m \geq 2$, $d_0 < 0$.

Proof Similarly to the proof of Theorem 4.3 in Chapter 1, we can obtain a bounded solution $u_0(z)$ of equation (2.6) in D satisfying the boundary condition (3.1) on Γ, which is a limit function of a subsequence $\{u_{n_k}(z)\}$ of $\{u_n(z)\}$. It remains to prove only that $u_0(z)$ satisfies the boundary condition (3.1) on γ. We choose an arbitrary point $x_0 \in [0, 1]$ and a neighborhood $U = \{|x - x_0| < \delta, 0 < y < 1/\eta\}$, $\eta > 1$, and introduce a function

$$W^{\pm}(z) = CV(z) + \beta \pm r(x_0) \mp Y(z), \ V(z) = (x - x_0)^2 + (\eta y)^s \text{ on } \overline{E_n}, \tag{3.36}$$

where $\overline{E_n} = \overline{U \cap D_n}$, $Y(z) = 2\text{Re}[\overline{\lambda(x_0)}u_{nz}] + \sigma(x_0)u_n$, c, η, s are unknown positive constants to be determined appropriately, $\lambda(x_0), \sigma(x_0)$ are as stated in (3.18). We shall prove that $W^{\pm}(z)$ cannot attain a negative minimum in E_n. Otherwise, $W^{\pm}(z)$ attains a negative minimum at a point $z^* \in E_n$. Consequently,

$$W^{\pm}_{z\bar{z}} - \text{Re}[QW^{\pm}_{zz} + A_1 W^{\pm}_z] - A_2 W^{\pm} \geq 0, \ W^{\pm}_z = W^{\pm}_{\bar{z}} = 0 \text{ at } z^*. \tag{3.37}$$

Moreover, we have also (3.21) and

$$\begin{aligned} W^{\pm}_{z\bar{z}} &- \text{Re}[QW^{\pm}_{zz} + A_1 W^{\pm}_z] - A_2 W^{\pm} \\[2mm] &\leq C\{1 + k_8[d_0 + (s-1)a_0 y^{m-1}]s\eta^s y^{s-1}\} + k_0(X_1 - X_2) \\[2mm] &+ k_0(\beta + M_1) + 8k_7 k_4[2c(2\delta y^{m+3} + s/\eta^{m+2}) \\[2mm] &+ 2(2M_1 + k_0)M_8 + k_0 M_7] + 8k_6(M_8 + M_6) < 0 \text{ at } z^*, \end{aligned} \tag{3.38}$$

where $k_4 = 1/(1 - |Q(z^*)|)$, $k_8 = \inf_{z \in \bar{D}} 1/2(a_1 + a_3 y^m)$, and $m = 1$, $d_0 < a_0$, $s < (a_0 - d_0)/a_0$, and for fixed positive number C, the constant η is chosen large enough, such that the last inequality in (3.38) holds. This is a contradiction to the second inequality in (3.37). Hence $W^{\pm}(z) \geq 0$ in E_n. The remaining proof is similar to the proof of Theorem 4.3 in Chapter 1.

Finally, we discuss the uniqueness of solutions to Problem P for (2.5) and to Problem Q for (2.6).

Theorem 3.6　*Let equation (2.5) satisfy Condition C_* and (2.32). Then the solution of its Problem P is unique.*

Proof　Suppose that $u_1(z)$, $u_2(z)$ are two solutions of Problem P for (2.5) and denote $u(z) = u_1(z) - u_2(z)$. It is clear that $u(z)$ is a solution of the equation

$$\begin{cases} u_{z\bar{z}} - \mathrm{Re}[q u_{zz}] = \mathrm{Re}[B_1 u_z] + B_2 u, \ B_2 > 0, \\ \mathrm{Re}[q u_{zz}] = F_1(z, u_1, u_{1z}, u_{1zz}) - F_2(z, u_1, u_{1z}, u_{2zz}), \end{cases} \tag{3.39}$$

and the boundary condition

$$l(u) = \frac{\partial u}{\partial \nu} + \sigma(t) u(t) = 0, \ t \in \partial D. \tag{3.40}$$

We arbitrarily select a positive number ε. Obviously, the function $U(z) = u(z)/\varepsilon$ is a solution of the boundary value problem

$$\begin{cases} U_{z\bar{z}} - \mathrm{Re}[q U_{zz} + B_1 U_z] - B_2 U = 0 \ \text{in} \ D, \\ l(u) = \frac{\partial U}{\partial \nu} + \sigma(t) U(t) = 0 \ \text{on} \ \partial D. \end{cases} \tag{3.41}$$

According to Lemma 3.1, $U(z)$ satisfies the estimate

$$|U(z)| \leq M_4, \ \text{i.e.} \ |u(z)| \leq M_4 \varepsilon \ \text{on} \ \overline{D_n}, \tag{3.42}$$

in which the constant M_4 is independent of n. Letting ε tend to 0, it shows $u(z) \equiv 0$, i.e. $u_1(z) \equiv u_2(z)$ in $\bar{D}$.

Similarly, we can also obtain the following result.

Theorem 3.7　*If equation (2.6) satisfies Condition C^* and (2.32), then the solution of Problem Q for (2.6) is unique.* (See [126]2)).

4　Boundary Value Problems for Another Class of Nonlinear Degenerate Elliptic Equations of Second Order

First of all, we discuss the following linear degenerate elliptic equation of second order

$$Lu = y^{m_1} u_{xx} + y^{m_2} u_{yy} + d_1 u_x + d_2 u_y + d_3 u = 0 \ \text{in} \ D, \tag{4.1}$$

where the domain D is as stated in Section 2, m_1 and m_2 are positive constants, $C_\mu[d_j, \bar{D}] \le k_0 < \infty$, $j = 1, 2, 3$, $0 < \mu < 1$, $d_3 \le 0$ in D. The complex form of (4.1) is

$$\begin{cases} u_{z\bar{z}} = F(z, u, u_z, u_{zz}), \ F = \text{Re}[Qu_{zz} + A_1 u_z] + A_2 u, \\[2mm] q = -\dfrac{y^{m_1} - y^{m_2}}{y^{m_1} + y^{m_2}}, \ A_1 = -\dfrac{d_1 + id_2}{y^{m_1} + y^{m_2}}, \ A_2 = -\dfrac{d_3}{y^{m_1} + y^{m_2}} \ge 0. \end{cases} \quad (4.2)$$

The so–called Problem D is to find a continuous solution $u(z)$ of (4.2) on $\bar{D}$ satisfying the Dirichlet boundary condition

$$u(t) = r(t), \ t \in \partial D, \quad (4.3)$$

in which $C_\mu^1[r, \partial D] \le l_0 < \infty$, $0 < \mu < 1$. If condition (4.3) is replaced by

$$u(t) = r(t), \ t \in \Gamma, \quad (4.4)$$

and the solution $u(z)$ of (4.2) is bounded on γ, then the boundary value problem is called Problem E.

Lemma 4.1 *If there exists a function $V(z)$ which is continuous in a neighborhood of any point $z_0 = x_0 \in \gamma$ and satisfies the conditions*

(A) $V(z_0) = 0$;

(B) $V(z) > 0$, for $z \ne x_0$;

(C) $LV \le -c_0 < 0$;

then Problem D for equation (3.1) or (3.2) has a unique classical solution $u(z)$.

Proof Following the proof of Theorem 2.1, we continuously extend $r(z)$ on the closure of the domain $\bar{D}$ such that (2.12) holds, and choose a sequence of domains $\{D_n\}$. It is clear that equation (4.1) or (4.2) is uniformly elliptic in D_n. According to Theorem 4.4 of Chapter 3 in [140], there exists a solution $u_n(z)$ of (4.1) in D_n, which satisfies the boundary condition: $u_n(z) = r(t)$, $t \in \partial D_n$. By the extremum principle as stated in Corollary 2.10 of Chapter 3 in [140], the solution $u_n(z)$ satisfies the estimate

$$C[u_n, \bar{D}_n] \le C[r, \partial D_n] \le M_1.$$

We then can prove that $u_n(z)$ satisfies

$$C_\alpha^1[u_n, \overline{D_n} \cap \{y \ge \eta\}] \le M_2,$$

where $\alpha = 1 - 2/p$, $p(> 2)$ and $\eta(> 0)$ are constants, M_2 is a nonnegative constant independent of n. Hence there exists a subsequence of $\{u_n(z)\}$, which uniformly converges to a continuous function $u_0(z)$. We can prove that $u_0(z)$ is a solution of equation (4.1) in D and satisfies the boundary condition (4.3) on Γ. It remains to prove that $u_0(z)$ satisfies the boundary condition (4.3) on γ. We choose an arbitrary point $x_0 \in [0, 1]$ such that $|r(z) - r(x_0)| < \beta$ in $D \cap U$, where β is an arbitrary positive number, $U = \{|x - x_0| < \delta, 0 < y < 1/\eta\}, \eta > 1$, and then introduce the function

$$W^{\pm}(z) = cV(z) + \beta \pm r(x_0) \mp u_n(z),$$

in which c and β are undetermined positive constants. It is evident that

$$LW^{\pm} = cLV - d_3[\beta \pm r(x_0)] < 0 \text{ in } D_n,$$

when the constant c is sufficiently large. This shows that $W^{\pm}(z)$ cannot attain a negative minimum in D_n. Moreover, it is not difficult to see that

$$W^{\pm}(z) \geq cV(z) + \beta - |r(z) - r(x_0)| > 0 \text{ in } \partial D_n \cap U,$$

and

$$W^{\pm}(z) \geq cV(z) + \beta \pm r(x_0) - M_1 > 0 \text{ in } \partial D_n \backslash U$$

for a large enough constant c. Hence

$$W^{\pm}(z) = cV(z) + \beta \pm r(x_0) \mp u_n(z) \geq 0 \text{ in } \overline{D_n}.$$

It follows that

$$|u_n(z) - r(x_0)| \leq cV(z) + \beta \text{ in } \overline{D_n}.$$

Letting n tend to ∞ and z tend to x_0, we obtain

$$\overline{\lim_{n \to \infty}} \ \overline{\lim_{z \to x_0}} |u_n(z) - r(x_0)| \leq \beta.$$

Noting the arbitrariness of β, implies $u_0(x_0) = r(x_0)$(see [75]).

The uniqueness of solutions of Problem D can be derived by the extremum principle.

Lemma 4.2   *If there exists a function $V(z) \in C(\bar{D}) \cap C^2(D)$, such that*

    (A') $V(z) \geq 0$ *in* $\bar{D}$;

    (B') $\lim_{y \to +0} V(z) = +\infty$ *uniformly for* $z \in \gamma$;

    (C') $LV < 0$ *in* D;

then Problem E for (4.1) or (4.2) has a unique solution.

Proof   According to the method in the proof of Theorem 2.1, we can prove the existence of solutions of Problem E for (4.1). To prove the uniqueness of solutions of Problem E, it is sufficient to prove that Problem E for (4.1) with homogeneous boundary condition $u(t) = 0$ on Γ has only the trivial solution. Since $L(cV - u) < 0$, where c is any positive number, it can be seen that $cV(z) - u(z)$ cannot attain a negative minimum in D. Thus we can see that

$$\lim_{z \to t} [cV(z) - u(z)] \geq 0, \quad \text{for } t \in \partial D.$$

Hence

$$cV(z) - u(z) \geq 0, \quad \text{i.e. } u(z) \leq cV(z) \text{ in } D.$$

Letting c tend to 0 we have $u(z) \leq 0$ in D. Similarly, we consider the function $-cV(z) - u(z)$. From $L(-cV - u) > 0$, it can be obtained

$$-cV(z) - u(z) \leq 0, \quad \text{i.e. } -cV(z) \leq u(z) \text{ in } D.$$

Thus $u(z) \geq 0$ in D. So $u(z) = 0$ in D(see [75]).

Theorem 4.3 *Suppose that one of the following conditions holds:*

(a) $m_2 < 1$;

(b) $m_2 = 1$, $d_2(x,0) < 1$, $x \in [0,1]$;

(c) $1 < m_2 < 2$, $d_2(x,y) \in C^1(\overline{D^*})$, $D^* = \{-\delta_0 < x < 1+\delta_0, 0 < y < \delta_0\}$, where δ_0 is a positive constant, $d_2(x,y) \leq 0$, $x \in (-\delta_0, 1+\delta_0)$;

(d) $m_2 \geq 2$, $d_2(x,y) < 0$, $x \in [0,1]$.

Then Problem D for (4.1) or (4.2) has a unique solution.

Proof We introduce a function

$$V(z) = (x - x_0)^2 + y^\beta, \quad 0 \leq x_0 \leq 1, \tag{4.5}$$

where $\beta(0 < \beta < 1)$ is an unknown constant to be determined appropriately. It is clear that $V(z)$ satisfies the condition (A) and (B) in Lemma 4.1, and

$$LV \leq 2y^{m_1} + \beta(\beta - 1)y^{m_2+\beta-2} + 2d_1(x - x_0) + \beta d_2 y^{\beta-1}. \tag{4.6}$$

(a) $m_2 < 1$. We choose $\beta = 1/2$ and denote $X = \max_{z \in \bar{D}} x$, $Y = \max_{z \in \bar{D}} y$. From (4.6), it follows that

$$LV \leq 2y^{m_1} - \frac{1}{4}y^{m_2-\frac{3}{2}} + 2d_1(x - x_0) + \frac{1}{2}d_2 y^{-\frac{1}{2}}$$

$$\leq 2Y^{m_1} + 2k_0(X + 1) - \frac{1}{4}y^{m_2-\frac{3}{2}}(1 - 2k_0 y^{1-m_2}) \to -\infty, \quad \text{as } y \to +0. \tag{4.7}$$

Hence there exists a positive constant δ, such that $L(V) \leq -1$, if $0 < y < \delta$. This shows that $V(z)$ satisfies condition (C) in Lemma 4.1.

(b) $m_2 = 1$, $d_2(x,0) < 1$, $x \in [0,1]$. Obviously, we can find a constant $\sigma(0 < \sigma < 1)$, so that

$$d_2(x,0) \leq 1 - 2\sigma < 1, \quad x \in [0,1].$$

Then there exists a positive number η, such that $d_2(x,y) \leq 1 - \sigma$, if $(x,y) \in \{0 \leq x \leq 1, 0 < y < \eta\}$. Choosing $\beta = \sigma/2$, we see that

$$LV \leq 2y^{m_1} + 2k_0(x+1) + \beta(\beta-1)y^{\beta-1} + \beta(1-\sigma)y^{\beta-1}$$

$$\leq 2Y^{m_1} + 2k_0(X+1) - \frac{\sigma^2}{4}y^{\sigma/2-1} \to -\infty \text{ as } y \to +0. \tag{4.8}$$

Consequently there is a positive number $\delta < \eta$, so that $LV \leq -1$, if $0 < y < \delta$. Hence, condition (C) is satisfied.

(c) $1 < m_1 < 2$, $d_2(x,y) \in C(\overline{D^*})$, $d_2(x,0) \leq 0$ for $-\delta_0 < x < 1 + \delta_0$. When $y < \delta_0$, we have

$$d_2(x,y) = d_2(x,0) + d_2'(x,\theta y)y \leq C^1[d_2,\bar{D}]y, \quad 0 < \theta < 1.$$

Assuming $C^1[d_2,\bar{D}] = k_1 < \infty$ and choosing $\beta = 1 - m_2/2 \, (0 < \beta < 1)$, we find

$$\begin{aligned} LV \quad \leq \quad & 2y^{m_1} + 2k_0(x+1) + \beta(\beta-1)y^{m_2+\beta-2} + \beta k_1 y^\beta \\[2mm] \leq \quad & 2Y^{m_1} + 2k_0(X+1) - \frac{m_2}{2}\left(1 - \frac{m_2}{2}\right)y^{m_2/2-1} \\[2mm] + \quad & \left(1 - \frac{m_2}{2}\right)k_1 y^{1-m_2/2} \to -\infty, \text{ as } y \to +0. \end{aligned} \tag{4.9}$$

Hence, we can derive condition (C).

(d) $m_1 \geq 2$, $d_2(x,0) < 0$, $x \in [0,1]$. Similarly to (b), there exist two positive constants σ and η, such that $d_2(x,y) \leq -\sigma < 0$, if $(x,y) \in \{0 \leq x \leq 1, 0 < y < \eta\}$. From (4.6), we can obtain

$$LV \leq 2Y^{m_1} + 2k_0(X+1) - \beta\sigma y^{\beta-1} \to -\infty, \text{ as } y \to +0. \tag{4.10}$$

Therefore we have condition (C).

On the basis of Lemma 4.1, Theorem 4.3 is proved.

Theorem 4.4   *Suppose that one of the following conditions holds:*

(a') $m_2 = 1$, $d_2(x,y) \in C^1(\overline{D^*})$, $d_2(x,0) \geq 1$, $x \in (-\delta_0, 1 + \delta_0)$, *where D^* and δ_0 are as stated in Theorem 3.3;*

(b') $m_2 > 1$, $d_2(x,0) > 0$, $x \in [0,1]$.

Then Problem E for (4.1) or (4.2) has a unique solution.

Proof (a') Denote $Y = \max_{z \in \bar{D}} y$ and $k_1 = C^1[d_2,\bar{D}]$, we introduce a function

$$V(y) = e^{nY} - e^{ny} + V_1(y), \quad V(y) = \int_y^Y \frac{e^{k_1 t}}{t} dt, \tag{4.11}$$

where n is an unknown positive constant. It is clear that $w(y)$ satisfies the condition (A') in Lemma 4.2. Due to

$$V_1(y) \geq \int_y^Y \frac{dt}{t} = \ln Y - \ln y \to +\infty, \quad \text{as } y \to +0,$$

the condition (B') in Lemma 4.2 is satisfied. In order to ensure condition (C'), we note that because of the continuity of $d_2(x,y)$, from $d_2(x,0) \geq 1$, $x \in (-\delta_0, 1 + \delta_0)$, there exists a positive constant $\delta < 1$, such that $d_2(x,y) \geq 1/2$ if $(x,y) \in \{0 \leq x \leq 1, 0 < y < \delta\}$. Moreover,

$$
\begin{aligned}
LV &= L(e^{nY} - e^{ny}) + LV_1 \leq -n^2 y e^{ny} - d_2(x,y)n e^{ny} \\
&+ yV_1'' + d_2V_1' \leq yV_1'' + [d_2(x,0) + d_2{}'(x,\theta y)y]V_1' \\
&\leq y\left(-k_1\frac{e^{k_1 y}}{y} + \frac{e^{k_1 y}}{y^2}\right) - \frac{e^{k_1 y}}{y} + k_1 e^{k_1 y} = 0,
\end{aligned}
\tag{4.12}
$$

if $0 < y < \delta$. If $\delta \leq y < Y$, there exists a positive constant M_3 independent of n, such that $LV_1 \leq M_3$. Let n be large enough, so that

$$n > \max\{(k_0 + 1)/\delta, M_3\}.$$

We immediately obtain

$$LV \leq -\delta n^2 e^{ny} + n k_0 e^{ny} + M_3 < 0. \tag{4.13}$$

So condition (C') in Lemma 4.2 holds. According to Lemma 4.2, under condition (a'), the result in Theorem 4.4 is true.

(b') $m_2 > 1$ and $d_2(x,0) > 0$, $x \in [0,1]$. Similarly to (d) in the proof of Theorem 4.3, there exist two positive constants σ and η, such that $d_2(x,y) \geq \sigma$, if $0 < y < \eta$. We introduce an auxiliary function

$$V(y) = e^{nY} - e^{ny} + V_2(y), \quad V_2(y) = \int_y^Y e^{\frac{\sigma}{m_2-1}t^{1-m_2}} dt. \tag{4.14}$$

Obviously, $V(y)$ satisfies condition (A') in Lemma 4.2. Let $j = [1/(m_2 - 1)] + 1$. It is clear that if $0 < t < 1$, then

$$
e^{\frac{\sigma}{m_2-1}t^{1-m_2}} \geq \frac{1}{j!}\left(\frac{\sigma}{m_2-1}t^{1-m_2}\right)^j
$$

$$
= \frac{1}{j!}\left(\frac{\sigma}{m_2-1}\right)^j \left(\frac{1}{t}\right)^{j(m_2-1)} \geq \frac{1}{j!}\left(\frac{\sigma}{m_2-1}\right)^j \frac{1}{t}.
$$

Consequently

$$V(y) \geq \int_y^{\min(1,Y)} \frac{1}{j!}\left(\frac{\sigma}{m_2-1}\right)^j \frac{dt}{t} \to +\infty, \quad \text{as } y \to +0.$$

This shows that the condition (B') is satisfied. If $0 < y < \delta$, then

$$\begin{aligned}
LV &\leq -n^2 y^{m_2} e^{ny} - n d_2 e^{ny} + y^{m_2} V_2'' + d_2 V_2' \\
&< y^{m_2}\left(\frac{\sigma}{y^{m_2}} e^{\frac{\sigma}{m_2-1} y^{1-m_2}}\right) - \sigma e^{\frac{\sigma}{m_2-1} y^{1-m_2}} = 0.
\end{aligned} \tag{4.15}$$

When $\delta \leq y \leq Y$, we choose n large enough such that

$$n > \max\{(k_0 + 1)/\delta^{m_2}, M_4\},$$

where $M_4 (\geq LV_2)$ is a positive constant independent of n. Thus

$$\begin{aligned}
LV &\leq -n^2 y^{m_2} e^{ny} + n k_0 e^{ny} + LV_2 \\
&\leq -n e^{ny}(n \delta^{m_2} - k_0) + M_4 < 0.
\end{aligned} \tag{4.16}$$

Therefore condition (C') holds. By Lemma 4.2, the result in Theorem 4.4 is justified.

Next, we consider the case when $d_j(x,y)(j = 1,2,3)$ satisfy $d_j(x,y) \in C_\alpha^1(\bar{D})(0 < \alpha < 1)$, $j = 1,2,3$, $d_3(x,y) \leq 0$ in D, and $m_1 > 0$, $m_2 > 0$, and discuss the mixed boundary value problem (Problem M) for (4.1) or (4.2) with the boundary condition

$$\begin{cases}
l\,u = \dfrac{\partial u}{\partial \nu} + \sigma(t)u(t) = r(t) \ \ t \in \Gamma, \\[2mm]
u(t) = r(t), \ \ t \in \gamma,
\end{cases} \tag{4.17}$$

where $\Gamma \in C_\alpha^2$, $\lambda = \cos(\nu,x) - i\cos(\nu,y) \in C_\alpha^1(\Gamma)$, $\sigma(x,y) \in C_\alpha^1(\Gamma)$, $\sigma(x,y) > 0$ on Γ, $r(t) \in C_\alpha^1(\Gamma \cup \gamma)$, $\cos(\nu,n) > 0$, in which n is the outer normal direction on Γ, and the angles between ν and the y–axis are not less than $\pi/2$ in the neighborhood of the points $(0,0)$ and $(1,0)$.

Theorem 4.5   *Under the foregoing conditions and the hypotheses in Theorem 4.3, Problem M for (4.1) has a unique solution.*

Proof   The uniqueness of solutions of Problem M for (4.1) can be derived by the extremum principle as stated in Corollary 2.10, Chapter 3, [140]. In order to prove the existence of solutions of problem M, from $\{u_n(z)\}$ we can choose a subsequence which converges to a function $u_0(z)$ satisfying equation (4.1) and the boundary condition $l(u) = r(t)$ on Γ. As in the proof of Theorem 4.3, we can prove that $u_0(z)$ satisfies the boundary condition $u_0(t) = r(t)$ on $\gamma\backslash\{0,1\}$. It remains to prove that $u_0(0) = r(0)$ and $u_0(1) = r(1)$. We choose an auxiliary function

$$V(x,y) = (x - x_0)^2 + y^\beta, \ \ x_0 = 0 \text{ or } 1, 0 < \beta < 1, \tag{4.18}$$

which satisfies the following conditions:

(A'') $V(x_0, 0) = 0$;

(B'') $V(x,y) > 0$, $(x,y) \neq (x_0,0)$;

(C'') $lV \leq -\sigma < 0$ in E_n;

(D'') $LV \leq -\sigma < 0$ in E_n.

Here σ is a positive constant. In fact, the conditions (A''), (B''), (D'') in $E_n = D_n \cap \{|x - x_0| < \delta, 0 < y < \delta\}$ are obviously satisfied. From the condition in this theorem, there exist two positive constants $\theta(< \pi/2)$ and δ, such that $\cos(\nu, y) \leq -\cos\theta < 0$, if $y < \delta$. Thus

$$lV \leq |V_x| + V_y \cos(\nu, y) \leq 2X - \beta \cos\theta \, y^{\beta-1} \to -\infty \text{ as } y \to +0.$$

This shows that condition (C'') holds. This completes the proof.

Finally, we consider the general degenerate elliptic equation of second order

$$\begin{cases} y^{m_1} b_1 u_{xx} + 2y^{m_2} b_2 u_{xy} + y^{m_3} b_3 u_{yy} + f(x,y,u,u_x,u_y) = 0, \\ f(x,y,u,u_x,u_y) = d_1 u_x + d_2 u_y + d_3, \, d_j = d_j(x,y,u,u_x,u_y), j = 1,2,3, \end{cases} \tag{4.19}$$

where $a_j = y^{m_j} b_j (j=1,2,3)$ satisfy (2.3) and (2.4)($\Delta_0 = 0$), $m_1, m_3, m_2(> (m_1+m_3)/2)$ are positive constants, $d_j \in C^1_\alpha(\bar{D})$, $C^1[d_j, \bar{D}] \leq k_1 < \infty \, (j = 1,2,3)$, $d_3 \leq 0$ in D for any twice continuously differentiable function $u(x,y)$ in D. With a similar method as used for Theorem 3.4, we can prove the following result.

Theorem 4.6 *If one of the conditions $(a),(b),(c),(d)$ in Theorem 4.3 holds, then Problem M for (4.19) with the boundary condition (4.17) is solvable (see [41]).*

IV   Boundary Value Problems for Several Complex Variables

In this chapter, we mainly consider some boundary value problems for analytic functions of several complex variables and systems of first order complex equations in the polycylinder and the ball. The stated boundary value problems include the Riemann boundary value problem, the Dirichlet problem and the Riemann–Hilbert problem. We establish the integral expressions of solutions for the foregoing boundary value problems and give their solvability conditions. In addition, some boundary value problems for harmonic functions of several complex variables and systems of complex equations of second order are discussed.

1   Riemann Boundary Value Problem for Several Complex Variables

For the sake of convenience, we discuss only the Riemann problem for two complex variables. By a similar way, it can be obtained the corresponding results for several complex variables.

1.1   Riemann boundary value problem for analytic functions

Let Γ_j be a simply closed curve in the z_j-plane and $\Gamma_j \in C_\mu^1$, $j = 1, 2$, where $\mu(0 < \mu < 1)$ is a constant. Without loss of generality, we may consider $\Gamma_j = \{|z| = 1\}$, $j = 1, 2$. Denote by D_j^+, D_j^- the inner domain and outer domain of Γ_j respectively, and $D^{++} = D_1^+ \times D_2^+$, $D^{+-} = D_1^+ \times D_2^-$, $D^{-+} = D_1^- \times D_2^+$, $D^{--} = D_1^- \times D_2^-$, $\Gamma = \Gamma_1 \times \Gamma_2$. There is no harm in assuming that $(0,0) \in D^{++}$.

Problem R   The Riemann boundary value problem for analytic functions is to find a sectionally analytic function $F(z_1, z_2)$ in $D^{++}, D^{+-}, D^{-+}, D^{--}$, such that $F(z_1, z_2)$ are continuous on $\overline{D^{++}}, \overline{D^{+-}}, \overline{D^{-+}}, \overline{D^{--}}$ and satisfies the boundary condition:

$$F^{++}(t_1, t_2) = G_1(t_1, t_2)F^{+-}(t_1, t_2) + G_2(t_1, t_2)F^{-+}(t_1, t_2)$$

$$+G_3(t_1, t_2)F^{--}(t_1, t_2) + g(t_1, t_2), \quad t = (t_1, t_2) \in \Gamma, \tag{1.1}$$

where $G_1(z_1, z_2)$, $G_2(z_1, z_2)$, $G_3(z_1, z_2)$ are analytic in D^{+-}, D^{-+}, D^{--} and are continuous on $\overline{D^{+-}}, \overline{D^{-+}}, \overline{D^{--}}$ respectively, which have no zero, $G_j(t_1, t_2)(j = 1, 2, 3)$, $g(t_1, t_2) \in C_\alpha$, $\alpha(0 < \alpha < 1)$ is a constant.

Introduce a so–called integral of Cauchy type

$$\Phi(z_1, z_2) = \frac{1}{(2\pi i)^2} \int_{\Gamma_1} \int_{\Gamma_2} \frac{g(\zeta_1, \zeta_2)}{(\zeta_1 - z_1)(\zeta_2 - z_2)} d\zeta_1 d\zeta_2. \tag{1.2}$$

It is clear that $\Phi(z_1, z_2)$ in $C^2\backslash\Gamma$ is analytic and $\Phi(z_1, \infty) = \Phi(\infty, z_2) = 0$. We first prove a lemma.

Lemma 1.1   *Denote by* $\Phi^{++}(t_1, t_2)$, $\Phi^{+-}(t_1, t_2)$, $\Phi^{-+}(t_1, t_2)$, $\Phi^{--}(t_1, t_2)$ *the boundary values of the integral* (1.2) *as* $z = (z_1, z_2)(\in D^{++}, D^{+-}, D^{-+}, D^{--})$ *tend to* $t \in \Gamma$ *respectively, then* $\Phi^{++}(t_1, t_2)$, $\Phi^{+-}(t_1, t_2)$, $\Phi^{-+}(t_1, t_2)$, $\Phi^{--}(t_1, t_2)$ *satisfy the boundary condition*

$$\Phi^{++}(t_1, t_2) + \Phi^{--}(t_1, t_2) - \Phi^{+-}(t_1, t_2) - \Phi^{-+}(t_1, t_2) = g(t_1, t_2), \ t \in \Gamma. \qquad (1.3)$$

Proof   Setting

$$\Psi(z_1, z_2) = \frac{1}{2\pi i} \int_\Gamma \frac{g(\zeta_1, z_2)}{\zeta_1 - z_1} d\zeta_1,$$

and according to the Plemelj formula, it can be obtained that when $t \in \Gamma$,

$$\frac{1}{(2\pi i)^2} \int_{\Gamma_1} \int_{\Gamma_2} \frac{g(\zeta_1, \zeta_2)}{(\zeta_1 - t_1)(\zeta_2 - t_2)} d\zeta_1 d\zeta_2$$

$$= \frac{1}{2\pi i} \int_{\Gamma_2} \frac{1}{\zeta_2 - t_2} [\frac{1}{2\pi i} \int_{\Gamma_1} \frac{g(\zeta_1, \zeta_2)}{\zeta_1 - t_1} d\zeta_1] d\zeta_2$$

$$= \frac{1}{2\pi i} \int_{\Gamma_2} \frac{1}{\zeta_2 - t_2} [\Psi^+(t_1, \zeta_2) - \frac{1}{2} g(t_1, \zeta_2)] d\zeta_2$$

$$= \Phi^{++}(t_1, t_2) - \frac{1}{2} \Psi^+(t_1, t_2) - \frac{1}{4\pi i} \int_{\Gamma_2} \frac{g(t_1, \zeta_2)}{\zeta_2 - t_2} d\zeta_2$$

$$= \Phi^{++}(t_1, t_2) - \frac{1}{4\pi i} \int_{\Gamma_1} \frac{g(\zeta_1, t_2)}{\zeta_1 - t_1} d\zeta_1$$

$$- \frac{1}{4} g(t_1, t_2) - \frac{1}{4\pi i} \int_{\Gamma_2} \frac{g(t_1, \zeta_2)}{\zeta_2 - t_2} d\zeta_2.$$

Proceeding similarly, we can derive the formulae

$$\begin{cases} \Phi^{++}(t_1, t_2) = \frac{1}{4} g(t_1, t_2) + \frac{1}{4\pi i} \int_{\Gamma_1} \frac{g(\zeta_1, t_2)}{\zeta_1 - t_1} d\zeta_1 \\[2mm] \quad + \frac{1}{4\pi i} \int_{\Gamma_2} \frac{g(t_1, \zeta_2)}{\zeta_2 - t_2} d\zeta_2 + \frac{1}{(2\pi i)^2} \int_{\Gamma_1} \int_{\Gamma_2} \frac{g(\zeta_1, \zeta_2)}{(\zeta - t_1)(\zeta_2 - t_2)} d\zeta_1 d\zeta_2, \\[3mm] \Phi^{+-}(t_1, t_2) = -\frac{1}{4} g(t_1, t_2) - \frac{1}{4\pi i} \int_{\Gamma_1} \frac{g(\zeta_1, t_2)}{\zeta_1 - t_1} d\zeta_1 \\[2mm] \quad + \frac{1}{4\pi i} \int_{\Gamma_2} \frac{g(t_1, \zeta_2)}{\zeta_2 - t_2} d\zeta_2 + \frac{1}{(2\pi i)^2} \int_{\Gamma_1} \int_{\Gamma_2} \frac{g(\zeta_1, \zeta_2)}{(\zeta_1 - t_1)(\zeta_2 - t_2)} d\zeta_1 d\zeta_2, \end{cases}$$

$$
\left\{
\begin{aligned}
&\Phi^{-+}(t_1, t_2) = -\frac{1}{4}g(t_1, t_2) + \frac{1}{4\pi i}\int_{\Gamma_1}\frac{g(\zeta_1, t_2)}{\zeta_1 - t_1}d\zeta_1 \\
&\quad -\frac{1}{4\pi i}\int_{\Gamma_2}\frac{g(t_1, \zeta_2)}{\zeta_2 - t_2}d\zeta_2 + \frac{1}{(2\pi i)^2}\int_{\Gamma_1}\int_{\Gamma_2}\frac{g(\zeta_1, \zeta_2)}{(\zeta_1 - t_1)(\zeta_2 - t_2)}d\zeta_1 d\zeta_2, \\
&\Phi^{--}(t_1, t_2) = \frac{1}{4}g(t_1, t_2) - \frac{1}{4\pi i}\int_{\Gamma_1}\frac{g(\zeta_1, t_2)}{\zeta_1 - t_1}d\zeta_1 \\
&\quad -\frac{1}{4\pi i}\int_{\Gamma_2}\frac{g(t_1, \zeta_2)}{\zeta_2 - t_2}d\zeta_2 + \frac{1}{(2\pi i)^2}\int_{\Gamma_1}\int_{\Gamma_2}\frac{g(\zeta_1, \zeta_2)}{(\zeta_1 - t_1)(\zeta_2 - t_2)}d\zeta_1 d\zeta_2.
\end{aligned}
\right.
\tag{1.4}
$$

Hence we have (1.3).

Let

$$
F(z_1, z_2) =
\begin{cases}
\Phi(z_1, z_2) & \text{for } z = (z_1, z_2) \in D^{++}, \\
\Phi(z_1, z_2)/G_1(z_1, z_2) & \text{for } z \in D^{+-}, \\
\Phi(z_1, z_2)/G_2(z_1, z_2) & \text{for } z \in D^{-+}, \\
-\Phi(z_1, z_2)/G_3(z_1, z_2) & \text{for } z \in D^{--}.
\end{cases}
\tag{1.5}
$$

On the basis of Lemma 1.1, it is easy to derive the following result.

Theorem 1.2 *The function $F(z_1, z_2)$ in (1.5) is a solution of Problem R for analytic functions.*

In the boundary condition (1.1), if

$$
G_j(t_1, t_2) = t_1^{K_{1j}}t_2^{K_{2j}}, \; j = 1, 2, 3,
\tag{1.6}
$$

where K_{ij} $(i = 1, 2, \; j = 1, 2, 3)$ are integers, and $G_j(z_1, z_2)$ $(j = 1, 2, 3)$ are admitted to possess zeroes, for instance when $K_{11} < 0$, $z_1^{K_{11}}z_2^{K_{21}}$ has a zero of order $|K_{11}|$ at $z_1 = \infty$. The above boundary value problem for analytic functions is called Problem R_1.

Theorem 1.3 (1) *If $K_{11} \le 0$, $K_{22} \le 0$, and*

$$
K_{21}, \; K_{12}, \; K_{13}, \; K_{23} \ge 0,
\tag{1.7}
$$

then Problem R_1 has a solution of the form

$$
F(z_1, z_2) =
\begin{cases}
F_1(z_1, z_2) + \Phi_1(z_1, z_2) & \text{for } z = (z_1, z_2) \in D^{++}, \\
z_1^{-K_{11}}z_2^{-K_{21}}[F_1(z_1, z_2) + \Phi_1(z_1, z_2)] & \text{for } z \in D^{+-}, \\
z_1^{-K_{12}}z_2^{-K_{22}}[F_1(z_1, z_2) + \Phi_1(z_1, z_2)] & \text{for } z \in D^{-+}, \\
-z_1^{-K_{13}}z_2^{-K_{23}}[F_1(z_1, z_2) + \Phi_1(z_1, z_2)] & \text{for } z \in D^{--},
\end{cases}
\tag{1.8}
$$

in which $\Phi_1(z_1, z_2)$ is the integral $\Phi(z_1, z_2)$ as stated in (1.2), and

$$F_1(z_1, z_2) = \sum_{m=0}^{\min(K_{12}, K_{13})} \sum_{n=0}^{\min(K_{21}, K_{23})} c_{mn} z_1^m z_2^n,$$

where c_{mn} are arbitrary complex constants.

(2) If $K_{11} \leq 0$, $K_{22} \leq 0$, and (1.7) is not satisfied, then under

$$I_1 = \begin{cases} m_1 + n_1 - 2 \ \text{ for } \ m_1 n_1 = 0, \\[2mm] m_1 = \max(|K_{12}| - K_{12}, |K_{13}| - K_{13}), \\[2mm] m_1 + n_1 - 4 \ \text{ for } \ m_1 n_1 \neq 0, \\[2mm] n_1 = \max(|K_{21}| - K_{21}, |K_{23}| - K_{23}) \end{cases} \tag{1.9}$$

solvability conditions, Problem R_1 possesses a bounded solution $F(z_1, z_2)$.

(3) If $K_{11} \geq 0$, $K_{22} \geq 0$, and

$$K_{21} \geq K_{22}, \ K_{12} \geq K_{11}, \ K_{13} \geq K_{11}, \ K_{23} \geq K_{22}, \tag{1.10}$$

then Problem R_1 has a solution of the form

$$F(z_1, z_2) = \begin{cases} z_1^{K_{11}} z_2^{K_{22}} [F_2(z_1, z_2) + \Phi_2(z_1, z_2)] \ \text{for} \ z \in D^{++}, \\[2mm] z_2^{-K_{21}+K_{22}} [F_2(z_1, z_2) + \Phi_2(z_1, z_2)] \ \text{for} \ z \in D^{+-}, \\[2mm] z_1^{-K_{12}+K_{11}} [F_2(z_1, z_2) + \Phi_2(z_1, z_2)] \ \text{for} \ z \in D^{-+}, \\[2mm] -z_1^{-K_{13}+K_{11}} z_2^{-K_{23}+K_{22}} [F_2(z_1, z_2) + \Phi_2(z_1, z_2)] \ \text{for} \ z \in D^{--}, \end{cases} \tag{1.11}$$

where

$$\Phi_2(z_1, z_2) = \frac{1}{(2\pi i)^2} \int_{\Gamma_1} \int_{\Gamma_2} \frac{\zeta^{-K_{11}} \zeta_2^{-K_{22}} g(\zeta_1, \zeta_2)}{(\zeta_1 - z_1)(\zeta_2 - z_2)} d\zeta_1 d\zeta_2,$$

$$F_2(z_1, z_2) = \sum_{m=0}^{\min(K_{12} - K_{11}, K_{13} - K_{11})} \sum_{m=0}^{\min(K_{21} - K_{22}, K_{23} - K_{22})} c_{mn} z_1^m z_2^n, \tag{1.12}$$

in which c_{mn} are arbitrary complex constants.

(4) If $K_{11} \geq 0$, $K_{22} \geq 0$, and (1.10) is not satisfied, then under

$$I_2 = \begin{cases} m_2 + n_2 - 2, \ m_2 n_2 = 0, \\[2mm] m_2 = \max(|K_{12} - K_{11}| - (K_{12} - K_{11}), |K_{13} - K_{11}| - (K_{13} - K_{11})), \\[2mm] m_2 + n_2 - 4, \ m_2 n_2 \neq 0, \\[2mm] n_2 = \max(|K_{21} - K_{22}| - (K_{21} - K_{22}), |K_{23} - K_{22}| - (K_{23} - K_{22})) \end{cases} \tag{1.13}$$

solvability conditions, Problem R_1 possesses a bounded solution $F(z_1, z_2)$.

Besides, for the cases $K_{11} > 0$, $K_{22} \leq 0$ and $K_{11} \leq 0$, $K_{22} > 0$, we can also write the result of solvability for Problem R_1 similar to the cases (1)-(4).

Proof   (1) It is easy to see that the function $F(z_1, z_2)$ in (1.8) is sectionally analytic and bounded in $D^{++}, D^{+-}, D^{-+}, D^{--}$, and

$$F_1(t_1, t_2) = F_1^{++}(t_1, t_2) = F_1^{+-}(t_1, t_2) = F^{-+}(t_1, t_2) = F_1^{--}(t_1, t_2) \text{ on } \Gamma. \qquad (1.14)$$

On the basis of (1.3), we know that the function $F(z_1, z_2)$ in (1.8) satisfies the boundary condition (1.1) with (1.6).

(2) Obviously, $m_1, n_1 \geq 0$, and the special solution $\Phi_1(z_1, z_2)$ is analytic in D^{+-}, D^{-+}, D^{--}, if and only if $\Phi_1(z_1, z_2)$ has a zero of order m_1 at $z_1 = \infty$ and a zero of order n_1 at $z_2 = \infty$. Noting that

$$\int_{\Gamma_1} \int_{\Gamma_2} \frac{g(\zeta_1, \zeta_2)}{(\zeta_1 - z_1)(\zeta_2 - z_2)} d\zeta_1 d\zeta_2 = -\frac{1}{z_1} \int_{\Gamma_1} \int_{\Gamma_2} \frac{g(\zeta_1, \zeta_2)}{(1 - \zeta_1/z_1)(\zeta_2 - z_2)} d\zeta_1 d\zeta_2$$

$$= -\sum_{m=0}^{\infty} \frac{1}{z_1^{m+1}} \int_{\Gamma_1} \int_{\Gamma_2} \frac{\zeta_1^m g(\zeta_1, \zeta_2)}{\zeta_2 - z_2} d\zeta_1 d\zeta_2, \ |z_1| > 1,$$

if $m_1 \geq 0$, and the following $2m_1 - 2$ conditions hold:

$$\int_{\Gamma_1} \int_{\Gamma_2} \frac{\zeta_1^m g(\zeta_1, \zeta_2)}{\zeta_2 - z_2} d\zeta_1 d\zeta_2 = 0, \ m = 0, 1, \cdots, m_1 - 2,$$

then $\Phi_1(z_1, z_2)$ is analytic in D^{-+}, when $m_1 = 0$, the number $2m_1 - 2$ is replaced by 0. Similarly, under the $2n_1 - 2$ conditions

$$\int_{\Gamma_1} \int_{\Gamma_2} \frac{\zeta_2^n g(\zeta_1, \zeta_2)}{\zeta_1 - z_1} d\zeta_1 d\zeta_2 = 0, \ n = 0, 1, \cdots, n_1 - 2,$$

the function $\Phi_1(z_1, z_2)$ is analytic in D^{+-}, when $n = 0$, the number $2n_1 - 2$ is replaced by 0. Hence Problem R_1 possesses I_1 solvability conditions.

(3) From the boundary condition (1.1) with (1.6),

$$\begin{aligned} t_1^{-K_{11}} t_2^{-K_{22}} F^{++}(t_1, t_2) &= t_2^{K_{21}-K_{22}} F^{+-}(t_1, t_2) + t_1^{K_{12}-K_{11}} F^{-+}(t_1, t_2) \\ &+ t_1^{K_{13}-K_{11}} t_2^{K_{23}-K_{22}} F^{--}(t_1, t_2) + t_1^{-K_{11}} t_2^{-K_{22}} g(t_1, t_2) \text{ on } \Gamma \end{aligned} \qquad (1.15)$$

can be obtained. Setting

$$\Phi_2(z_1, z_2) = \frac{1}{(2\pi i)^2} \int_{\Gamma_1} \int_{\Gamma_2} \frac{\zeta_1^{-K_{11}} \zeta_2^{-K_{22}} g(\zeta_1, \zeta_2)}{(\zeta_1 - z_1)(\zeta_2 - z_2)} d\zeta_1 d\zeta_2,$$

according to the proof in (1) and (2), we can derive that Problem R_1 possesses the solution (1.11).

The remaining part of the proof is not difficult. It can be given by a similar method as before.

1.2   Riemann problem of inhomogeneous Cauchy–Riemann systems

Next, we discuss the system of first order complex equations

$$w_{\overline{z_1}} = f_1(z_1, z_2), \quad w_{\overline{z_2}} = f_2(z_1, z_2), \tag{1.16}$$

where $f_j(z_1, z_2) \in C^1(\overline{E_1} \times \overline{E_2})$, $E_j = \{|z| < R_j\}$, $R_j(j = 1, 2)$ are positive constants, and $f_j(z_1, z_2)\,(j = 1, 2)$ satisfy the compatibility condition

$$f_{1\overline{z_2}} = f_{2\overline{z_1}}, \quad z = (z_1, z_2) \in E = E_1 \times E_2. \tag{1.17}$$

Lemma 1.4   *The general solution $w(z_1, z_2)$ of system (1.16) possesses the form*

$$w(z_1, z_2) = \Phi(z_1, z_2) + w_0(z_1, z_2), \tag{1.18}$$

where $\Phi(z_1, z_2)$ is an arbitrary analytic function in $E_1 \times E_2$, and

$$\begin{cases} w_0(z_1, z_2) = T_{E_1} f_1 + T_{E_2} \Phi_0, \quad T_{E_1} = -\dfrac{1}{\pi} \displaystyle\iint_{E_1} \dfrac{f_1(\zeta_1, z_2)}{\zeta_1 - z_1} d\sigma_{\zeta_1}, \\[4mm] T_{E_2} \Phi_0 = -\dfrac{1}{\pi} \displaystyle\iint_{E_2} \dfrac{\Phi_0(z_1, \zeta_2)}{\zeta_2 - z_2} d\sigma_{\zeta_2}, \quad \Phi_0(z_1, z_2) = \dfrac{1}{2\pi i} \displaystyle\int_{\partial E_1} \dfrac{f_2(\zeta_1, z_2)}{\zeta_1 - z_1} d\zeta_1. \end{cases} \tag{1.19}$$

Proof   It is clear that for a fixed point $z_2 \in \overline{E_2}$, $\Phi_0(z_1, z_2)$ is analytic in $z_1 \in E_1$, consequently

$$(T_{E_2} \Phi_0)_{\overline{z_1}} = T_{E_2}(\Phi_{0\overline{z_1}}) = 0.$$

Noting that $(T_{E_1} f_1)_{\overline{z_1}} = f_1(z_1, z_2)$, so $w_0(z_1, z_2)$ satisfies the first equation of (1.16). By the condition (1.17) and the Pompeiu formula, we have

$$(T_{E_1} f_1)_{\overline{z_2}} = -\frac{1}{\pi} \iint_{E_1} \frac{[f_1(\zeta_1, z_2)]_{\overline{z_2}}}{\zeta_1 - z_1} d\sigma_{\zeta_1}$$

$$= -\frac{1}{\pi} \iint_{E_1} \frac{[f_2(\zeta_1, z_2)]_{\overline{\zeta_1}}}{\zeta_1 - z_1} d\sigma_{\zeta_1} = f_2(z_1, z_2) - \frac{1}{2\pi i} \int_{\partial E_1} \frac{f_2(\zeta_1, z_2)}{\zeta_1 - z_1} d\zeta_1$$

$$= f_2(z_1, z_2) - \Phi_0(z_1, z_2) = f_2(z_1, z_2) - (T_{E_2} \Phi_0)_{\overline{z_2}}.$$

It shows that $w_0(z_1, z_2)$ satisfies the second equation of (1.16). Hence $w_0(z_1, z_2)$ is a special solution of system (1.16). Taking into account

$$[w(z_1, z_2) - w_0(z_1, z_2)]_{\overline{z_j}} = [\Phi(z_1, z_2)]_{\overline{z_j}} = 0, \quad j = 1, 2,$$

Lemma 1.4 is proved.

Theorem 1.5 *Problem R for system (1.16) with the boundary condition (1.1) is solvable, and its solution $w(z_1, z_2)$ can be written as (1.18), where $\Phi(z_1, z_2)$ satisfies the boundary condition:*

$$\Phi^{++}(t_1, t_2) = G_1(t_1, t_2)\Phi^{+-}(t_1, t_2) + G_2(t_1, t_2)\Phi^{-+}(t_1, t_2)$$
$$+ G_3(t_1, t_3)\Phi^{--}(t_1, t_2) + g_0(t_1, t_2), \quad t \in \Gamma, \tag{1.20}$$

in which

$$g_0(t_1, t_2) = g(t_1, t_2) + (G_1 + G_2 + G_3 + 1)w_0(t_1, t_2) \in C_\alpha(\Gamma), \tag{1.21}$$

where $\alpha\,(0 < \alpha < 1)$ is a constant.

Proof From Lemma 1.4, it follows that

$$w_0(t_1, t_2) = w_0^{++}(t_1, t_2) = w_0^{+-}(t_1, t_2) = w_0^{-+}(t_1, t_2) = w_0^{--}(t_1, t_2), \quad t \in \Gamma.$$

In accordance with Lemma 1.1, the Riemann problem (1.20) for analytic functions has a solution $\Phi(z_1, z_2)$. Let $\Phi(t_1, t_2) = w(t_1, t_2) - w_0(t_1, t_2)$ be substituted into (1.20), it can be derived

$$w^{++}(t_1, t_2) = G_1 w^{+-} + G_2 w^{-+} + G_3 w^{--} + g(t_1, t_2), \quad t \in \Gamma. \tag{1.22}$$

Hence $w(z_1, z_2) = \Phi(z_1, z_2) + w_0(z_1, z_2)$ in (1.18) is a solution of Problem R for (1.16). (See [166])

2 Riemann–Hilbert Problem for Analytic Functions in a Polycylinder

First of all, we consider the simple case, where the domain D is the unit bicylinder $D = D_1 \times D_2(D_1 = \{|z_1| < 1, D_2 = \{|z_2| < 1\})$ and $\Gamma = \Gamma_1 \times \Gamma_2(\Gamma_1 = \{|t_1| = 1\}, \Gamma_2 = \{|t_2| = 1\})$.

2.1 A special boundary value problem for analytic functions

Problem A The so–called Riemann–Hilbert boundary value problem is to find an analytic function $\Phi(z_1, z_2)$ in D continuous on $\bar{D}$ satisfying the boundary condition

$$\text{Re}[\bar{t_1}^{K_1}\bar{t_2}^{K_2}\Phi(t_1, t_2)] = r(t_1, t_2), \quad t = (t_1, t_2) \in \Gamma, \tag{2.1}$$

where K_1, K_2 are integers and $r(z_1, z_2)$ satisfies the condition

$$C_\alpha^1[r(t_1, t_2), \Gamma] \le l, \tag{2.2}$$

in which $\alpha\,(0 < \alpha < 1)$, $l\,(0 \le l < \infty)$ are real constants. (K_1, K_2) is called the index of Problem A. Problem A with $K_1 = K_2 = 0$ is called the Dirichlet problem (Problem D).

It is known that the solution of the Dirichlet problem for analytic functions of one complex variable can be given by the Schwarz formula. However, Problem A for analytic functions of several complex variables in D usually is not solvable. Hence, we propose the modified Riemann–Hilbert problem B for analytic functions of two complex variables with the boundary condition

$$\mathrm{Re}[\overline{t_1}^{-K_1}\overline{t_2}^{-K_2}\Phi(t_1, t_2)] = r(t_1, t_2) + H(t_1, t_2) + h(t_1, t_2),\ t = (t_1, t_2) \in \Gamma, \tag{2.3}$$

where

$$H(t_1, t_2) = \mathrm{Re}\sum_{j=1}^{\infty}\sum_{k=1}^{\infty} H_{jk}\overline{t_1}^{j}t_2^{k},\ t \in \Gamma, \tag{2.4}$$

and

$$h(t_1, t_2) = \begin{cases} -\mathrm{Re}\sum_{j=1}^{K_1}\sum_{K=1}^{K_2} H_{jk}\overline{t_1}^{j}t_2^{k}\ \text{for}\ K_1 \ge 0,\ K_2 \ge 0, \\[2mm] \mathrm{Re}\left(\sum_{m=0}^{-K_1-1}\sum_{n=0}^{\infty} + \sum_{m=-K_1}^{\infty}\sum_{n=0}^{-K_2-1}\right) h_{mn}t_1^{m}t_2^{n}, \\[2mm] \qquad\qquad\qquad \text{for}\ K_1 < 0,\ K_2 < 0, \\[2mm] \mathrm{Re}\sum_{m=0}^{-K_1-1} h_m^2(t_2)t_1^{m}\ \text{for}\ K_1 < 0,\ K_2 \ge 0, \\[2mm] \mathrm{Re}\sum_{n=0}^{-K_2-1} h_n^1(t_1)t_2^{n}\ \text{for}\ K_1 \ge 0,\ K_2 < 0, \end{cases} \tag{2.5}$$

in which $H_{jk}(j, k = 1, 2, \cdots)$, $h_{mn}(m = 1, \cdots, -K_1 - 1$, or $n = 1, \cdots, -K_2 - 1)$ are unknown complex constants to be determined appropriately, h_{00} is an unknown real constant, $h_n^1(t_1)\,(n = 1, 2, \cdots)$ and $h_m^2(t_2)\,(m = 1, 2, \cdots)$ are unknown complex functions on Γ_1 and Γ_2 respectively, $h_0^1(t_1)$ and $h_0^2(t_2)$ are unknown real functions. Problem B with $K_1 = K_2 = 0$ is called the modified Dirichlet problem (Problem E).

We need to establish some lemmas.

Lemma 2.1　*Problem E for analytic functions in D has a solution $\Phi_1(z_1, z_2)$ which can be expressed as the Schwarz formula:*

$$\Phi_1(z_1, z_2) = \frac{1}{(2\pi i)^2}\int_{\Gamma_1}\int_{\Gamma_2} r(t_1, t_2)\left[\frac{2}{(t_1 - z_1)(t_2 - z_2)} - \frac{1}{t_1 t_2}\right] dt_1 dt_2 + ic_0 = Pr + ic_0, \tag{2.6}$$

where c_0 is an arbitrary real constant. Moreover, Problem D is solvable if and only if the following conditions hold:

$$\int_0^{2\pi}\int_0^{2\pi} r(t_1, t_2)\frac{t_1^{m+1}}{t_2^{n+1}}d\phi_1 d\phi_2 = 0,\ m, n = 0, 1, 2, \cdots, \tag{2.7}$$

in which $t_1 = e^{i\phi_1}$, $t_2 = e^{i\phi_2}$.

Proof (1) We first use the Cauchy formula. Let $\Phi_1(z_1, z_2)$ be a solution of Problem D. Then for $z \in D$, there are

$$\Phi_1(z_1, z_2) = \frac{1}{(2\pi i)^2} \int_{\Gamma_1} \int_{\Gamma_2} \frac{\Phi_1(\zeta_1, \zeta_2)}{(\zeta_1 - z_1)(\zeta_2 - z_2)} d\zeta_1 d\zeta_2,$$

$$0 = \frac{1}{(2\pi i)^2} \int_{\Gamma_1} \int_{\Gamma_2} \frac{\Phi_1(\zeta_1, \zeta_2)}{(\zeta_1 - 1/\overline{z_1})(\zeta_2 - 1/\overline{z_2})} d\zeta_1 d\zeta_2.$$

From the above formulae, we obtain

$$\Phi_1(z_1, z_2) = \frac{1}{(2\pi i)^2} \int_{\Gamma_1} \int_{\Gamma_2} \left[\frac{\Phi_1(\zeta_1, \zeta_2) d\zeta_1 d\zeta_2}{(\zeta_1 - z_1)(\zeta_2 - z_2)} + \frac{\overline{\Phi_1(\zeta_1, \zeta_2) d\zeta_1 d\zeta_2}}{(\overline{\zeta_1} - 1/z_1)(\overline{\zeta_2} - 1/z_2)} \right]$$

$$= \frac{1}{(2\pi i)^2} \int_{\Gamma_1} \int_{\Gamma_2} \left[\frac{\Phi_1(\zeta_1, \zeta_2)}{(\zeta_1 - z_1)(\zeta_2 - z_2)} + \frac{\overline{\Phi_1(\zeta_1, \zeta_2) \zeta_1 \zeta_2}}{(\overline{\zeta_1} - 1/z_1)(\overline{\zeta_2} - 1/z_2)\zeta_1\zeta_2} \right] d\zeta_1 d\zeta_2$$

$$= \frac{1}{(2\pi i)^2} \int_{\Gamma_1} \int_{\Gamma_2} \left[\frac{\Phi_1(\zeta_1, \zeta_2)}{(\zeta_1 - z_1)(\zeta_2 - z_2)} + \overline{\Phi_1(\zeta_1, \zeta_2)} \right.$$

$$\left. \times \left(\frac{1}{(\zeta_1 - z_1)(\zeta_2 - z_2)} - \frac{1}{(\zeta_1 - z_1)\zeta_2} - \frac{1}{\zeta_1(\zeta_2 - z_2)} + \frac{1}{\zeta_1 \zeta_2} \right) \right] d\zeta_1 d\zeta_2.$$

Noting that

$$\frac{1}{(2\pi i)^2} \int_{\Gamma_1} \int_{\Gamma_2} \frac{\overline{\Phi_1(\zeta_1, \zeta_2)}}{(\zeta_1 - z_1)\zeta_2} d\zeta_1 d\zeta_2 = -\frac{1}{2\pi i} \int_{\Gamma_2} \frac{d\zeta_2}{\zeta_2} \frac{1}{2\pi i} \int_{\Gamma_1} \frac{\overline{\Phi_1(\zeta_1, \zeta_2)}\zeta_1}{(\zeta_1 - z_1)\overline{\zeta_1}} d\overline{\zeta_1}$$

$$= \frac{1}{2\pi i} \int_{\Gamma_2} \frac{d\zeta_2}{\zeta_2} \overline{\frac{1}{2\pi i} \int_{\Gamma_1} \frac{\Phi_1(\zeta_1, \zeta_2) d\zeta_1}{(1 - \zeta_1 \overline{z_1})\zeta_1}} = \frac{1}{2\pi i} \int_{\Gamma_2} \frac{\overline{\Phi_1(0, \zeta_2)}}{\zeta_2} d\zeta_2 =$$

$$= \overline{\frac{1}{2\pi i} \int_{\Gamma_2} \frac{\Phi_1(0, \zeta_2)}{\zeta_2} d\zeta_2} = \overline{\Phi_1(0, 0)},$$

and

$$\frac{1}{(2\pi i)^2} \int_{\Gamma_1} \int_{\Gamma_2} \frac{\overline{\Phi_1(\zeta_1, \zeta_2)}}{\zeta_1(\zeta_2 - z_2)} d\zeta_1 d\zeta_2 = \frac{1}{(2\pi i)^2} \int_{\Gamma_1} \int_{\Gamma_2} \frac{\overline{\Phi_1(\zeta_1, \zeta_2)}}{\zeta_1 \zeta_2} d\zeta_1 d\zeta_2$$

$$= \overline{\Phi_1(0, 0)} = \frac{1}{(2\pi i)^2} \int_{\Gamma_1} \int_{\Gamma_2} \frac{r(\zeta_1, \zeta_2)}{\zeta_1 \zeta_2} d\zeta_1 d\zeta_2 - i c_0,$$

where

$$c_0 = \frac{1}{(2\pi)^2} \int_{\Gamma_1} \int_{\Gamma_2} \mathrm{Im}\Phi_1(\zeta_1, \zeta_2) d\phi_1 d\phi_2, \quad 2r(\zeta_1, \zeta_2) = \Phi_1(\zeta_1, \zeta_2) + \overline{\Phi_1(\zeta_1, \zeta_2)},$$

it can be derived

$$\Phi_1(z_1, z_2) = \frac{1}{(2\pi i)^2} \int_{\Gamma_1} \int_{\Gamma_2} r(\zeta_1, \zeta_2) \left[\frac{2}{(\zeta_1 - z_1)(\zeta_2 - z_2)} - \frac{1}{\zeta_1 \zeta_2} \right] d\zeta_1 d\zeta_2 + i c_0.$$

(2) Next, we use another method to prove the Schwarz formula (2.6). By the condition (2.2), the function $r(t_1, t_2)$ can be expanded as a Fourier series in complex form

$$r(t_1, t_2) = \sum_{m=-\infty}^{\infty} \sum_{n=-\infty}^{\infty} a_{mn} t_1^m t_2^n, \tag{2.8}$$

where the Fourier coefficients

$$a_{mn} = \frac{1}{(2\pi i)^2} \int_{\Gamma_1} \int_{\Gamma_2} \frac{r(t_1, t_2)}{t_1^{m+1} t_2^{n+1}} dt_1 dt_2 = \overline{a_{-m,-n}}, \; m, n = 0, \pm 1, \pm 2, \cdots.$$

It is clear that

$$a_{mn} = \frac{1}{(2\pi)^2} \int_0^{2\pi} \int_0^{2\pi} \frac{r(t_1, t_2)}{t_1^m t_2^n} d\phi_1 d\phi_2 = \overline{a_{-m,-n}}, \; m, n = 0, \pm 1, \pm 2, \cdots. \tag{2.9}$$

If Problem D for analytic functions has a solution $\Phi_1(z_1, z_2)$, then $\Phi_1(z_1, z_2)$ satisfies the boundary condition

$$\operatorname{Re} \Phi_1(t_1, t_2) = r(t_1, t_2), \; t \in \Gamma, \tag{2.10}$$

and $\Phi_1(z_1, z_2)$ can be expanded as a Taylor series

$$\Phi_1(z_1, z_2) = \sum_{m=0}^{\infty} \sum_{n=0}^{\infty} c_{mn} z_1^m z_2^n \text{ in } D. \tag{2.11}$$

According to the Abel theorem, we can conclude

$$\operatorname{Re} \Phi_1(t_1, t_2) = \frac{1}{2} \Big[\sum_{m=0}^{\infty} \sum_{n=0}^{\infty} c_{mn} t_1^m t_2^n$$

$$+ \sum_{m=-\infty}^{0} \sum_{n=-\infty}^{0} \overline{c_{-m,-n}} t_1^m t_2^n \Big] = r(t_1, t_2), \; t \in \Gamma. \tag{2.12}$$

In comparison with (2.8),

$$\begin{cases} \operatorname{Re} c_{00} = a_{00} = \dfrac{1}{(2\pi)^2} \int_0^{2\pi} \int_0^{2\pi} r(t_1, t_2) d\phi_1 d\phi_2, \\[2mm] c_{mn} = 2a_{mm} = \dfrac{\cdot 2}{(2\pi)^2} \int_0^{2\pi} \int_0^{2\pi} \dfrac{r(t_1, t_2)}{t_1^m t_2^n} d\phi_1 d\phi_2, \\[2mm] \qquad m, n = 0, 1, 2, \cdots, \; m + n \neq 0, \end{cases} \tag{2.13}$$

can be obtained and

$$a_{mn} = \frac{1}{(2\pi)^2} \int_0^{2\pi} \int_0^{2\pi} \frac{r(t_1, t_2)}{t_1^m t_2^n} d\phi_1 d\phi_2 = 0, \; m < 0, n > 0 \text{ or } m > 0, n < 0. \tag{2.14}$$

Obviously, (2.14) is a necessary condition that Problem D is solvable, which can be written in the form (2.7). Under the conditions, from (2.11) and (2.13) it follows the Schwarz formula

$$
\begin{aligned}
\Phi_1(t_1, t_2) &= \sum_{m=0}^{\infty} \sum_{n=0}^{\infty} 2a_{mn} z_1^m z_2^n - a_{00} + i \operatorname{Im} \Phi_1(0,0) \\
&= \frac{1}{(2\pi i)^2} \int_{\Gamma_1} \int_{\Gamma_2} r(t_1, t_2) \left[\sum_{m=0}^{\infty} \sum_{n=0}^{\infty} \frac{2 z_1^m z_2^n}{t_1^{m+1} t_2^{n+1}} - \frac{1}{t_1 t_2} \right] dt_1 dt_2 + i c_0 \\
&= \frac{1}{(2\pi i)^2} \int_{\Gamma_1} \int_{\Gamma_2} r(t_1, t_2) \left[\frac{2}{(t_1 - z_1)(t_2 - z_2)} - \frac{1}{t_1 t_2} \right] dt_1 dt_2 + i c_0,
\end{aligned}
\tag{2.15}
$$

where $c_0 = \operatorname{Im} \Phi_1(0,0)$.

From (2.7)–(2.9) for $(z_1, z_2) \in \Gamma$

$$
\begin{aligned}
r(z_1, z_2) &= \sum_{m=0}^{\infty} \sum_{n=0}^{\infty} a_{mn} z_1^m z_2^n + \sum_{m=-\infty}^{0} \sum_{n=-\infty}^{0} a_{mn} z_1^m z_2^n - a_{00} \\
&= 2\operatorname{Re} \sum_{m=0}^{\infty} \sum_{n=0}^{\infty} a_{mn} z_1^m z_2^n - a_{00},
\end{aligned}
\tag{2.16}
$$

can be derived. On the basis of the Abel theorem, the series in (2.16) absolutely converges in D, and then

$$
\begin{aligned}
\Phi_1(z_1, z_2) &= 2 \sum_{m=0}^{\infty} \sum_{n=0}^{\infty} a_{mn} z_1^m t_2^n - a_{00} + i c_0 \\
&= \frac{1}{(2\pi i)^2} \int_{\Gamma_1} \int_{\Gamma_2} r(t_1, t_2) \left[\frac{2}{(t_1 - z_1)(t_2 - z_2)} - \frac{1}{t_1 t_2} \right] dt_1 t_2 + i c_0,
\end{aligned}
\tag{2.17}
$$

where c_0 is an arbitrary real constant. This shows that (2.7) is a sufficient condition for the solvability of Problem D.

(3) Finally, we discuss Problem E. Due to $h(t_1, t_2) \doteq 0$ in this case, let $r(t_1, t_2) + H(t_1, t_2)$ in (2.3) be substituted into the position of $r(t_1, t_2)$ in (2.7), we have

$$
\begin{aligned}
0 &= \int_0^{2\pi} \int_0^{2\pi} [r(t_1, t_2) + H(t_1, t_2)] \frac{t_1^{m+1}}{t_2^{n+1}} d\phi_1 d\phi_2 \\
&= \int_0^{2\pi} \int_0^{2\pi} r(t_1, t_2) \frac{t_1^{m+1}}{t_2^{n+1}} d\phi_1 d\phi_2 + \frac{1}{2} \int_0^{2\pi} \int_0^{2\pi} \sum_{j=1}^{\infty} \sum_{k=1}^{\infty} (H_{jk} \overline{t_1}^{j} t_2^k \\
&\quad + \overline{H_{jk}} t_1^j \overline{t_2}^k) \frac{t_1^{m+1}}{t_2^{n+1}} d\phi_1 d\phi_2 = \int_0^{2\pi} \int_0^{2\pi} r(t_1, t_2) \frac{t_1^{m+1}}{t_2^{n+1}} d\phi_1 d\phi_2 + \frac{(2\pi)^2}{2} H_{m+1, n+1},
\end{aligned}
$$

i.e.

$$H_{m+1,n+1} = \frac{-2}{(2\pi)^2} \int_0^{2\pi} \int_0^{2\pi} r(t_1, t_2) \frac{t_1^{m+1}}{t_2^{n+1}} d\phi_1 d\phi_2, \quad m, n = 0, 1, 2, \cdots. \tag{2.18}$$

Under the above conditions, Problem E is solvable. Noting that

$$\int_{\Gamma_1} \int_{\Gamma_2} H(t_1, t_2) \left[\frac{2}{(t_1 - z_1)(t_2 - z_2)} - \frac{1}{t_1 t_2} \right] dt_1 dt_2$$

$$= \int_{\Gamma_1} \int_{\Gamma_2} H(t_1, t_2) \frac{1}{t_1 t_2} [2 \sum_{m=0}^{\infty} \sum_{n=0}^{\infty} (\frac{z_1}{t_1})^m (\frac{z_2}{t_2})^n - 1] dt_1 dt_2 = 0,$$

hence the solution $\Phi_1(z_1, z_2)$ of Problem E can be expressed as (2.6), where c_0 is an arbitrary real constant.

Lemma 2.2 *If $K_1 \geq 0$, $K_2 \geq 0$, Problem B for analytic functions is solvable, and its general solution $\Phi(z_1, z_2)$ can be expressed as*

$$\Phi(z_1, z_2) = z_1^{K_1} z_2^{K_2} \Phi_1(z_1, z_2) + \Phi_0(z_1, z_2), \tag{2.19}$$

where $\Phi_1(z_1, z_2) = Pr$, Pr is as stated in (2.6), and

$$\Phi_0(z_1, z_2) = \sum_{m=0}^{2K_1} \sum_{n=0}^{2K_2} c_{mn} z_1^m z_2^n, \tag{2.20}$$

in which $c_{mn}(0 \leq m \leq 2K_1, 0 \leq n \leq 2K_2)$ are arbitrary complex constants satisfying the condition

$$c_{m+K_1, n+K_2} + \overline{c_{-m+K_1, -n+K_2}} = \begin{cases} 2a_{mn}, \ 0 < m \leq K_1, \ -K_2 \leq n < 0 \\[4pt] \text{and} \ -K_1 \leq m < 0, \ 0 < n \leq K_2, \\[4pt] 0, \ 0 \leq m \leq K_1, \ 0 \leq n \leq K_2 \\[4pt] \text{and} \ -K_1 \leq m \leq 0, \ -K_2 \leq n \leq 0, \end{cases} \tag{2.21}$$

in which $a_{mn} = \overline{a_{-m,-n}} \ (0 < m \leq K_1, \ -K_2 \leq n < 0)$ are as stated in (2.8). Under the conditions

$$\int_{\Gamma_1} \int_{\Gamma_2} \frac{r(t_1, t_2)}{t_1^{m+1} t_2^{n+1}} dt_1 dt_2 = 0, \ \text{for} \begin{cases} 0 < m \leq K_1, \ -\infty < n < -K_2 \\[4pt] \text{and} \ K_1 < m < \infty, \ -\infty < n < 0, \end{cases} \tag{2.22}$$

the general solution $\Phi(z_1, z_2)$ of Problem A also possesses the form (2.19).

Proof We first discuss Problem B. Noting (2.8), the coefficients of $H(t_1, t_2) + h(t_1, t_2)$ in (2.3) can be determined as follows:

$$H_{-mn} = -2a_{mn} = \frac{-2}{(2\pi i)^2} \int_{\Gamma_1} \int_{\Gamma_2} \frac{r(t_1, t_2)}{t_1^{m+1} t_2^{n+1}} dt_1 dt_2,$$

for $0 < m \leq K_1, \ -\infty < n < -K_2$ and $K_1 < m < \infty, \ -\infty < n < 0$.

Moreover, due to

$$\text{Re}\,[\overline{t_1}^{-K_1}\overline{t_2}^{-K_2}\Phi_0(t_1,t_2)] = \text{Re}\,[\sum_{m=0}^{\infty}\sum_{n=0}^{\infty} c_{mn} t_1^{m-K_1} t_2^{n-K_2}]$$

$$= \frac{1}{2}\sum_{m=0}^{2K_1}\sum_{n=0}^{2K_2}(c_{mn} + \overline{c_{2K_1-m,2K_2-n}})t_1^{m-K_1} t_2^{n-K_2},$$

we can choose the complex constants $c_{mn}(0 \le m \le 2K_1,\, 0 \le n \le 2K_2)$ such that (2.21) holds. Consequently, the function $\Phi(z_1,z_2)$ in (2.19) is the general solution of Problem B. If we require that the solution $\Phi(z_1,z_2)$ satisfies the $K = 4(K_1 + 1)(K_2 + 1) - 3$ point conditions

$$\text{Im}\,[\overline{a_{1j}}^{-K_1}\overline{a_{2j}}^{-K_2}\Phi(a_{1j},a_{2j})] = b_j \quad j = 1,\cdots,K,$$

where $(a_{1j},a_{2j}) \in \Gamma\,(j = 1,\cdots,K)$ are distinct points, $b_j\,(j = 1,\cdots,K)$ are real constants, then the solution of Problem B is unique. If $H(t_1,t_2) + h(t_1,t_2) = 0$ in (2.3), we can derive that (2.22) is a necessary and sufficient condition of solvability for Problem A.

Lemma 2.3 *When $K_1 < 0$, $K_2 < 0$, Problem B is solvable, and its solution $\Phi(z_1,z_2)$ possesses the representation*

$$\Phi(z_1,z_2) = z_1^{K_1} z_2^{K_2}\Phi_1(z_1,z_2), \tag{2.23}$$

where $\Phi_1(z_1,z_2) = P(r + h)$, the operator P is as stated in (2.6). Moreover, Problem A is solvable if and only if (2.7) holds and

$$\int_{\Gamma_1}\int_{\Gamma_2}\frac{r(t_1,t_2)}{t_1^{m+1}t_2^{n+1}}dt_1 dt_2 = 0 \text{ for } \begin{cases} 0 \le m \le -K_1 - 1,\, 0 \le n < \infty, \text{ and} \\[2mm] -K_1 \le m < \infty,\, 0 \le n \le -K_2 - 1. \end{cases} \tag{2.24}$$

Proof Because $z_j^{K_j}$ has a pole of order $|K_j|$ at $z_j = 0\,(j = 1,2)$ and the solution $z_1^{K_1}z_2^{K_2}\Phi_1(z_1,z_2)$ is continuous at the points $(0,z_2)$ and $(z_1,0)$, we must require that $\Phi_1(z_1,z_2)$ has at least a zero of the order $|K_j|$ at the point $z_j = 0\,(j = 1,2)$, namely

$$\Phi_1(z_1,z_2) = \frac{1}{(2\pi i)^2}\int_{\Gamma_1}\int_{\Gamma_2}[r(t_1,t_2) + h(t_1,t_2)][\frac{2}{(t_1 - z_1)(t_2 - z_2)}$$

$$-\frac{1}{t_1 t_2}]dt_1 t_2 = \frac{1}{(2\pi i)^2}\int_{\Gamma_1}\int_{\Gamma_2}[r(t_1,t_2) + \frac{1}{2}(\sum_{m=0}^{-K_1-1}\sum_{n=0}^{\infty} + \sum_{m=-K_1}^{\infty}\sum_{n=0}^{-K_2-1})$$

$$\times(h_{mn}t_1^m t_2^n + \overline{h_{mn}t_1^m t_2^n})][2\sum_{m=0}^{\infty}\frac{z_1^m}{t_1^{m+1}}\sum_{n=0}^{\infty}\frac{z_2^n}{t_2^{n+1}} - \frac{1}{t_1 t_2}]dt_1 dt_2$$

$$= \frac{2}{(2\pi i)^2}\sum_{m=-K_1}^{\infty}\sum_{n=-K_2}^{\infty}\int_{\Gamma_1}\int_{\Gamma_2}\frac{r(t_1,t_2)}{t_1^{m+1}t_2^{n+1}}dt_1 dt_2\, z_1^m z_2^n.$$

118   *Heinrich Begehr and Guo Chun Wen*

Hence

$$\begin{cases} h_{00} = -\dfrac{1}{(2\pi i)^2} \displaystyle\int_{\Gamma_1} \int_{\Gamma_2} \dfrac{r(t_1 t_2)}{t_1 t_2} dt_1 dt_2, \\[3mm] h_{mn} = -\dfrac{2}{(2\pi i)^2} \displaystyle\int_{\Gamma_1} \int_{\Gamma_2} \dfrac{r(t_1, t_2)}{t_1^{m+1} t_2^{n+1}} dt_1 dt_2, \\[3mm] \text{for } 0 \le m \le -K_1 - 1, \ 0 \le n < \infty, \ (m,n) \ne (0,0) \\[2mm] \text{and } -K_1 \le m < \infty, \ 0 \le n \le -K_2 - 1. \end{cases} \tag{2.25}$$

Besides, it is easy to see that

$$\int_0^{2\pi} \int_0^{2\pi} h(t_1, t_2) \frac{t_1^{m+1}}{t_2^{n+1}} d\phi_1 d\phi_2 = 0, \quad m, n = 0, 1, \cdots. \tag{2.26}$$

Similarly to Lemma 2.2, taking $H(t_1, t_2) = 0$, $h(t_1, t_2) = 0$, the necessary and sufficient condition for solvability of Problem A can be derived.

Lemma 2.4   *When $K_1 \ge 0$, $K_2 < 0$, Problem B is solvable, and its solution $\Phi(z_1, z_2)$ has the representation (2.23), but $c_0 = 0$ and $r(t_1, t_2)$ has to be replaced by $r(t_1, t_2) + h(t_1, t_2)$. In addition, Problem A is solvable if and only if (2.7) holds and*

$$\begin{aligned} &\int_{\Gamma_1} \int_{\Gamma_2} \frac{(t_1 + z_1) r(t_1, t_2)}{(t_1 - z_1) t_1 t_2} dt_1 t_2 = 0, \ z_1 \in D_1, \\[2mm] &\int_{\Gamma_1} \int_{\Gamma_2} \frac{r(t_1, t_2) dt_1 dt_2}{(t_1 - z_1) t_2^{n+1}} = 0, \ z_1 \in D_1, \ n = 1, \cdots, -K_2 - 1. \end{aligned} \tag{2.27}$$

When $K_1 < 0$, $K_2 \ge 0$, a similar statement holds.

Proof   Because the solution of Problem B is continuous on $\bar{D}$, we must require that $\Phi_1(z_1, z_2)$ has a zero of the order $|K_2|$ at $z_2 = 0$, i.e.

$$\Phi_1(z_1, z_2) = \frac{1}{(2\pi i)^2} \int_{\Gamma_1} \int_{\Gamma_2} [r(t_1, t_2) + h(t_1, t_2)][\frac{2}{(t_1 - z_1)(t_2 - z_2)} - \frac{1}{t_1 t_2}]$$

$$\times dt_1 dt_2 = \frac{1}{(2\pi i)^2} \int_{\Gamma_1} \int_{\Gamma_2} [r(t_1, t_2) + \frac{1}{2} \sum_{n=0}^{-K_2-1} (h_n^1(t_1) t_2^n + \overline{h_n^1(t_1) t_2^n})][\frac{2}{t_1 - z_1}$$

$$\times \sum_{n=0}^{\infty} \frac{z_2^n}{t_2^{n+1}} - \frac{1}{t_1 t_2}] dt_1 dt_2 = \frac{2}{(2\pi i)^2} \sum_{n=-K_2}^{\infty} \int_{\Gamma_1} \int_{\Gamma_2} \frac{r(t_1, t_2)}{(t_1 - z_1) t_2^{n+1}} dt_1 dt_2 \, z_2^n. \tag{2.28}$$

Consequently, if

$$\begin{cases} \dfrac{1}{2\pi i} \displaystyle\int_{\Gamma_1} \dfrac{t_1 + z_1}{(t_1 - z_1) t_1} h_0(t_1) dt_1 = \dfrac{-1}{(2\pi i)^2} \displaystyle\int_{\Gamma_1} \int_{\Gamma_2} \dfrac{(t_1 + z_1) r(t_1, t_2)}{(t_1 - z_1) t_1 t_2} dt_1 dt_2, \\[3mm] \dfrac{1}{2\pi i} \displaystyle\int_{\Gamma_1} \dfrac{h_n(t_1)}{t_1 - z_1} dt_1 = \dfrac{-2}{(2\pi i)^2} \displaystyle\int_{\Gamma_1} \int_{\Gamma_2} \dfrac{r(t_1, t_2) dt_1 dt_2}{(t_1 - z_1) t_2^{n+1}}, \ n = 1, \cdots, -K_2 - 1, \end{cases} \tag{2.29}$$

then Problem B has a solution $z_1^{K_1} z_2^{K_2} \Phi_1(z_1 z_2)$. Let $\Phi(z_1, z_2)$ be the general solution and denote $\Phi_0(z_1, z_2) = \Phi(z_1, z_2) - z_1^{K_1} z_2^{K_2} \Phi_1(z_1, z_2)$. Similarly to Lemma 2.3, we can obtain

$$\Phi_0(z_1, z_2) = \sum_{m=0}^{2K_1} c_m^1(z_2) z_1^m, \tag{2.30}$$

where $c_m^1(z_2)$ are analytic functions in D_2, and

$$z_2^{|K_2|} c_m^1(z_2) = -z_2^{|K_2|} \overline{c_{2K_1-m}^1(z_2)}$$

are also analytic in D_2. This shows that $z_2^{|K_2|} c_m^1(z_2)$ are constants, hence $c_m^1(z_2) = 0$, $z \in D_2$. So the solution of Problem B possesses the representation (2.23). Similarly as before, we can obtain the remaining result of this lemma.

Combining Lemma 2.2 – Lemma 2.4, we have the following theorem.

Theorem 2.5 *Problem B for analytic functions is solvable, and its solution $\Phi(z_1, z_2)$ can be expressed by (2.19), where*

$$\Phi_1(z_1, z_2) = \begin{cases} Pr, & \text{for } K_1 \geq 0, \ K_2 \geq 0, \\ P(r+h), & \text{for } K_1 < 0 \text{ or } K_2 < 0, \end{cases} \tag{2.31}$$

and

$$\Phi_0(z_1, z_2) = \begin{cases} \displaystyle\sum_{m=0}^{2K_1} \sum_{n=0}^{2K_2} c_{mn} z_1^m z_2^n, & \text{for } K_1 \geq 0, \ K_2 \geq 0, \\ 0, & \text{for } K_1 < 0, \text{ or } K_2 < 0, \end{cases} \tag{2.32}$$

in which the operator P and c_{mn} are as stated in (2.6) and (2.20) respectively. Moreover, Problem A is solvable if and only if (2.22)($K_1 \geq 0$, $K_2 \geq 0$); (2.7), (2.24)($K_1 < 0$, $K_2 < 0$), (2.7), (2.27)($K_1 \geq 0$, $K_2 < 0$), and (2.7) and

$$\int_{\Gamma_1} \int_{\Gamma_2} \frac{(t_2 + z_2) r(t_1, t_2)}{(t_2 - z_2) t_1 t_2} dt_1 dt_2 = 0, \quad \int_{\Gamma_1} \int_{\Gamma_2} \frac{r(t_1, t_2) dt_1 dt_2}{(t_2 - z_2) t_1^{m+1}} = 0,$$

$$z_2 \in D_2, \ m = 1, \cdots, -K_1 - 2 (K_1 < 0, K_2 \geq 0) \tag{2.33}$$

hold.

2.2 General boundary value problem for analytic functions

Secondly , we consider the general boundary condition for analytic functions:

$$\text{Re}[\overline{\lambda(t_1, t_2)} \Phi(t_1, t_2)] = r(t_1, t_2), \ t = (t_1, t_2) \in \Gamma, \tag{2.34}$$

where $r(t_1, t_2)$ satisfies the condition (2.2), and $|\lambda(t_1, t_2)| = 1$, $C_\alpha^1[\lambda, \Gamma] \leq l < \infty$, $0 < \alpha < 1$. Denote that

$$K_1 = \frac{1}{2\pi}\Delta_{\Gamma_1} \arg \lambda(t_1, t_2), \ K_2 = \frac{1}{2\pi}\Delta_{\Gamma_2} \arg \lambda(t_1, t_2) \tag{2.35}$$

and (K_1, K_2) is called the index of $\lambda(t_1, t_2)$. We take $t_1 = e^{i\phi_1}$, $t_2 = e^{i\phi_2}$, $0 \leq \phi_1, \phi_2 \leq 2\pi$. It is clear that $\lambda(e^{i\phi_1}, e^{i\phi_2})$ is uniformly continuous on the closed set: $0 \leq \phi_1 \leq 2\pi$, $0 \leq \phi_2 \leq 2\pi$, hence K_1 and K_2 are independent of $\phi_2 \, (0 \leq \phi_2 \leq 2\pi)$ and $\phi_1 \, (0 \leq \phi_1 \leq 2\pi)$ respectively. It is not difficult to see that

$$S_1(t_1, t_2) = \arg \lambda(t_1, t_2) - K_1 \arg t_1 - K_2 \arg t_2$$

is continuous on Γ. If $S_1(t_1, t_2)$ satisfies

$$\int_0^{2\pi} \int_0^{2\pi} S_1(t_1, t_2)\frac{t_1^{m+1}}{t_2^{n+1}} d\phi_1 d\phi_2 = 0, \ m, n = 0, 1, 2, \cdots, \tag{2.36}$$

then by Lemma 2.1, there exists an analytic function

$$S(z_1, z_2) = \frac{1}{(2\pi i)^2} \int_{\Gamma_1} \int_{\Gamma_2} S_1(t_1, t_2) \left[\frac{2}{(t_1 - z_1)(t_2 - z_2)} - \frac{1}{t_1 t_2} \right] dt_1 dt_2 \ \text{in } D, \quad \tag{2.37}$$

which is continuous on $\bar{D}$ and satisfies the Dirichlet boundary condition

$$\mathrm{Re} S(t_1, t_2) = S_1(t_1, t_2), \ t \in \Gamma.$$

As before, we need the modified boundary value problem(Problem B^*) for analytic functions with the boundary condition

$$\mathrm{Re}[\overline{\lambda(t_1, t_2)}\Phi(t_1, t_2)] = r(t_1, t_2) + [H(t_1, t_2) + h(t_1, t_2)]e^{-S_2(t_1, t_2)}, \tag{2.38}$$

where $r(t_1, t_2)$, $H(t_1, t_2)$ and $h(t_1, t_2)$ are stated as in (2.3), and $S_2(t_1, t_2) = \mathrm{Im} S(t_1, t_2)$. Similarly to Section 1, Chapter 1 in [140], the boundary condition (2.38) can be reduced to the following boundary condition for analytic functions $\Psi(z_1, z_2) = \Phi(z_1, z_2)e^{-iS(z_1, z_2)}$:

$$\mathrm{Re}[\overline{t_1}^{-K_1}\overline{t_2}^{-K_2}\Psi(t_1, t_2)] = R(t_1, t_2) + H(t_1, t_2) + h(t_1, t_2), \ t \in \Gamma, \tag{2.39}$$

in which $R(t_1, t_2) = r(t_1, t_2)e^{S_2(t_1, t_2)}$. According to the following Lemma 3.1, we see that $R(t_1, t_2) \in C_\alpha^1(\Gamma)$. On the basis of Theorem 2.5, we can find a solution $\Psi(z_1, z_2)$ of Problem B with the boundary condition (2.39), and then

$$\Phi(z_1, z_2) = \Psi(z_1, z_2)e^{iS(z_1, z_2)} \tag{2.40}$$

is just a solution of Problem B^*. Thus we have

Theorem 2.6 *Under the condition (2.36), Problem B^* for analytic functions is solvable, and its solution $\Phi(z_1, z_2)$ is the solution of Problem B with the boundary condition (2.39), and $S(z_1, z_2)$ is as stated in (2.37). Besides, we can give the solvability condition of the boundary value problem A^* for analytic functions with the boundary condition (2.34) (see [91]2),[150]2)).*

Finally, we introduce the Riemann–Hilbert boundary value problem for analytic functions of several complex variables in the polycylinder $D = D_1 \times \cdots \times D_n$, $D_j = \{|z_j| < 1\}$, $j = 1, \cdots, n$, the boundary condition of which is as follows

$$\mathrm{Re}[\overline{\lambda(t_1, \cdots, t_n)}\Phi(t_1, \cdots, t_n)] = r(t_1, \cdots, t_n),\ t = (t_1, \cdots, t_n) \in \Gamma, \qquad (2.41)$$

where $\Gamma = \Gamma_1 \times \cdots \times \Gamma_n$, $\Gamma_j = \{|z_j| = 1\}$, $j = 1, \cdots, n$, $|\lambda(t_1, \cdots, t_n)| = 1$, $r(t_1, \cdots, t_n) \in C_\alpha^{n-1}(\Gamma)$, $0 < \alpha < 1$. The Riemann–Hilbert problem will be called Problem A', and the corresponding modified problem (Problem B') is to find a continuous solution $\Phi(z_1, \cdots, z_n)$ on $\bar{D}$ satisfying the boundary condition

$$\mathrm{Re}[\overline{\lambda(t_1, \cdots, t_n)}\Phi(t_1, \cdots, t_n)] = r(t_1, \cdots, t_n)$$
$$+h(t_1, \cdots, t_n)e^{-S_2(t_1, \cdots, t_n)},\ t = (t_1, \cdots, t_n) \in \Gamma, \qquad (2.42)$$

in which $S_2(z_1, \cdots, z_n) = \mathrm{Im}\, S(z_1, \cdots, z_n)$, and

$$\begin{cases} S(z_1, \cdots, z_n) = \dfrac{1}{(2\pi i)^n} \displaystyle\int_{\Gamma_1} \cdots \int_{\Gamma_n} S_1(t_1, \cdots, t_n) \left[\dfrac{2}{\Pi_{j=1}^n (t_j - z_j)} - \dfrac{1}{\Pi_{j=1}^n t_j} \right] \\ \times dt_1 \cdots dt_n = PS_1,\ S_1(t_1, \cdots, t_n) = \arg \lambda(t_1, \cdots, t_n) - \displaystyle\sum_{j=1}^n K_j \arg t_j, \end{cases} \qquad (2.43)$$

we assume that $S_1(t_1, \cdots, t_n)$ satisfies

$$\int_{\Gamma_1} \cdots \int_{\Gamma_n} \frac{S_1(t_1, \cdots, t_n)}{t_1^{\beta_1+1} \cdots t_n^{\beta_n+1}} dt_1 \cdots dt_n = 0,\ (\beta_1, \cdots, \beta_n) \in E_0, \qquad (2.44)$$

where $\beta_j (j = 1, \cdots, n)$ are nonnegative integers and $E_0 = \{(-\infty, \infty), \cdots, (-\infty, \infty)\}\setminus \{[0, \infty), \cdots, [0, \infty)\}\setminus\{(-\infty, 0], \cdots, (-\infty, 0]\}$. Moreover, $h(t_1, \cdots, t_n)$ is an unknown real function to be determined appropriately:

$$h(t_1, \cdots, t_n) \in \begin{cases} B \setminus B_0 \setminus B_1 & \text{for } K_j \geq 0,\ j = 1, \cdots, n, \\ B \setminus B_2 & \text{for } K_j < 0,\ j = 1, \cdots, n, \end{cases} \qquad (2.45)$$

in which B_k $(k = 0, 1, 2)$ are sets of real functions on Γ, such that $h(t_1, \cdots, t_n) \in B =$

$C_\alpha^{n-1}(\Gamma)$, $0 < \alpha < 1$, and

$$\int_{\Gamma_1} \cdots \int_{\Gamma_n} \frac{h(t_1, \cdots, t_n)}{t_1^{\beta_1+1} \cdots t_n^{\beta_n+1}} dt_1 \cdots dt_n = 0, \ \beta_j \geq 0, \ j = 1, \cdots, n,$$

$$\int_{\Gamma_1} \cdots \int_{\Gamma_n} \frac{h(t_1, \cdots, t_n)}{t_1^{\beta_1+1} \cdots t_n^{\beta_n+1}} dt_1 \cdots dt_n = 0, \ |\beta_j| \leq K_j, \ j = 1, \cdots, n, \qquad (2.46)$$

$$\int_{\Gamma_1} \cdots \int_{\Gamma_n} \frac{h(t_1, \cdots, t_n)}{t_1^{\beta_1+1} \cdots t_n^{\beta_n+1}} dt_1 \cdots dt_n = 0, \ \beta_j \geq -K_j, \ j = 1, \cdots, n,$$

respectively. For other cases, the form of $h(t_1, \cdots, t_n)$ can be given too.

Using a method similar as before, we can obtain the result on the solvability of Problem A' and Problem B'. In the following, we only state the case of $K_j \geq 0, j = 1, \cdots, n$.

Theorem 2.7 *Problem B' for analytic functions of several complex variables has a solution, and Problem A' ($K_j \geq 0$, $j = 1, \cdots, n$) is solvable if and only if the following conditions hold:*

$$\int_{\Gamma_1} \cdots \int_{\Gamma_n} \frac{r(t_1, \cdots, t_n)}{t_1^{\beta_1+1} \cdots t_n^{\beta_n+1}} dt_1 \cdots dt_n = 0, \ (\beta_1, \cdots, \beta_n) \in E_0 \backslash E_1, \qquad (2.47)$$

where E_0 is defined as in (2.44), $E_1 = \{(\beta_1, \cdots, \beta_n) | (\beta_1, \cdots, \beta_n) \in E_0, |\beta_j| \leq K_j, j = 1, \cdots, n\}$. Under the above conditions, the general solution of Problem A' and Problem B' can be expressed as

$$\Phi(z_1, \cdots, z_n) = z_1^{K_1} \cdots z_n^{K_n} P(r + h) + \Phi_0(z_1, \cdots, z_n), \ z \in D, \qquad (2.48)$$

where the operator P is as stated in (2.43), and $\Phi_0(z_1, \cdots, z_n)$ is the solution of the corresponding homogeneous problem (see [135]).

3 Riemann–Hilbert Problem for Elliptic Systems of Complex Equations in a Polycylinder

3.1 Riemann–Hilbert problem for elliptic complex systems and a priori estimates of its solutions

We first give two lemmas on estimates of solutions for Problem E and Problem B^* for analytic functions.

Lemma 3.1 *If the solution $\Phi_1(z_1, z_2)$ of Problem E for analytic functions satisfies the condition $|\operatorname{Im} \Phi_1(0, 0)| \leq l < \infty$, then $\Phi_1(z_1, z_2)$ satisfies the estimate*

$$C_\alpha^1[\Phi_1, \bar{D}] \leq M_1 = M_1(\alpha, l), \ 0 < \alpha < 1. \qquad (3.1)$$

Proof From the Schwarz formula (2.6), $c_0 = \text{Im}\Phi_1(0,0)$, and by using the Plemelj formula, we have

$$\Phi_1(z_1, \zeta_2) = \frac{1}{2\pi i} \int_{\Gamma_1} \frac{dt_1}{t_1 - z_1} \left\{ \frac{1}{2\pi i} \int_{\Gamma_2} \left[\frac{2r(t_1, t_2)}{t_2 - \zeta_2} \right.\right.$$
$$\left.\left. - \frac{1}{t_1 t_2} \right] dt_2 + r(t_1, \zeta_2) \right\} + ic_0, \quad z \in D_1, \ \zeta_2 \in \Gamma_2. \tag{3.2}$$

It can be seen that $\Phi_1(z_1, \zeta_2)$ is an analytic function in D_1, and by Theorem 3.1, Chapter 5 in [139]17), $\Phi_1(z_1, \zeta_2)$ satisfies the estimate

$$C_\alpha^1[\Phi_1(z_1, \zeta_2), \overline{D_1}] \le M_2 = M_2(\alpha, l). \tag{3.3}$$

Similarly, it can be obtained

$$C_\alpha^1[\Phi_1(\zeta_1, z_2), \overline{D_2}] \le M_3 = M_3(\alpha, l). \tag{3.4}$$

On the basis of the maximum modulus principle for analytic functions, the estimate (3.1) is derived.

Lemma 3.2 *The solution $\Phi(z_1, z_2)$ of Problem B^* for analytic functions satisfies the estimate*

$$C_\alpha^1[\Phi, \overline{D}] \le M_4 = M_4(\alpha, l, c, K), \tag{3.5}$$

where c represents the constants $c_{mn}(0 \le m \le 2K_1, 0 \le n \le 2K_2$, for $K_1 \ge 0$, $K_2 \ge 0)$, and $K = (K_1, K_2)$.

Proof By Lemma 3.1, we see that the function $S(z_1, z_2)$ in (2.37) satisfies

$$C_\alpha^1[S, \overline{D}] \le M_5 = M_5(\alpha, l). \tag{3.6}$$

Hence $R(t_1, t_2) = r(t_1, t_2)e^{S_2(z_1, z_2)}$ satisfies

$$C_\alpha^1[R, \Gamma] \le M_6 = M_6(\alpha, l, K). \tag{3.7}$$

Noting that

$$\begin{cases} \Psi(z_1, z_2) = z_1^{K_1} z_2^{K_2} \Psi_1(z_1, z_2) + \Psi_0(z_1, z_2), \\[2mm] \Psi_1(z_1, z_2) = \dfrac{1}{(2\pi i)^2} \int_{\Gamma_1} \int_{\Gamma_2} \left[\dfrac{2R(t_1, t_2)}{(t_1 - z_1)(t_2 - z_2)} - \dfrac{1}{t_1 t_2} \right] dt_1 dt_2, \\[4mm] \Psi_0(z_1, z_2) = \displaystyle\sum_{m=0}^{2K_1} \sum_{n=0}^{2K_2} c_{mn} z_1^m z_2^n, \end{cases} \tag{3.8}$$

124   *Heinrich Begehr and Guo Chun Wen*

where $c_{mn}(0 \leq m \leq 2K_1, 0 \leq n \leq 2K_2)$ are arbitrary complex constants satisfying the conditions

$$c_{m+K_1,n+K_2} + \overline{c_{-m+K_1,-n+K_2}} = \begin{cases} 2a_{mn}, \text{ for } \begin{cases} 0 < m \leq K_1, \; -K_2 \leq n < 0; \\ -K_1 \leq m < 0, \; 0 < n \leq K_2, \end{cases} \\ 0, \text{ for } \begin{cases} 0 \leq m \leq K_1, \; 0 \leq n \leq K_2; \\ -K_1 \leq m \leq 0, \; -K_2 \leq n \leq 0, \end{cases} \end{cases}$$

and $\Psi_1(z_1, z_2)$ satisfies

$$C_\alpha^1[\Psi_1, \overline{D}] \leq M_7 = M_7(\alpha, l). \tag{3.9}$$

So $\Psi(z_1, z_2)$ satisfies the estimate

$$C_\alpha^1[\Psi, \overline{D}] \leq M_8 = M_8(\alpha, l, c, K). \tag{3.10}$$

Thus $\Phi(z_1, z_2) = \Psi(z_1, z_2)e^{iS(z_1, z_2)}$ satisfies the estimate (3.5).

Now we consider a quasilinear system of first order complex equations

$$w_{\overline{z_j}} = F_j(z_1, z_2, w), \; F_j = A_j(z_1, z_2, w)w + A_{j+2}(z_1, z_2, w), \; j = 1, 2, \tag{3.11}$$

and suppose that (3.11) satisfies the following conditions which will be denoted by Condition C.

(1) $A_j(z_1, z_2, w)(j = 1, \cdots, 4)$, $F_j(z_1, z_2, w)(j = 1, 2)$ are continuously differentiable in $z \in D$ and $w \in \mathbb{C}$, and satisfy

$$C_\alpha^1[A_j(z_1, z_2, w), \overline{D}] \leq L, \; j = 1, \cdots, 4 \tag{3.12}$$

for any continuously differentiable function $w(z_1, z_2) \in C_\alpha^1(\overline{D})$, where $L(\geq 0)$, $\alpha\,(0 < \alpha < 1)$ are real constants.

(2) $A_j(z_1, z_2, w)$ and $F_j(z_1, z_2, w)$ satisfy

$$\begin{cases} A_{1\overline{z_2}} = A_{2\overline{z_1}}, \; A_{3\overline{z_1}} = A_{4\overline{z_2}}, \; A_1 A_4 = A_2 A_3, \\ C_\alpha^1[F_j(z_1, z_2, w_1) - F_j(z_1, z_2, w_2), \overline{D}] \leq LC_\alpha^1[w_1 - w_2, \overline{D}], \; j = 1, 2, \end{cases} \tag{3.13}$$

where $w_j(z_1, z_2) \in C_\alpha^1(\overline{D})$, $j = 1, 2$.

Problem C   Find out a solution $w(z_1, z_2) \in C_\alpha^1(\overline{D})$ for the complex system (3.11), which satisfies the boundary condition

$$\text{Re}[\overline{\lambda(t_1, t_2)}w(t_1, t_2)] = r(t_1, t_2) + [H(t_1, t_2) + h(t_1, t_2)]e^{-S_2(t_1, t_2)}, \; t \in \Gamma, \tag{3.14}$$

where $r(t_1, t_2)$, $H(t_1, t_2)$, $h(t_1, t_2)$, $S_2(t_1, t_2)$ and $\lambda(t_1, t_2)$ are as stated in (2.3), (2.38) and (2.34). $K = (K_1, K_2)$ is called the index of Problem C.

Lemma 3.3 *The solution $w(z_1, z_2)$ of Problem B^* for the system (3.11) can be expressed as*

$$w(z_1, z_2) = [\Phi(z_1, z_2) + T_1 G_1 + T_2 G_2 + T_3 G_3] e^{-iS((z_1, z_2))}, \qquad (3.15)$$

where $G_j = F_j e^{-iS}$, $j = 1, 2$, $G_3 = G_{1\overline{z_2}} = G_{2\overline{z_1}}$, and

$$\begin{cases} T_1 G_1 = -\dfrac{1}{\pi} \displaystyle\iint_{D_1} \dfrac{G_1(\zeta_1, z_2)}{\zeta_1 - z_1} d\sigma_{\zeta_1}, \quad T_2 G_2 = -\dfrac{1}{\pi} \displaystyle\iint_{D_2} \dfrac{G_2(z_1, \zeta_2)}{\zeta_2 - z_2} d\sigma_{\zeta_2}, \\[4mm] T_3 G_3 = -\dfrac{1}{\pi^2} \displaystyle\iint_{D_1} \iint_{D_2} \dfrac{G_3(\zeta_1, \zeta_2)}{(\zeta_1 - z_1)(\zeta_2 - z_2)} d\sigma_{\zeta_1} d\sigma_{\zeta_2}, \end{cases} \qquad (3.16)$$

and $\Phi(z_1, z_2)$ is an analytic function, $S(z_1, z_2)$ is as stated in (2.37).

Proof Similarly to the proof from (2.38) to (2.39), we can see that the function

$$W(z_1, z_2) = w(z_1, z_2) e^{-iS(z_1, z_2)} \qquad (3.17)$$

satisfies the system

$$W_{\overline{z_j}} = F_j(z_1, z_2, W e^{iS(z_1, z_2)}) e^{-iS(z_1, z_2)} = G_j, \ j = 1, 2 \qquad (3.18)$$

and the boundary condition

$$\mathrm{Re}[\overline{t_1}^{K_1} \overline{t_2}^{K_2} W(t_1, t_2)] = R(t_1, t_2) + H(t_1, t_2) + h(t_1, t_2), \ t \in \Gamma, \qquad (3.19)$$

where $R(t_1, t_2) = r(t_1, t_2) e^{S_2(t_1, t_2)}$, and the coefficients of (3.18) $B_j(z_1, z_2, W) = A_j(z_1, z_2, w) e^{-iS(z_1, z_2)}$ $(j = 1, \cdots, 4)$ and $G_j(z_1, z_2, W)$ $(j = 1, 2)$ satisfy a similar Condition C', where the constant L is replaced by another constant L'. Because the function $\Phi(z_1, z_2)$ satisfies the system

$$\Phi_{\overline{z_j}} = (W - T_1 G_1 - T_2 G_2 - T_3 G_3)_{\overline{z_j}} = 0, \ j = 1, 2, \qquad (3.20)$$

we know that $\Phi(z_1, z_2)$ is an analytic function in D. This completes the proof.

Theorem 3.4 *The solution $w(z_1, z_2)$ of Problem C satisfies the estimate*

$$C_\alpha^1[w, \overline{D}] \leq M_9 = M_9(\alpha, l, L, c, K), \qquad (3.21)$$

where α, l, L, c, K are as stated in (3.10) and (3.12).

Proof Let the solution $w(z_1, z_2)$ be substituted into the system (3.11) and the boundary condition (3.14). Setting

$$\begin{cases} g_j(z_1, z_2) = A_j(z_1, z_2, w), \ j = 1, 2, \\[3mm] \phi(z_1, z_2) = -\dfrac{1}{\pi} \displaystyle\iint_{D_1} \dfrac{g_1(\zeta_1, z_2)}{\zeta_1 - z_1} d\sigma_{\zeta_1} - \dfrac{1}{\pi} \displaystyle\iint_{D_2} \dfrac{g_2(z_1, \zeta_2)}{\zeta_2 - z_2} d\sigma_{\zeta_2} \\[4mm] \qquad\qquad -\dfrac{1}{\pi} \displaystyle\iint_{D_1} \iint_{D_2} \dfrac{g_{1\overline{\zeta_2}}(\zeta_1, \zeta_2)}{(\zeta_1 - z_1)(\zeta_2 - z_2)} d\sigma_{\zeta_1} d\sigma_{\zeta_2} \end{cases} \qquad (3.22)$$

and

$$\begin{cases} f_j(z_1, z_2) = A_{j+2}(z_1, z_2, w)e^{-\phi(z_1, z_2)}, \ j = 1, 2, \\[2mm] \psi(z_1, z_2) = -\dfrac{1}{\pi} \iint_{D_1} \dfrac{f_1(\zeta_1, z_2)}{\zeta_1 - z_1} d\sigma_{\zeta_1} - \dfrac{1}{\pi} \iint_{D_2} \dfrac{f_2(z_1, \zeta_2)}{\zeta_2 - z_2} d\sigma_{\zeta_2} \\[2mm] \qquad\qquad -\dfrac{1}{\pi} \iint_{D_1} \iint_{D_2} \dfrac{f_{1\overline{\zeta_2}}(\zeta_1, \zeta_2)}{(\zeta_1 - z_1)(\zeta_2 - z_2)} d\sigma_{\zeta_1} d\sigma_{\zeta_2}, \end{cases} \tag{3.23}$$

where

$$f_{1\overline{z_2}} = A_{3\overline{z_2}}e^{-\phi} - A_3 e^{-\phi} A_2 = A_{4\overline{z_1}}e^{-\phi} - A_4 e^{-\phi} A_1 = f_{2\overline{z_1}}.$$

We see that the function

$$\Phi(z_1, z_2) = w(z_1, z_2)e^{-\phi(z_1, z_2)} - \psi(z_1, z_2) \tag{3.24}$$

satisfies the complex system

$$\Phi_{\overline{z_j}} = w_{\overline{z_j}}e^{-\phi} - we^{-\phi}A_j - A_{j+2}e^{-\phi} = 0 \text{ in } D, \ j = 1, 2, \tag{3.25}$$

and hence $\Phi(z_1, z_2)$ is an analytic function in D. So $w(z_1, z_2)$ can be expressed as

$$w(z_1, z_2) = [\Phi(z_1, z_2) + \psi(z_1, z_2)]e^{\phi(z_1, z_2)}. \tag{3.26}$$

By Condition C, it can be derived that $\phi(z_1, z_2)$ and $\psi(z_1, z_2)$ satisfy

$$C_\alpha^1[\phi, \overline{D}] \le M_{10}, \ C_\alpha^1[\psi, \overline{D}] \le M_{11}, \tag{3.27}$$

where $M_j = M_j(\alpha, l), \ j = 10, 11$. Moreover, the analytic function $\Phi(z_1, z_2)$ satisfies the boundary condition

$$\begin{aligned} \operatorname{Re}[\overline{\lambda(t_1, t_2)}e^{\phi(t_1, t_2)}\Phi(t_1, t_2)] &= r(t_1, t_2) - \operatorname{Re}[\overline{\lambda(t_1, t_2)} \\ \times \psi(t_1, t_2)e^{\phi(t_1, t_2)}] &+ [H(t_1, t_2) + h(t_1, t_2)]e^{-S_2(t_1, t_2)}, \ t \in \Gamma. \end{aligned} \tag{3.28}$$

On the basis of Lemma 3.2, we have the estimate

$$C_\alpha^1[\Phi, \overline{D}] \le M_{12} = M_{12}(\alpha, l, L, c, K). \tag{3.29}$$

Combining (3.27) and (3.29), the estimate (3.21) in obtained.

3.2 Existence of solutions to the Riemann–Hilbert problem for elliptic complex systems of first order

In the following, by using the preceding estimate and the Schauder fixed–point theorem, we shall prove the existence of solutions of Problem B^* for the system (3.11).

Theorem 3.5 *Suppose that the system* (3.11) *satisfies Condition C. Then Problem B^* for* (3.11) *has a solution.*

Proof   We introduce the complex system with the parameter $k \in [0, 1]$

$$w_{j\overline{z_j}} - kF_j(z_1, z_2, w) = B_j(z_1, z_2), \ j = 1, 2, \tag{3.30}$$

where

$$B_{1\overline{z_2}} = B_{2\overline{z_1}}, \ A_1 B_2 = A_2 B_1, \ B_j(z_1, z_2) \in C_\alpha^1(\overline{D}), \ j = 1, 2.$$

When $k = 0$, it is easy to see that Problem C for (3.30) has a solution of the form

$$w_*(z_1, z_2) = [\Phi_*(z_1, z_2) + \tilde{T}B]e^{-iS(z_1, z_2)}, \ \tilde{T}B = T_1 B_1 + T_2 B_2 + T_3 B_3, \tag{3.31}$$

in which $B_3 = B_{1\overline{z_2}}$, and $\Phi_*(z_1, z_2)$ is an analytic function in D satisfying the boundary condition

$$\mathrm{Re}[\overline{t_1}^{K_1} \overline{t_2}^{K_2} \Phi_*(t_1, t_2)] = R_*(t_1, t_2) + H(t_1, t_2) + h(t_1, t_2), \ t \in \Gamma, \tag{3.32}$$

which possesses the form

$$\Phi_*(z_1, z_2) = z_1^{K_1} z_2^{K_2} P(R_*), \ R_* = R(t_1, t_2) - \mathrm{Re}[\overline{t_1}^{K_1} \overline{t_2}^{K_2} \tilde{T}B]. \tag{3.33}$$

Let $k_0(0 \le k_0 < 1)$ be a number such that Problem C for (3.30) with k_0 and any of the above functions $B_j(z_1, z_2)(j = 1, 2)$ is solvable. The system (3.30) can be rewritten as

$$w_{\overline{z_j}} - k_0 F_j(z_1, z_2, w) = (k - k_0)F_j(z_1, z_2, w) + B_j(z_1, z_2), \ j = 1, 2. \tag{3.34}$$

We choose an arbitrary function $w_0(z_1, z_2) \in C_\alpha^1(\overline{D})$. There is no harm in assuming $w_0(z_1, z_2) = 0$. Substituting $w_0(z_1, z_2)$ into the position of w in the right–hand side of (3.34), it is clear that Problem C for (3.34) has a solution $w_1(z_1, z_2)$ similar to $w_*(z_1, z_2)$. Thus by successive iteration, we obtain a sequence of solutions $w_n(z_1, z_2)$, $n = 1, 2, \cdots$, which satisfy

$$w_{n+1\overline{z_j}} - k_0 F_j(z_1, z_2, w_{n+1}) = (k - k_0)F_j(z_1, z_2, w_n) + B_j(z_1, z_2),$$
$$j = 1, 2, \ n = 0, 1, 2, \cdots \tag{3.35}$$

and the boundary condition

$$\mathrm{Re}[\overline{\lambda(t_1, t_2)} w_{n+1}(t_1, t_2)] = r(t_1, t_2) + [H(t_1, t_2) + h(t_1, t_2)]e^{-iS(t_1, t_2)}, \ t \in \Gamma. \tag{3.36}$$

By using Theorem 3.4, it can be proved that there exists a function $w^*(z_1, z_2) \in C_\alpha^1(\overline{D})$, such that $C_\alpha^1[w_n - w^*, \overline{D}] \to 0$ as $n \to \infty$, and $w^*(z_1, z_2)$ is a solution of Problem C for (3.30) in a neighborhood of k_0. Moreover, we can derive that Problem C for (3.30) with $k = 1$ and $B_j(z_1, z_2) = 0(j = 1, 2)$ is solvable and the solution is just a solution of Problem C for (3.11)(see [150]2)).

Finally, we mention that by using a similar method we can also discuss the Riemann–Hilbert boundary value problem for a quasilinear system of n complex equations of first order

$$w_{\overline{z_j}} = F_j(z_1, \cdots, z_n, w), \ F_j = A_j(z_1, \cdots, z_n, w)w + B_j(z_1, \cdots, z_n, w)$$

in the polycylinder $D = D_1 \times \cdots \times D_n$, $D_j = \{|z_j| < 1\}$, $j = 1, \cdots, n$ with the boundary condition

$$\mathrm{Re}[\overline{\lambda(t_1, \cdots, t_n)}w(t_1, \cdots, t_n)] = r(t_1, \cdots, t_n), \ t \in \Gamma,$$

where $\Gamma = \{|z_1| = 1\} \times \cdots \times \{|z_n| = 1\}$.

4   Some Boundary Value Problems in the Ball

In this section, we mainly discuss the Riemann–Hilbert boundary value problem for analytic functions of several complex variables in the unit ball $G = \{(z_1, \cdots, z_n) \in \mathbb{C}^n, \sum_{k=1}^n |z_k|^2 < 1\}$ with the boundary condition

$$\mathrm{Re}[\overline{t_1}^{-K_1} \cdots \overline{t_n}^{-K_n}\Phi(t_1, \cdots, t_n)] = r(t_1, \cdots, t_n), \ t = (t_1, \cdots, t_n) \in \partial G, \qquad (4.1)$$

where $r(t_1, \cdots, t_n) \in C_\alpha^{n-1}(\partial G)$, ∂G is the boundary of the domain G, $\alpha \, (0 < \alpha < 1)$ is a constant, $K_1, \cdots, K_n$ are integers, and $(K_1, \cdots, K_n)$ is called the index of the above boundary value problem. Without loss of generality, we only consider the case of $n = 2$, and the above problem in the case will be called Problem A, we require that the solution $\Phi(z_1, z_2)$ is continuous on $\overline{G}$, and Problem A with $K_1 = K_2 = 0$ is called Problem D, the boundary condition of Problem D is as follows:

$$\mathrm{Re}\,\Phi(t_1, t_2) = r(t_1, t_2), \ t = (t_1, t_2) \in \partial G. \qquad (4.2)$$

     We first give two lemmas.

Lemma 4.1   *Let $\Phi(z_1, z_2)$ be a solution of Problem D for analytic functions. Then $\Phi(z_1, z_2)$ can be expressed by the following Schwarz formula:*

$$\Phi(z) = \int_{\partial G} r(t)[2H(z, \overline{t}) - 1]d\sigma(t) + ic = Pr + ic \text{ in } G, \qquad (4.3)$$

where

$$H(z, \overline{t}) = \frac{1}{(1 - z_1\overline{t_1} - z_2\overline{t_2})^2}, \ z = (z_1, z_2), \ t = (t_1, t_2),$$

and $\sigma(t)$ is the rotation–invariant positive Borel measure on ∂G for which $\sigma(\partial G) = 1$ (see [115]).

Proof   From the Cauchy formula, it is not difficult to prove that

$$\Phi(z) = \int_{\partial G} 2r(t)H(z, \overline{t})d\sigma(t) - \overline{\Phi(0)}. \qquad (4.4)$$

In fact, the right–hand side in (4.4) can be written as follows:

$$\int_{\partial G} 2r(t)H(z, \overline{t})d\sigma(t) - \overline{\Phi(0)} = F(\Phi) + F(\overline{\Phi}) - \overline{\Phi(0)},$$

in which

$$F(\Phi) = \int_{\partial G} \Phi(t) H(z, \bar{t}) d\sigma(t). \tag{4.5}$$

Setting $\Psi(\zeta) = \Phi(\zeta) H(\zeta, \bar{z})$, $z = (z_1, z_2) \in G$, $\zeta = (\zeta_1, \zeta_2) \in \overline{G}$, it is clear that $\Psi(\zeta)$ is an analytic function in G continuous on $\overline{G}$ and $\Psi(0) = \Phi(0)$. Consequently,

$$\int_{\partial G} \Psi(\zeta) d\sigma(\zeta) = \Phi(0), \ F(\overline{\Phi}) = \int_{\partial G} \overline{\Psi d\sigma(\zeta)} = \overline{\Phi(0)}.$$

Thus from (4.5) and the Cauchy formula (see [115])

$$\Phi(z) = \int_{\partial G} \Phi(t) H(z, \bar{t}) d\sigma(t),$$

formula (4.4) is derived.

Taking $z_1 = z_2 = 0$ in (4.4), we have

$$\Phi(0) = \int_{\partial G} 2r(t) d\sigma(t) - \overline{\Phi(0)},$$

and

$$\Phi(z) - \frac{1}{2}\Phi(0) = \int_{\partial G} r(t)[2H(z, \bar{t}) - 1]d\sigma(t) - \overline{\Phi(0)} + \frac{1}{2}\overline{\Phi(0)}.$$

Hence (4.3) is true, where $c = \operatorname{Im}\Phi(0)$.

Generally speaking, the Dirichlet problem D for analytic functions may not be sovable. Next, using Lemma 4.1 and the Plemelj formula, we shall obtain the solvability condition of Problem D.

Lemma 4.2 *Suppose that $r(t) \in C_\alpha(\partial G)$, $0 < \alpha < 1$. Then Problem D for analytic functions is solvable if and only if*

$$\int_{\partial G} r(t)[H(t, \bar{\tau}) + H(\bar{t}, \tau) - 1]d\sigma(\tau) = 0, \tag{4.6}$$

where $t \in \partial G$. The above integral is determined by the Cauchy principle value, i.e.

$$\int_{\partial G} r(\tau) H(t, \bar{\tau}) d\sigma(\tau) = \lim_{\varepsilon \to 0} \int_{|1 - t\bar{\tau}'| \geq \varepsilon} r(\tau) H(t, \bar{\tau}) d\sigma(\tau),$$

where $\tau \overline{\tau'} = 1$.

Proof From (4.3), we see that $\operatorname{Re} Pr = r(t)$, $t \in \partial G$ is a necessary and sufficient condition of Problem D to be solvable. Secondly, we shall consider the limit of $\operatorname{Re} Pr$ on the boundary ∂G.

When $z = (z_1, z_2)\,(\in G)$ tends to $t = (t_1, t_2)\,(\in \partial G)$, according to the Plemelj formula, we have

$$
\begin{aligned}
\mathrm{Re}Pr|_{z=t} &= \lim_{z\to t} \int_{\partial G} r(\tau)\mathrm{Re}[2H(z,\bar\tau) - 1]d\sigma(\tau) \\
&= \lim_{z\to t} \int_{\partial G} r(\tau)[H(z,\bar\tau) + H(\bar z,\tau) - 1]d\sigma(\tau) \\
&= \int_{\partial G} r(\tau)H(t,\bar\tau)d\sigma(\tau) + \tfrac{1}{2}r(t) + \\
&\quad + \int_{\partial G} \overline{r(\tau)H(t,\bar\tau)d\sigma(\tau)} + \tfrac{1}{2}r(t) - \int_{\partial G} r(\tau)d\sigma(\tau) \\
&= \int_{\partial G} r(\tau)[H(t,\bar\tau) + H(\bar t,\tau) - 1]d\sigma(\tau) + r(t).
\end{aligned}
$$

This shows that (4.6) is a necessary and sufficient condition that Problem D is solvable.

In the following, we consider the solvability of Problem A separating three cases.

(1) $K_1 \geq 0$, $K_2 \geq 0$. We introduce a function

$$
\Psi(z_1, z_2) = z_1^{-K_1} z_2^{-K_2} \int_{\partial G} r(t_1, t_2)[2H(z,\bar t) - 1]d\sigma(t). \tag{4.7}
$$

If $r(t_1, t_2) \in C_\alpha(\partial G)$, $0 < \alpha < 1$, and (4.6) holds. Obviously, $\Psi(z_1, z_2)$ is a solution of Problem A. Setting

$$
F(z_1, z_2) = z_1^{K_1} z_2^{K_2}[\Phi(z_1, z_2) - \Psi(z_1, z_2)],
$$

it is clear $F(z_1, z_2)$ is analytic in $G\backslash\{0,0\}$, and has a pole of the order $\leq (K_1, K_2)$ at $(0,0)$. Hence $F(z_1, z_2)$ can be expressed as

$$
F(z_1, z_2) = \sum_{m=-K_1}^{\infty} \sum_{n=-K_2}^{\infty} c_{mn} z_1^m z_2^n. \tag{4.8}
$$

Noting that $F(z_1, z_2)$ satifies the boundary condition

$$
\mathrm{Re}\,F(t_1, t_2) = 0, \quad (t_1, t_2) \in \partial G, \tag{4.9}
$$

we denote

$$
c_{mn} = \alpha_{mn} + i\beta_{mn}, \quad t_1 = re^{i\theta_1}, \quad 0 < r < 1, \quad \text{and } t_2 = \sqrt{1 - r^2}\,e^{i\theta_2},
$$

then

$$
\mathrm{Re}\,F(z_1, z_2) = \sum_{m=-K_1}^{\infty} \sum_{n=-K_2}^{\infty} r^m (1 - r^2)^{n/2}[\alpha_{mn}\cos(m\theta_1 + n\theta_2) -
$$

$$
- \beta_{mn}\sin(m\theta_1 + n\theta_2)] = 0, \quad t = (t_1, t_2) \in \partial G.
$$

According to the uniqueness of the Fourier series expansion and the arbitrariness of $r(0 < r < 1)$, it can be derived that

$$
\alpha_{mn} = 0, \quad \beta_{mn} = 0, \quad (m, n) \neq (0, 0), \quad \alpha_{00} = 0,
$$

thus $F(z_1, z_2) = i\beta_{00}$. Therefore, the general solution of Problem A possesses the form

$$\Phi(z_1, z_2) = z_1^{K_1} z_2^{K_2} \{ \int_{\partial G} r(t_1, t_2)[2H(z, \bar{t}) - 1]d\sigma(t) + i\beta_{00}\}, \qquad (4.10)$$

where β_{00} is an arbitrary real constant.

(2) $K_1 < 0$, $K_2 < 0$. By Lemma 4.1 and Lemma 4.2, the solution of Problem A can be expressed as

$$\Phi(z_1, z_2) = z_1^{K_1} z_2^{K_2} \{ \int_{\partial G} r(t_1, t_2)[2H(z, \bar{t}) - 1]d\sigma(t) + ic\}.$$

We require that $\Phi(z_1, z_2)$ is continuous at $(0, 0)$, hence the function between the braces of above formula has to have a zero of degree $\geq (-K_1, -K_2)$ at the origin $(0, 0)$. Indeed, if

$$\int_{\partial G} \bar{t_1}^{j} \bar{t_2}^{k} r(t_1, t_2)d\sigma(t) = 0, \ j = 0, 1, \cdots, -K_1 - 1 \text{ or } k = 0, 1, \cdots, -K_2 - 1 \qquad (4.11)$$

hold, then

$$Pr + ic = 2 \int_{\partial G} \frac{r(t_1, t_2)}{(1 - z_1\bar{t_1} - z_2\bar{t_2})^2} d\sigma(t) - \int_{\partial G} r(t_1, t_2)d\sigma(t) + ic$$

$$= 2 \int_{\partial G} r(t_1, t_2)[\sum_{n=1}^{\infty}(n + 1)(z_1\bar{t_1} + z_2\bar{t_2})^n]d\sigma(t) - \int_{\partial G} r(t_1, t_2)d\sigma(t) + ic$$

$$= 2 \sum_{n=1}^{\infty}(n + 1) \int_{\partial G} r(t_1, t_2)(z_1\bar{t_1} + z_2\bar{t_2})^n d\sigma(t) + \int_{\partial G} r(t_1, t_2)d\sigma(t) + ic$$

$$= 2 \sum_{n=1}^{\infty} \sum_{k=0}^{n}(n + 1)C_n^k \int_{\partial G} r(t_1, t_2)\bar{t_1}^{n-k}\bar{t_2}^{k} d\sigma(t) z_1^{n-k} z_2^{k} + \int_{\partial G} r(t_1, t_2)d\sigma(t) + ic,$$

in which we take $c = 0$. Under the conditions (4.6) and (4.11), the solution of Problem A can be written in the form

$$\Phi(z_1, z_2) = z_1^{K_1} z_2^{K_2} \int_{\partial G} r(t_1, t_2)[2H(z, \bar{t}) - 1]d\sigma(t). \qquad (4.12)$$

(3) $K_1 < 0$, $K_2 \geq 0$ (or $K_1 \geq 0$, $K_2 < 0$). Similarly to (2), a necessary and sufficient condition under which Problem A is solvable can be derived. We have

$$\int_{\partial G} r(t_1, t_2)\bar{t_1}^{m}\bar{t_2}^{n} d\sigma(t) = 0, \ m = 0, 1, \cdots, -K_1 - 1,$$

$$\qquad (4.13)$$

$$n = 0, 1, 2, \cdots \text{ (or } m = 0, 1, 2, \cdots, \ n = 0, 1, \cdots, -K_2 - 1).$$

The above results are collected in the following theorem.

Theorem 4.3   *Suppose that condition (4.6) holds. Then we have:*

(1) *When $K_1 \geq 0$, $K_2 \geq 0$, Problem A is solvable, and its solution $\Phi(z_1, z_2)$ possesses the form (4.10).*

(2) *When $K_1 < 0$, $K_2 < 0$, Problem A is solvable if and only if that (4.11) holds, and its solution $\Phi(z_1, z_2)$ possesses the form (4.12).*

(3) *When $K_1 < 0$, $K_2 \geq 0$ (or $K_1 \geq 0$, $K_2 < 0$), Problem A is solvable if and only if the condition (4.13) holds, and its solution $\Phi(z_1, z_2)$ can be written in the form (4.12).*

Finally, we consider the system of complex equations

$$w_{\overline{z_1}} = A_1(z_1, z_2), \ w_{\overline{z_2}} = A_2(z_1, z_2), \ z = (z_1, z_2) \in G, \tag{4.14}$$

where $A_j(z_1, z_2)(j = 1, 2)$ possess continuous partial derivatives up to second order with respect to $z_1, z_2((z_1, z_2) \in G)$ satisfying the compatibility condition $A_{1\overline{z_2}} = A_{2\overline{z_1}}$, $z \in G$. We may extend $A_j(z_1, z_2)\,(j = 1, 2)$ into $G_1 \times G_2\,(G_j = \{|z_j| < 1\}, j = 1, 2)$.

The Riemann–Hilbert boundary value problem A^* is to find a continuous solution of (4.14) satisfying the boundary condition

$$\mathrm{Re}[\overline{t_1}^{K_1}\overline{t_2}^{K_2} w(t_1, t_2)] = r(t_1, t_2), \ t = (t_1, t_2) \in \partial G, \tag{4.15}$$

in which K_1, K_2 are integers and $r(t_1, t_2) \in C_\alpha(\partial G)$, $0 < \alpha < 1$.

Setting

$$\Phi(z_1, z_2) = w(z_1, z_2) - T_1 A_1 - T_2 \tilde{A}_2,$$

where

$$\tilde{A}_2 = A_2 - (T_1 A_1)_{\overline{z_2}}, \ T_1 A_1 = -\frac{1}{\pi} \iint_{G_1} \frac{A_1(\zeta_1, z_2)}{\zeta_1 - z_1} d\sigma_{\zeta_1},$$

$$T_2 \tilde{A}_2 = -\frac{1}{\pi} \iint_{G_2} \frac{\tilde{A}_2(z_1, \zeta_2)}{\zeta_2 - z_2} d\sigma_{\zeta_2},$$

it is easy to see that $\Phi(z_1, z_2)$ is an analytic function in G. Thus Problem A^* can be transformed into the boundary value problem A for analytic functions with the boundary condition

$$\mathrm{Re}[\overline{t_1}^{K_1}\overline{t_2}^{K_2} \Phi(t_1, t_2)] = R(t_1, t_2), \ t = (t_1, t_2) \in \partial G,$$

$$R(t_1, t_2) = r(t_1, t_2) - \mathrm{Re}[\overline{t_1}^{K_1}\overline{t_2}^{K_2}(T_1 A_1 + T_2 \tilde{A}_2)]. \tag{4.16}$$

If $R(t_1, t_2)$ satisfies some conditions similar to those for $r(t_1, t_2)$ as stated in $(4.6), (4.11)$ and (4.13), then Problem A with the boundary condition (4.16) is solvable. Consequently, the result of solvability of Problem A^* with the boundary condition (4.15) is obtained (see [164]2)).

5 Dirichlet Boundary Value Problem for Overdetermined Systems of Second Order Complex Equations

In this section, we first introduce the linear Dirichlet problem for two classes of overdetermined systems of second order complex equations in a polycylinder, and then discuss the nonlinear Dirichlet problem for quasilinear overdetermined systems of second order complex equations (see [92]2)).

5.1 Linear Dirichlet problem for a class of overdetermined complex systems

Denote $D = D_1 \times D_2$, where $D_j = \{|z_j| < 1\}, j = 1, 2$, and $\partial D = \partial D_1 \times \partial D_2$, $\partial D_j = \{|z_j| = 1\}, j = 1, 2$. We consider the overdetermined complex system of second order equations

$$w_{z_i \bar{z_j}} = F_{ij}(z_1, z_2), \ i, j = 1, 2, \ z = (z_1, z_2) \in D, \tag{5.1}$$

where $F_{ij}(z_1, z_2)(j = 1, 2)$ are continuously differentiable up to the first order in the variables $(z_1, z_2) \in \bar{D}$ and satisfy the conditions

$$F_{ij\bar{z_k}} = F_{ik\bar{z_j}}, \ F_{ijz_k} = F_{kjz_i}, \ i, j, k = 1, 2. \tag{5.2}$$

Problem D The linear Dirichlet problem for the complex system of second order equations (5.1) is to find a solution $w(z_1, z_2) \in C_\alpha^1(\bar{D})$ satisfying the boundary condition

$$w(z_1, z_2) = r(z_1, z_2), \ z = (z_1, z_2) \in \partial D, \tag{5.3}$$

where $r(t_1, t_2) \in C_\alpha^1(\partial D)$, $\alpha(0 < \alpha < 1)$ is a constant.

In order to give the solvability conditions of Problem D for (5.1), we need the following two lemmas.

Lemma 5.1 *The function*

$$
\begin{aligned}
w^*(z_1, z_2) \ = \ & \overline{T_1}[T_1 F_{11} + T_2 F_{12} - T_1 T_2 F_{11\bar{z_2}} + \textstyle\int_0^{z_2} G(z_1, \zeta_2)d\zeta_2] \\
& + \ \overline{T_2}[T_1 F_{21} + T_2 F_{22} - T_1 T_2 F_{22\bar{z_1}}] \\
& - \ \overline{T_1 T_2}[\Pi_1 F_{21} + T_2 F_{22z_1} - \Pi_1 T_2 F_{22\bar{z_1}}]
\end{aligned}
\tag{5.4}
$$

is a special solution of (5.1), *where*

$$T_1 F_{11} = -\frac{1}{\pi} \iint_{D_1} \frac{F_{11}(\zeta_1, z_2)}{\zeta_1 - z_1} d\sigma_{\zeta_1}, \ \Pi_1 F_{11} = (T_1 F_{11})_{z_1},$$

$$T_2 F_{12} = -\frac{1}{\pi} \iint_{D_2} \frac{F_{12}(z_1, \zeta_2)}{\zeta_2 - z_2} d\sigma_{\zeta_2}, \ \Pi_2 F_{12} = (T_2 F_{12})_{z_2},$$

and

$$G(z_1, z_2) = \Pi_1 F_{21} + T_2 F_{22z_1} - \Pi_1 T_2 F_{22\overline{z_1}} - T_1 F_{11z_2} - \Pi_2 F_{12} + T_1 \Pi_2 F_{11\overline{z_2}}$$

is analytic in D.

Proof   Since the function $F_{ij}(i, j = 1, 2)$ satisfy condition (5.2) and

$$-G(z_1, z_2) + \Pi_1 F_{21} + T_2 F_{22z_1} - \Pi_1 T_2 F_{22\overline{z_1}} = T_1 F_{11z_2} + \Pi_2 F_{12} - T_1 \Pi_2 F_{12\overline{z_2}},$$

we have

$$w^*_{z_1} = T_1 F_{11} + T_2 F_{12} - T_1 T_2 F_{11\overline{z_2}} + \int_0^{z_2} G(z_1, \zeta_2) d\zeta_2$$

and

$$w^*_{z_2} = T_1 F_{21} + T_2 F_{22} - T_1 T_2 F_{22\overline{z_1}}.$$

So it can be calculated directly that $w^*_{z_i \overline{z_j}} = F_{ij}$, $i, j = 1, 2$.

Lemma 5.2   *If $F_{ij}(z_1, z_2) = 0$, $i, j = 1, 2$ in $\bar{D}$, then the solutions of the complex system (5.1) have the representation*

$$w(z_1, z_2) = \Phi(z_1, z_2) + \overline{\Psi(z_1, z_2)}, \tag{5.5}$$

where $\Phi(z_1, z_2)$ and $\Psi(z_1, z_2)$ are analytic functions in D.

Proof   From the system (5.1) and some results on functions of several complex variables, it can be obtained that

$$w_{z_1} = \Phi_1(z_1, z_2), \; w_{z_2} = \Phi_2(z_1, z_2), \tag{5.6}$$

in which $\Phi_1(z_1, z_2)$ and $\Phi_2(z_1, z_2)$ are analytic functions in D. From the system (5.6), we see that $\Phi_j(z_1, z_2)(j = 1, 2)$ satisfy the condition

$$\Phi_{1z_2} = \Phi_{2z_1}. \tag{5.7}$$

Setting $\phi(z_1, z_2) = \Phi_{1z_2}$, we derive that

$$\Phi_1(z_1, z_2) = \int_0^{z_2} \phi(z_1, \zeta_2) d\zeta_2 + \psi_1(z_1),$$

$$\Phi_2(z_1, z_2) = \int_0^{z_1} \phi(\zeta_1, z_2) d\zeta_1 + \psi_2(z_2),$$

where $\phi_j(z_1, z_2)\,(j = 1, 2)$ are analytic functions in $z_j \in D_j, j = 1, 2$. When the condition (5.7) is satisfied, it can be concluded that

$$w(z_1, z_2) = \int_0^{z_1} \int_0^{z_2} \phi(\zeta_1, \zeta_2) d\zeta_1 d\zeta_2 + \int_0^{z_1} \psi_1(\zeta_1) d\zeta_1$$

$$+ \int_0^{z_2} \psi_2(\zeta_2) d\zeta_2 + \overline{\Psi(z_1, z_2)}.$$

This completes the proof.

Next, we introduce a new function

$$\tilde{w}(z_1, z_2) = w(z_1, z_2) - w^*(z_1, z_2), \tag{5.8}$$

where $w^*(z_1, z_2)$ is a special solution of (5.1) as stated in (5.4), and then Problem D for (5.1) is reduced to the following boundary value problem(Problem $\tilde{D}$) :

$$\tilde{w}_{z_i \overline{z_j}} = 0, \ i, j = 1, 2 \text{ in } D \tag{5.9}$$

$$\tilde{w}(z_1, z_2) = r(z_1, z_2) - w^*(z_1, z_2) = \tilde{r}(z_1, z_2) \text{ on } \partial D, \tag{5.10}$$

where $\operatorname{Re} \tilde{r} = \tilde{r}_1, \operatorname{Im} \tilde{r} = \tilde{r}_2$.

Theorem 5.3 *A necessary and sufficient condition of Problem $\tilde{D}$ to be solvable is that*

$$S_{mn} \tilde{r}_1 = \iint_{\partial D} \tilde{r}_1(\zeta_1, \zeta_2) \zeta_1^{m+1} / \zeta_2^{n+1} d\theta_1 d\theta_2, \ S_{mn} \tilde{r}_2 = 0, \ m, n = 0, 1, 2, \cdots, \tag{5.11}$$

where $\zeta_1 = e^{i\theta_1}$, $\zeta_2 = e^{i\theta_2}$. When these conditions are satisfied, the solution of Problem $\tilde{D}$ can be expressed as

$$\begin{cases} \tilde{w}(z_1, z_2) = P\tilde{r} + \overline{P\tilde{r}}, \ z = (z_1, z_2) \in \bar{D}, \\ P\tilde{r} = \dfrac{1}{(2\pi i)^2} \iint_{\partial D} \tilde{r}(\zeta_1, \zeta_2) \left[\dfrac{2}{(\zeta_1 - z_1)(\zeta_2 - z_2)} - \dfrac{1}{\zeta_1 \zeta_2} \right] d\zeta_1 d\zeta_2. \end{cases} \tag{5.12}$$

Proof By Lemma 5.2, we see that it is sufficient to find two analytic functions $\Phi(z_1, z_2)$, $\Psi(z_1, z_2)$ satisfying the boundary condition

$$\Phi(z_1, z_2) + \overline{\Psi(z_1, z_2)} = \tilde{r}(z_1, z_2) \text{ on } \partial D.$$

The above boundary condition is equivalent to

$$\begin{cases} \operatorname{Re}[\Phi(z_1, z_2) + \Psi(z_1, z_2)] = \tilde{r}_1(z_1, z_2) \text{ on } \partial D, \\ \operatorname{Re}[i(\Phi(z_1, z_2) - \Psi(z_1, z_2))] = -\tilde{r}_2(z_1, z_2) \text{ on } \partial D. \end{cases} \tag{5.13}$$

On the basis of Lemma 2.1, if (5.11) holds, then the harmonic functions $\operatorname{Re}(\Phi + \Psi)$ and $\operatorname{Re} i(\Phi - \Psi)$ can be expressed as

$$\operatorname{Re}(\Phi + \Psi) = \operatorname{Re} P\tilde{r}_1, \ \operatorname{Re} i(\Phi - \Psi) = -\operatorname{Re} P\tilde{r}_2 \text{ in } D$$

and

$$\Phi(z_1, z_2) + \overline{\Psi(z_1, z_2)} = \operatorname{Re} P\tilde{r}_1 + i\operatorname{Re} P\tilde{r}_2 \text{ in } D.$$

Thus the function

$$\tilde{w}(z_1, z_2) = \Phi(z_1, z_2) + \overline{\Psi(z_1, z_2)} = \frac{1}{2}(P\tilde{r} + \overline{P}\tilde{r}) \tag{5.14}$$

is a solution of Problem $\tilde{D}$. Moreover, we also see that (5.11) is a necessary and sufficient condition of the solvability of Problem $\tilde{D}$.

From Theorem 5.3, we immediately obtain

Corollary 5.4 *If condition* (5.11) *holds, then there exists the solution* $w(z_1, z_2)$ *of Problem D for* (5.1), *which can be expressed as*

$$w(z_1, z_2) = w^*(z_1, z_2) + \tilde{w}(z_1, z_2), \tag{5.15}$$

where $w^*(z_1, z_2)$ *and* $\tilde{w}(z_1, z_2)$ *are functions as stated in* (5.4) *and* (5.12) *respectively.*

5.2 Linear Dirichlet Problem of another class of overdetermined complex systems

Now, we discuss another class of complex system of second order equations

$$w_{\overline{z_i z_j}} = F_{ij}(z_1, z_2), \ i, j = 1, 2, \ z = (z_1, z_2) \in D, \tag{5.16}$$

where $F_{ij}(z_1, z_2) \in C^1(\bar{D})(i, j = 1, 2)$ and satisfy the conditions

$$F_{ij} = F_{ji}, \ F_{ij\overline{z_k}} = F_{ik\overline{z_j}} = F_{kj\overline{z_i}}, \ i, j, k = 1, 2. \tag{5.17}$$

Problem D' The linear Dirichlet problem for the complex system (5.16) is to find a solution $w(z_1, z_2) \in C_\alpha^1(\bar{D})$ satisfying the boundary condition

$$w(z_1, z_2) = r(z_1, z_2), \ (z_1, z_2) \in \partial D, \tag{5.18}$$

where $r(z_1, z_2) \in C_\alpha^1(\partial D)$, $\alpha(0 < \alpha < 1)$ is a constant.

Lemma 5.5 *The function*

$$\begin{aligned} w^*(z_1, z_2) &= T_1 T_1 F_{11} + T_1 T_2 F_{12} + T_2 T_2 F_{22} \\ &\quad - T_1 T_2 (T_1 F_{11\overline{z_2}} + T_2 F_{22\overline{z_1}}) \end{aligned} \tag{5.19}$$

is a special solution of the complex system (5.16).

Proof It is easy to see that

$$w^*_{\overline{z_1}} = T_1 F_{11} + T_2 F_{12} - T_2 T_1 F_{11\overline{z_2}},$$

$$w^*_{\overline{z_2}} = T_1 F_{12} + T_2 F_{22} - T_1 T_2 F_{22\overline{z_1}},$$

and then we have that $(w^*)_{\overline{z_i z_j}} = (F_{ij}(z_1, z_2))$, $j = 1, 2$.

Lemma 5.6 *If $F_{ij}(z_1, z_2) = 0$, $i, j = 1, 2$ on $\bar{D}$, then the solution of (5.16) can be written as*

$$w(z_1, z_2) = \Phi(z_1, z_2) + \overline{z_1}\Phi_1(z_1, z_2) + \overline{z_2}\Phi_2(z_1, z_2), \tag{5.20}$$

in which $\Phi(z_1, z_2)$, $\Phi_1(z_1, z_2)$, $\Phi_2(z_1, z_2)$ are analytic functions in D.

Proof By (5.16) and some results on functions of several complex variables, it can be obtained that

$$w_{\overline{z_1}} = \Phi_1(z_1, z_2), \ w_{\overline{z_2}} = \Phi_2(z_1, z_2).$$

From the above discussion, it is not difficult to derive the representation (5.20), where $\Phi(z_1, z_2)$, $\Phi_1(z_1, z_2)$, $\Phi_2(z_1, z_2)$ are three analytic functions in D.

Due to Lemma 5.6, Problem D' for (5.16) can be transformed into that to find three analytic functions $\Phi(z_1, z_2)$, $\Phi_1(z_1, z_2)$, $\Phi_2(z_1, z_2)$ in D, which are continuous on $\bar{D}$ and satisfy the boundary condition

$$\Phi(z_1, z_2) + \overline{z_1}\Phi_1(z_1, z_2) + \overline{z_2}\Phi_2(z_1, z_2) = r(z_1, z_2) \text{ on } \partial D. \tag{5.21}$$

By using the above method, we can also discuss the solvability of Problem D' for (5.16). Here we mention that Problem D' for (5.16) is of Hausdorff type. There are infinite solutions of the homogeneous Dirichlet problem (Problem D_0') with $F_{ij}(z_1, z_2) = 0$, $i, j = 1, 2$, $(z_1, z_2) \in D$ and $r(z_1, z_2) = 0$, $(z_1, z_2) \in \partial D$, for instance, we choose that

$$\Phi(z_1, z_2) = z_1^n + z_2^n, \ \Phi_1(z_1, z_2) = -z_1^{n+1}, \ \Phi_2(z_1, z_2) = -z_2^{n+1}, \ n = 1, 2, \cdots,$$

then

$$w_n(z_1, z_2) = \Phi(z_1, z_2) + \overline{z_1}\Phi_1(z_1, z_2) + \overline{z_2}\Phi_2(z_1, z_2), \ n = 1, 2, \cdots \tag{5.22}$$

are solutions of Problem D_0'.

5.3 Nonlinear Dirichlet problem for quasilinear systems of second order complex equations

In the following, we consider the quasilinear overdetermined complex system

$$w_{z_i \overline{z_j}} = F_{ij}(z_1, z_2, w), \ i, j = 1, 2, \ (z_1, z_2) \in D. \tag{5.23}$$

Suppose that (5.1) satisfies the following conditions.

Condition C

1) $F_{ij}(z_1, z_2, w) \, (i, j = 1, 2)$ are continuously differentiable up to second order in the variables $(z_1, z_2) \in D$, and are analytic with respect to $w \in G_R = \{|w| < R\}$, where R is any positive number.

2) For an arbitrary complex function $w(z_1, z_2) \in C_\alpha^1(\bar{D})$, $F_{ij}(z_1, z_2, w)\,(i, j = 1, 2)$ satisfy the compatibility conditions

$$F_{ij\overline{z_k}} = F_{ik\overline{z_j}}, \ F_{ijz_k} = F_{kjz_i}, \ i, j, k = 1, 2, \tag{5.24}$$

where $\alpha\,(0 < \alpha < 1)$ is a constant.

3) $F_{ij}(z_1, z_2, w)\,(i, j = 1, 2)$ and their first order derivatives are uniformly bounded in $\bar{D} \times \bar{G_R}$ by a nonnegative constant K_R, and the uniform Lipschitz conditions

$$|F_{ijz_k}(z_1, z_2, w) - F_{ijz_k}(z_1, z_2, \hat{w})|, \ |F_{ij\overline{z_k}}(z_1, z_2, w) - F_{ij\overline{z_k}}(z_1, z_2, \hat{w})| \le L_R|w - \hat{w}| \tag{5.25}$$

hold for arbitrary numbers $w, \hat{w} \in G_R$, where L_R is a nonnegative constant.

Problem D. The nonlinear Dirichlet problem for the system (5.23) is to find a solution $w(z_1, z_2) \in C_\alpha^1(\bar{D})$ satisfying the boundary condition

$$w(z_1, z_2) = r(z_1, z_2, w), \ (z_1, z_2) \in \partial D, \tag{5.26}$$

in which $r(z_1, z_2, w)$ satisfies the following condition. For any functions $w(z_1, z_2)$, $\hat{w}(z_1, z_2) \in C_\alpha^1(\bar{D})$, $r(z_1, z_2, w(z_1, z_2)) \in C_\alpha^1(\bar{D})$, and

$$\| r(z_1, z_2, w(z_1, z_2)) - r(z_1, z_2, \hat{w}(z_1, z_2)) \|_{C_\alpha^1(\bar{D})} \le M_1 \| w(z_1, z_2) - \hat{w}(z_1, z_2) \|_{C_\alpha^1(\bar{D})} \tag{5.27}$$

holds, where $\alpha\,(1/2 < \alpha < 1)$, M_1 are nonnegative constants.

In order to find the solutions of Problem D for (5.23), we introduce a modified problem (Problem D^*) with the boundary condition

$$w(z_1, z_2) = r(z_1, z_2, w) + H(z_1, z_2) \text{ on } \partial D, \tag{5.28}$$

where

$$H(z_1, z_2) = \mathrm{Re} \sum_{j=1}^{\infty} \sum_{k=1}^{\infty} H_{jk}^1 \overline{z_1^j} z_2^k + i\,\mathrm{Re} \sum_{j=1}^{\infty} \sum_{k=1}^{\infty} H_{jk}^2 \overline{z_1^j} z_2^k \text{ on } \partial D. \tag{5.29}$$

Here H_{jk}^1, $H_{jk}^2 (j, k = 1, 2, \cdots)$ are unknown complex constants to be determined appropriately.

We are free to choose a function $w_0(z_1, z_2) \in C_\alpha^1(\bar{D})$ and substitute it into the system (5.23) and the boundary condition (5.26). According to the method stated in Section 1, i.e. from Lemma 5.1, we can find a function $w_0^*(z_1, z_2)$ satisfying the following complex system

$$w_{0z_i\overline{z_j}}^* = F_{ij}(z_1, z_2, w_0), \ i, j = 1, 2 \text{ in } D. \tag{5.30}$$

Then as follows from Theorem 5.3, there exists a solution $\tilde{w}_0(z_1, z_2)$ of the boundary value problem

$$\tilde{w}_{0z_i\overline{z_j}} = 0, \ i, j = 1, 2 \text{ in } D, \tag{5.31}$$

$$\begin{aligned}
\tilde{w}_0(z_1, z_2) &= r(z_1, z_2, w_0) - w_0^*(z_1, z_2) + h(z_1, z_2) \\
&= \tilde{r}(z_1, z_2, w_0) + H(z_1, z_2) \text{ on } \partial D,
\end{aligned} \tag{5.32}$$

in which, according to (2.18), we choose the complex constants $H_{jk}^l (l = 1, 2, j, k = 1, 2, \cdots)$, such that

$$H_{m+1,n+1}^l = \frac{-2}{(2\pi)^2} \int_0^{2\pi} \int_0^{2\pi} \tilde{r}(\zeta_1, \zeta_2) \frac{\zeta_1^{m+1}}{\zeta_2^{n+1}} d\zeta_1 d\zeta_2 = 0, \, l = 1, 2, \, m, n = 1, 2, \cdots.$$

Denote that

$$w_1(z_1, z_2) = w_0^*(z_1, z_2) + \tilde{w}(z_1, z_2), \tag{5.33}$$

and substitute it into the positions of w in the right–hand sides of (5.23) and (5.28), by using the above method, there is a solution of the boundary value problem

$$w_{2z_i \overline{z_j}} = F_{ij}(z_1, z_2, w_1), \, i, j = 1, 2 \text{ in } D, \tag{5.34}$$

$$w_2(z_1, z_2) = r(z_1, z_2, w_1) + H(z_1, z_2) \text{ on } \partial D. \tag{5.35}$$

Applying successive iteration, we obtain a sequence of solutions $\{w_n(z_1, z_2)\}$, which satisfy the complex systems and the boundary conditions

$$w_{n+1 z_i \overline{z_j}} = F_{ij}(z_1, z_2, w_n), \, i, j = 1, 2, \text{ in } D, \tag{5.36}$$

$$w_{n+1}(z_1, z_2) = r(z_1, z_2, w_n) \text{ on } \partial D. \tag{5.37}$$

Noting that the functions $w_{n+1}(z_1, z_2) \, (n = 0, 1, 2, \cdots)$ possess the representations

$$w_{n+1}(z_1, z_2) = w_n^*(z_1, z_2) + \tilde{w}_n(z_1, z_2), \tag{5.38}$$

where $w_n^*(z_1, z_2)$ admits the form (5.4) with $F_{ij} = F_{ij}(z_1, z_2, w_n) \, (i, j = 1, 2)$, and $\tilde{w}_n(z_1, z_2)$ can be written as

$$\tilde{w}_n(z_1, z_2) = \Phi_n(z_1, z_2) + \overline{\Psi_n(z_1, z_2)} = \frac{1}{2}(P\tilde{r}_n + \overline{P}\tilde{r}_n),$$

with $\tilde{r}_n = r(z_1, z_2, w_n) - w_n^*(z_1, z_2)$ on ∂D. From (5.38), we have

$$w_{n+1} - w_n = w_n^* - w_{n-1}^* + \tilde{w}_n - \tilde{w}_{n-1}, \tag{5.39}$$

in which $w_n^* - w_{n-1}^*$ possesses the form (5.4), where $F_{ij} = F_{ij}(z_1, z_2, w_n) - F_{ij}(z_1, z_2, w_{n-1}), \, i, j = 1, 2$, and

$$\tilde{w}_n - \tilde{w}_{n-1} = \frac{1}{2}[P(\tilde{r}_n - \tilde{r}_{n-1}) + \overline{P}(\tilde{r}_n - r_{n-1})],$$

where $\tilde{r}_n - \tilde{r}_{n-1} = r(z_1, z_2, w_n) - r(z_1, z_2, w_{n-1}) - w_n^* + w_{n-1}^*$. Taking Condition C and (5.37) into account, from (5.39) it can be obtained that

$$\| w_n^* - w_{n-1}^* \|_{C_\alpha^1(\bar{D})} \leq L_R M_2 \| w_n - w_{n-1} \|_{C_\alpha^1(\bar{D})},$$

$$\| \tilde{w}_n - \tilde{w}_{n-1} \|_{C_\alpha^1(\bar{D})} \leq M_1 M_2 \| w_n - w_{n-1} \|_{C_\alpha^1(\bar{D})},$$

where M_2 is a nonnegative constant depending on K_R, α, D. Thus we have

$$\| w_{n+1} - w_n \|_{C^1_\alpha(\bar{D})} \leq (L_R + M_1) M_3 \| w_n - w_{n-1} \|_{C^1_\alpha(\bar{D})}, \qquad (5.40)$$

in which M_3 is a nonnegative constant. If we choose the constants L_R, M_1 so small that $(L_R + M_1)(1 + M_3) \leq 1/2$, then from (5.40) for $n \geq m \geq N + 1 > 1$,

$$\| w_n - w_m \|_{C^1_\alpha(\bar{D})} \leq 2^{-N+1} \| w_1 - w_0 \|_{C^1_\alpha(\bar{D})}$$

can be derived. Due to the completeness of the Banach space $C^1_\alpha(\bar{D})$, there exists a function $w_*(z_1, z_2) \in C^1_\alpha(\bar{D})$, so that

$$\| w_n - w_* \|_{C^1_\alpha(\bar{D})} \to 0 \text{ as } n \to \infty.$$

It is not difficult to derive that $w_*(z_1, z_2)$ is a solution of Problem D for (5.23).

Theorem 5.7 *Suppose that Condition C and (5.27) hold, and the constants L_R, M_1 are sufficiently small, then Problem D for (5.23) has a solution $w(z_1, z_2) \in C^1_\alpha(\bar{D})$.*

V Boundary Value Problems of Functions in Clifford Analysis

In [27],[38],[52] and [30], some basic results in the theory of regular functions and generalized regular functions in Clifford analysis and their applications are introduced. But research on boundary value problems of functions with values in a Clifford algebra is done only in recent years. In this chapter, we mainly discuss some boundary value problems of regular functions and generalized regular functions in Clifford analysis, for instance, the Riemann problem, the Riemann–Hilbert problem and the oblique derivative problem.

1 Integral of Cauchy Type and Plemelj Formula in Clifford Analysis

The Clifford algebra A_n over the space $I\!\!R^n$ is defined as follows: Let $e_1 = 1, e_2, \cdots, e_n$ be the standard orthonormal basis in $I\!\!R^n$, and denote by e_s the general basis element of A_n, where s is any subset of $\{1, 2, 3, \cdots, n\}$, i.e.

$$\begin{cases} e_1 = 1, \ e_j^2 = -1, \ 2 \leq j \leq n, \\ e_j e_k = -e_k e_j, \ 2 \leq j < k \leq n, \\ e_s = e_{j_1} e_{j_2} \cdots e_{j_s}, \ 2 \leq j_1 < j_2 < \cdots < j_s \leq n, \\ s = \{j_1, j_2, \cdots, j_s\}. \end{cases} \tag{1.1}$$

It is not difficult to see that A_n is a 2^{n-1}–dimensional algebra space. An arbitrary element of the Clifford algebra $A_n(I\!\!R)$ can be written as

$$x = x_1 e_1 + \cdots + x_n e_n + \cdots + x_{x_{j_1 \cdots j_n}} e_{j_1} \cdots e_{j_n} + \cdots + x_{2 \ldots n} e_2 \cdots e_n = \sum_s x_s e_s, \tag{1.2}$$

in which $x_1, \cdots, x_n, \cdots, x_{j_1 \cdots j_s}, \cdots x_{2 \ldots n} \in I\!\!R$.

Let $x = \sum x_s e_s, \ y = \sum y_s e_s \in A_n(I\!\!R)$, where $x_s, y_s \in I\!\!R$. Then $xy = \sum_{s,k} x_s y_k e_s e_k$, in which $e_s e_k$ is a basis element of $A_n(I\!\!R)$ and $x_s y_k \in I\!\!R$. For $x = x_1 e_1 + x_2 e_2 + \cdots + x_n e_n \in I\!\!R^n$, its conjugate element is defined by $\bar{x} = x_1 e_1 - x_2 e_2 - \cdots - x_n e_n$. It is evident that

$$x\bar{x} = \bar{x}x = |x|^2. \tag{1.3}$$

If $x \neq 0$, then x is invertible, and $x^{-1} = \bar{x}/|x|^2$.

Denote by D is a connected open set in $I\!\!R$, and by

$$F_D^{(r)} = \{f \mid f : D \to A_n(I\!\!R), \ f(x) = \sum_s f_s(x)e_s, \ f_s \in C^r(D)\} \tag{1.4}$$

the set of continuously differentiable functions up to degree r, the values of which belong to $A_n(I\!\!R)$. Define the differential operators

$$\bar{\partial} = e_1\frac{\partial}{\partial x_1} + e_2\frac{\partial}{\partial x_2} + \cdots + e_n\frac{\partial}{\partial x_n}, \ \partial = e_1\frac{\partial}{\partial x_1} - \cdots - e_n\frac{\partial}{\partial x_n}, \tag{1.5}$$

we have

$$\bar{\partial}\partial = \partial\bar{\partial} = \sum_{j=1}^{n} \frac{\partial^2}{\partial x_j^2} = \Delta. \tag{1.6}$$

If $f(x) \in F_D^{(r)} (r \geq 1)$ and $\bar{\partial}f = 0$, then $f(x)$ is called a (left)–regular function, and then $f_s(x)$ is harmonic in D.

The so–called integral of Cauchy type is the following integral

$$F(x) = \frac{1}{\omega_{n-1}} \int_\Sigma \frac{\bar{y} - \bar{x}}{|y - x|^n} n(y)f(y)dS_y, \tag{1.7}$$

where ω_{n-1} is the area of unit ball in $I\!\!R^n$, Σ is an $(n-1)$–dimensional smooth and orientated compact surface in $I\!\!R^n$, $f(y)$ is called the density of the integral (1.7), $n(y)$ is a unit normal vector at $y \in \Sigma$. If Σ is a closed surface, then $n(y)$ is always considered as the outward normal vector and dS_y is the surface element of Σ. Suppose that $f(y)$ is a Hölder continuous function, i.e. $f(y) \in C_\alpha(\Sigma)$, in which $\alpha \, (0 < \alpha < 1)$ is a real constant. It is not difficult to see that the function $F(x)$ in (1.7) is regular in $I\!\!R^n \backslash \Sigma$. When Σ is a closed surface, the space $I\!\!R^n$ is separated by Σ into the inner domain D^+ and the outer domain D^-. In general, the integral in (1.7) determines two different regular functions in D^+ and D^-, for instance, the integral

$$\frac{1}{\omega_{n-1}} \int_{|y|=1} \frac{\bar{y} - \bar{x}}{|y - x|^n} n(y)dS_y \tag{1.8}$$

equals 1 in $D^+ = \{|x| < 1\}$ and 0 in $D^- = \{|x| > 1\}$.

Next, we discuss the limit value of the function $F(x)$ in (1.7) as $x \to x^0 \in \Sigma$. For this, it is necessary to consider the sense of the singular integral

$$\frac{1}{\omega_{n-1}} \int_\Sigma \frac{\bar{y} - \bar{x}^0}{|y - x^0|^n} n(y)dS_y, \ x^0 \in \Sigma.$$

Generally speaking, the integral is divergent. However, we add some restriction, namely choose a sphere $S(x^0, r)$ of radius r and centre at x^0 and denote by Σ_0 the part of Σ in $S(x^0, r)$. If there exists the limit value

$$\lim_{r \to 0} \int_{\Sigma \backslash \Sigma_0} \frac{\bar{y} - \bar{x}^0}{|y - x^0|^n} n(y)f(y)dS_y,$$

then we define

$$\int_\Sigma \frac{\bar{y} - \bar{x}^0}{|y - x^0|^n} n(y) f(y) dS_y = \lim_{r \to 0} \int_{\Sigma \backslash \Sigma_0} \frac{\bar{y} - \bar{x}}{|y - x^0|^n} n(y) f(y) dS_y \qquad (1.9)$$

as the principle value of the integral

$$\int_\Sigma \frac{\bar{y} - \bar{x}^0}{|y - x^0|^n} n(y) f(y) dS_y.$$

Theorem 1.1 *Let Σ be an n–dimensional smooth and orientated compact surface in $I\!\!R^n$ and $f(y) \in C_\alpha(\Sigma), 0 < \alpha < 1$. Then the following formula holds at $x^0 \in \Sigma$,*

$$\frac{1}{\omega_{n-1}} \int_\Sigma \frac{\bar{y} - \bar{x}^0}{|y - x^0|^n} n(y) f(y) dS_y$$
$$= \frac{1}{\omega_{n-1}} \int_\Sigma \frac{\bar{y} - \bar{x}^0}{|y - x^0|^n} n(y) [f(y) - f(x^0)] dS_y + \frac{1}{2} f(x^0). \qquad (1.10)$$

Proof As before we denote by Σ_0 the part of Σ in $S(x^0, r)$. Obviously

$$\int_{\Sigma \backslash \Sigma_0} \frac{\bar{y} - \bar{x}^0}{|y - x^0|^n} n(y) f(y) dS_y - \int_{\Sigma \backslash \Sigma_0} \frac{\bar{y} - \bar{x}^0}{|y - x^0|^n} n(y) f(x^0) dS_y$$
$$= \int_{\Sigma \backslash \Sigma_0} \frac{\bar{y} - \bar{x}^0}{|y - x^0|^n} n(y) [f(y) - f(x^0)] dS_y. \qquad (1.11)$$

From $f(y) \in C_\alpha(\Sigma)$, there exists a positive constant M such that

$$|f(y) - f(x^0)| \leq M |y - x^0|^\alpha, \ 0 < \alpha < 1.$$

Consequently

$$|\frac{\bar{y} - \bar{x}^0}{|y - x^0|^n} n(y) [f(y) - f(x^0)]| \leq \frac{M}{|y - x^0|^{n-1-\alpha}}.$$

Because the surface Σ is smooth, the following integral is a proper integral:

$$\int_\Sigma \frac{\bar{y} - \bar{x}^0}{|y - x^0|^n} n(y) [f(y) - f(x^0)] dS_y$$
$$= \lim_{r \to 0} \int_{\Sigma \backslash \Sigma_0} \frac{\bar{y} - \bar{x}^0}{|y - x^0|^n} n(y) [f(y) - f(x^0)] dS_y. \qquad (1.12)$$

Denote by $S_1(x^0, r)$ the part of $S(x^0, r)$ outside of Σ. It can be seen that $(\Sigma \backslash \Sigma_0) \cup S_1$ is a closed surface and sectionally smooth. By the Cauchy formula, we have

$$\int_{(\Sigma \backslash \Sigma_0) \cup S_1} \frac{\bar{y} - \bar{x}^0}{|y - x^0|^n} n(y) f(x^0) dS_y = \omega_{n-1} f(x^0),$$

and

$$\int_{S_1} \frac{\bar{y} - \bar{x}^0}{|y - x^0|^n} n(y) f(x^0) dS_y = \int_{S_1} \frac{\bar{y} - \bar{x}^0}{|y - x^0|^n} \frac{y - x^0}{|y - x^0|} f(x^0) dS_y$$

$$= \int_{S_1} \frac{1}{|y - x^0|^{n-1}} f(x^0) dS_y = \frac{s_0}{r^{n-1}} f(x^0),$$

where s_0 is the area of S_1. Because Σ is a smooth surface, it is not difficult to obtain

$$\lim_{r \to 0} \int_{S_1} \frac{\bar{y} - \bar{x}^0}{|y - x^0|^n} n(y) f(x^0) dS_y = \frac{1}{2} \omega_{n-1} f(x^0). \tag{1.13}$$

Hence

$$\lim_{r \to 0} \int_{\Sigma \backslash \Sigma_0} \frac{\bar{y} - \bar{x}^0}{|y - x^0|^n} n(y) f(x^0) dS_y$$

$$= \lim_{r \to 0} \int_{(\Sigma \backslash \Sigma_0) \cup S_1} \frac{\bar{y} - \bar{x}^0}{|y - x^0|^n} n(y) f(x^0) dS_y$$

$$- \lim_{r \to 0} \int_{S_1} \frac{\bar{y} - \bar{x}^0}{|y - x^0|^n} n(y) f(x^0) dS_y = \frac{1}{2} \omega_{n-1} f(x^0).$$

From (1.11) as $r \to 0$ and (1.12),(1.13), it follows that (1.10) holds.

Now we rewrite (1.7) in the following form

$$F(x) = \frac{1}{\omega_{n-1}} \int_{\Sigma} \frac{\bar{y} - \bar{x}}{|y - x|^n} n(y) f(y) dS_y$$

$$= \frac{1}{\omega_{n-1}} \int_{\Sigma} \frac{\bar{y} - \bar{x}}{|y - x|^n} n(y)[f(y) - f(x^0)] dS_y \tag{1.14}$$

$$+ \frac{1}{\omega_{n-1}} \int_{\Sigma} \frac{\bar{y} - \bar{x}}{|y - x|^n} n(y) f(x^0) dS_y.$$

On the basis of the Cauchy formula:

$$\frac{1}{\omega_{n-1}} \int_{\Sigma} \frac{\bar{y} - \bar{x}}{|y - x|^n} n(y) f(x^0) dS_y = \begin{cases} f(x^0), & x \in D^+, \\ 0, & x \in D^-. \end{cases} \tag{1.15}$$

If we prove that the integral

$$E(x) = \frac{1}{\omega_{n-1}} \int_{\Sigma} \frac{\bar{y} - \bar{x}}{|y - x|^n} n(y)[f(y) - f(x^0)] dS_y \tag{1.16}$$

has the definite limit as $x(\notin \Sigma) \to x^0(\in \Sigma)$, then there exists a limit of the integral $F(x)$ of Cauchy type as $x(\in D^+) \to x^0(\in \Sigma)$ or $x(\in D_-) \to x_0(\in \Sigma)$.

Lemma 1.2 *Suppose that Σ is a smooth and closed surface and $f(y) \in C_\alpha(\Sigma)$, $0 < \alpha < 1$. Then $E(x)$ has a limit as $x(\notin \Sigma) \to x^0(\in E)$, i.e.*

$$\lim_{x \to x^0} E(x) = E(x^0). \tag{1.17}$$

Proof First of all, we consider that $x\,(\notin \Sigma) \to x_0\,(\in \Sigma)$ and x is not along with the tangent direction of Σ, in this case, we may assume that the angle between the direction of $x \to x_0$ and the tangent direction of Σ is greater than $2\beta_0 > 0$. We write that

$$E(x) - E(x^0) = \frac{1}{\omega_{n-1}} \int_\Sigma \frac{\bar{y} - \bar{x}}{|y - x|^n} n(y)[f(y) - f(x^0)]dS_y$$

$$-\frac{1}{\omega_{n-1}} \int_\Sigma \frac{\bar{y} - \bar{x}_0}{|y - x^0|^n} n(y)[f(y) - f(x^0)]dS_y$$

$$= \frac{1}{\omega_{n-1}} \int_\Sigma \left[\frac{\bar{y} - \bar{x}}{|y - x|^n} - \frac{\bar{y} - \bar{x}^0}{|y - x^0|^n} \right] n(y)[f(y) - f(x^0)]dS_y \qquad (1.18)$$

$$= \frac{1}{\omega_{n-1}} \int_{\Sigma_0} \left[\frac{\bar{y} - \bar{x}}{|y - x|^n} - \frac{\bar{y} - \bar{x}^0}{|y - x^0|^n} \right] n(y)[f(y) - f(x^0)]dS_y$$

$$+\frac{1}{\omega_{n-1}} \int_{\Sigma \setminus \Sigma_0} \left[\frac{\bar{y} - \bar{x}}{|y - x|^n} - \frac{\bar{y} - \bar{x}^0}{|y - x^0|^n} \right] n(y)[f(y) - f(x^0)]dS_y = I_1 + I_2.$$

Similarly to the plane case (see [105]1),[139]17)), there is

$$|x - x^0|/|y - x| \leq M_1 = \text{constant},$$

and

$$\left| \frac{x}{|x|^{m+2}} - \frac{t}{|t|^{m+2}} \right| \leq \frac{P_m(x,t)}{|x|^{m+1}|t|^{m+1}} |x - t|,$$

where $P_m(x,t) = \sum_{k=0}^m |x|^{m-k}|t|^k$, m is a nonnegative integer. Hence

$$\left| \frac{\bar{y} - \bar{x}}{|y - x|^n} - \frac{\bar{y} - \bar{x}^0}{|y - x^0|^n} \right| \leq \sum_{k=1}^{n-1} \frac{|x - x^0|}{|y - x|^k} \frac{1}{|y - x^0|^{n-k}}$$

$$= \sum_{k=1}^{n-1} \frac{|x - x^0|}{|y - x|} \frac{|y - x^0|^{k-1}}{|y - x|^{k-1}} \frac{1}{|y - x^0|^{n-1}} \leq \sum_{k=1}^{n-1} M_1^k \frac{1}{|y - x^0|^{n-1}}. \qquad (1.19)$$

The condition $f(y) \in C_\alpha(\Sigma)$ implies

$$|f(y) - f(x^0)| \leq M_2|y - x^0|^\alpha, \qquad (1.20)$$

where M_2 is a real constant. By (1.19) and (1.20), we have

$$|I_1| = \frac{1}{\omega_{n-1}} \int_{\Sigma_0} \left| \frac{\bar{y} - \bar{x}}{|y - x|^n} - \frac{\bar{y} - \bar{x}^0}{|y - x^0|^n} \right| |n(y)||f(y) - f(x^0)|dS_y$$

$$\leq \frac{1}{\omega_{n-1}} \sum_{k=1}^{n-1} M_1^k M_2 \int_{\Sigma_0} \frac{1}{|y - x^0|^{n-\alpha-1}} dS_y. \qquad (1.21)$$

Because Σ is a smooth surface and M_1, M_2 are indendent of x^0, we can find a positive constant ρ independent of x^0 so that $|I_1| \leq \varepsilon/2$, where $r < \rho$. For a fixed Σ_0, if x is sufficiently near to x^0, such that $|y - x^0| > r/2 = d$, $|y - x| > r/2$, then

$$|I_2| \leq |x - x^0| \sum_{k=1}^{n-1} \int_{\Sigma \backslash \Sigma_0} \frac{|f(y) - f(x^0)|}{|y - x|^k |y - x^0|^{n-k}} dS_y$$

$$\leq |x - x^0| \int_\Sigma \frac{n-1}{d^n} |f(y) - f(x^0)| dS_y.$$

It can be seen that there exists a positive constant δ independent of x^0 and when $|x - x^0| < \delta$, we have $|I_2| < \varepsilon/2$. Thus

$$|I| \leq |I_1| + |I_2| < \varepsilon, \text{ if } |x - x^0| < \delta.$$

Secondly, we consider that $x(\notin \Sigma) \to x^0 (\in \Sigma)$, where x is tending to x^0 along the tangent direction of Σ. Noting that the constant δ is independent of $x^0 \in \Sigma$ in the foregoing case, so there is uniformly $E(x) \to E(x^0)$ as $x \to x^0 \in \Sigma$. Let $x^0, x^1 \in \Sigma$, we have

$$|E(x^1) - E(x^0)| \leq |E(x^1) - E(x)| + |E(x) - E(x^0)|. \tag{1.22}$$

From (1.22) it is easy to see that $E(x)$ is uniformly continuous on Σ. We draw a straight line l through a point x perpendicular to Σ and denote by y the point of intersection of l and Σ, it is evident that l is not the tangent direction of Σ and if $|x - x^0|$ is small enough, then $|x - y|$ and $|y - x^0|$ are sufficiently small. According to the uniform continuity of $E(y)$ on Σ and the result in the first part, it can be seen that for any positive number ε, provided that x, y are near enough with x^0, so that $|E(y) - E(x^0)| < \varepsilon/2$, $|E(x) - E(y)| < \varepsilon/2$, thus

$$|E(x) - E(x^0)| \leq |E(x) - E(y)| + |E(y) - E(x^0)| < \varepsilon.$$

This completes the proof.

Theorem 1.3   *Let Σ be a smooth and closed surface in $\mathbb{R}^n$, D^+ and D^- be the inner domain and outer domain of Σ and $f(y) \in C_\alpha(\Sigma), 0 < \alpha < 1$. Then there exisits the limit of the integral (1.7) as $x\,(\in D^+$ or $D^-)$ tends to $x_0\,(\in \Sigma)$, and*

$$\begin{cases} F^+(x^0) = \dfrac{1}{\omega_{n-1}} \displaystyle\int_\Sigma \frac{\bar{y} - \bar{x}^0}{|y - x^0|^n} n(y) f(y) dS_y + \frac{1}{2} f(x^0), \\[4mm] F^-(x^0) = \dfrac{1}{\omega_{n-1}} \displaystyle\int_\Sigma \frac{\bar{y} - \bar{x}^0}{|y - x^0|^n} n(y) f(y) dS_y - \frac{1}{2} f(x^0), \end{cases} \tag{1.23}$$

where $F^+(x^0)$ and $F^-(x^0)$ are the limits of $F(x)$ as $x\,(\in D^+) \to x^0\,(\in \Sigma)$ and $x\,(\in D^-) \to x^0\,(\in \Sigma)$ respectively. Furthermore

$$\begin{cases} F^+(x^0) - F^-(x^0) = f(x^0), \\[4mm] F^+(x^0) + F^-(x^0) = \dfrac{2}{\omega^{n-1}} \displaystyle\int_\Sigma \frac{\bar{y} - \bar{x}^0}{|y - x^0|^n} n(y) f(y) dS_y. \end{cases} \tag{1.24}$$

The above formula is called the Plemelj formula. It gives the solution of a simple Riemann boundary value problem.

Proof We may write the integral (1.7) in the form (1.14). By (1.15) and Lemma 1.2, letting $x\,(\in G^+)$ tend to $x^0\,(\in \Sigma)$ and $x\,(\in D^-)$ tend to $x^0\,(\in \Sigma)$, we have

$$\begin{cases} F^+(x^0) = \dfrac{1}{\omega_{n-1}} \displaystyle\int_\Sigma \dfrac{\bar{y} - \bar{x}^0}{|y - x^0|^n} n(y)[f(y) - f(x^0)]dS_y + f(x^0), \\[3mm] F^-(x^0) = \dfrac{1}{\omega_{n-1}} \displaystyle\int_\Sigma \dfrac{\bar{y} - \bar{x}^0}{|y - x^0|^n} n(y)[f(y) - f(x^0)]dS_y. \end{cases} \tag{1.25}$$

By (1.10), it follows that (1.23) holds. As for (1.24), it can be easily derived from (1.23).

Corollary 1.4 *Under the same condition as in Theorem 1.3, the integral (1.7) of the Cauchy type is a Cauchy integral if and only if the following formula holds:*

$$F^-(\zeta) = 0 \ \text{ on } \Sigma. \tag{1.26}$$

Proof If (1.7) is the integral of Cauchy type, then $F(x) \to F^+(\zeta) = F(\zeta)$ as $x\,(\in D^+) \to \zeta\,(\in \Sigma)$. From (1.24), it follows (1.26). Inversely, by (1.24) and (1.26), we see that

$$F^+(\zeta) - F^-(\zeta) = F^+(\zeta) = f(\zeta) \ \text{ for } \zeta \in \Sigma.$$

Hence, the integral $F(x)$ in (1.7) is a Cauchy integral.

Finally, we mention that if $n = 2$, the regular function $f(x)$ in a domain D in $I\!\!R^2$ is an analytic function, and if $n = 3$, the regular function $f(x) = \sum_s f_s(x)e_s = f_1(x) + f_2(x)e_2 + f_3(x)e_3 + f_{23}(x)e_2e_3$ in a domain in $I\!\!R^3$ is a solution (f_1, f_2, f_3, f_{23}) of the system of partial differential equations of first order

$$\begin{pmatrix} \frac{\partial}{\partial x_1} & -\frac{\partial}{\partial x_2} & -\frac{\partial}{\partial x_3} & 0 \\[2mm] \frac{\partial}{\partial x_2} & \frac{\partial}{\partial x_1} & 0 & \frac{\partial}{\partial x_3} \\[2mm] \frac{\partial}{\partial x_3} & 0 & \frac{\partial}{\partial x_1} & -\frac{\partial}{\partial x_2} \\[2mm] 0 & -\frac{\partial}{\partial x_3} & \frac{\partial}{\partial x_2} & \frac{\partial}{\partial x_1} \end{pmatrix} \begin{pmatrix} f_1 \\ f_2 \\ f_3 \\ f_{23} \end{pmatrix} = \begin{pmatrix} 0 \\ 0 \\ 0 \\ 0 \end{pmatrix}. \tag{1.27}$$

It can be seen that it is an analytic function in the 3–dimensional domain. Hence, the general regular functions possess some properties similar to those of analytic functions in the planar domain(see [60]).

2 Riemann Problem for Regular Functions in Clifford Analysis

In this section, we discuss the solvability of the Riemann boundary value problem for regular functions with values in the Clifford algebra by the Plemelj formula as stated in Section 1.

Problem R   Let D^+ be a bounded domain in $I\!\!R$ with the boundary Σ, which is a Liapunov surface and D^- be the complement of $D^+ \cup \Sigma$ in $I\!\!R^n$. The so–called Riemann boundary value problem is to find a sectionally regular function

$$F(x) = \begin{cases} F^+(x), & x \in D^+, \\ F^-(x), & x \in D^-, \end{cases} \tag{2.1}$$

satisfying the boundary condition

$$F^+(x) = G(x)F^-(x) + g(x), \ x \in \Sigma, \tag{2.2}$$

and $F^-(\infty) = \lim_{|x| \to \infty} F^-(x) = 0$, where $G(x)(\neq 0)$, $g(x) \in C_\alpha(\Sigma), 0 < \alpha < 1$.

We first discuss the simple case, where $G(x) = G_0$ is a constant and has a inverse G_0^{-1}. This special Problem R is called Problem R'.

Lemma 2.1   *Problem R' for regular functions has a unique solution $F(x)$, and $F(x)$ can be expressed by*

$$F(x) = \frac{X(x)}{\omega_{n-1}} \int_\Sigma \frac{\bar{y} - \bar{x}}{|y - x|^n} G_0^{-1} n(y)f(y)dS_y, \tag{2.3}$$

where $X(x) = G_0, \ x \in D^+$ and $X(x) = 1, \ x \in D^-$.

Proof   It is easy to see that the function $F(x)$ in (2.3) is sectionally regular in D^+, D^-, and $F(\infty) = 0$. According to the Plemelj formula (1.23), we have

$$F^+(x) - G(x)F^-(x) = \frac{G_0}{\omega_{n-1}} \int_\Sigma \frac{\bar{y} - \bar{x}}{|y - x|^n} G_0^{-1} n(y)f(y)dS_y + \frac{X^+(x)}{2} G_0^{-1} f(x) -$$

$$-G_0 \Big[\frac{1}{\omega_{n-1}} \int_\Sigma \frac{\bar{y} - \bar{x}}{|y - x|^n} G_0^{-1} n(y)f(y)dS_y - \frac{X^-(x)}{2} G_0^{-1} f(x) \Big] = f(x), \ x \in \Sigma.$$

Let $F_1(x), F_2(x)$ be two solutions of Problem R' for regular functions. Then

$$F(x) = X^{-1}(x)[F_1(x) - F_2(x)]$$

is regular in D^+, D^- and is continuous on Σ. Noting that $F(\infty) = 0$, and applying the Liouville theorem, $F(x) \equiv 0$, i.e. $F_1(x) \equiv F_2(x)$ can be derived.

Lemma 2.2   *The integral operator*

$$Kf(x) = \frac{2}{\omega_{n-1}} \int_\Sigma \frac{\bar{y} - \bar{x}}{|y - x|^n} n(y)f(y)dS_y \tag{2.4}$$

is a bounded linear operator, which maps $C_\alpha(\Sigma)\,(0 < \alpha < 1)$ into itself, i.e. for any $f(x) \in C_\alpha(\Sigma)$ there exists a positive constant M_1 independent of $f(x)$, such that

$$C_\alpha[Kf, \Sigma] \le M_1 C_\alpha[f, \Sigma]. \tag{2.5}$$

Proof Since the Cauchy principle value integral

$$\frac{2}{\omega_{n-1}} \int_\Sigma \frac{\bar{y} - \bar{x}}{|y - x|^n} n(y) f(y) dS_y = 1, \ x \in \Sigma,$$

it can be obtained

$$|Kf(x)| = |\frac{2}{\omega_{n-1}} \int_\Sigma \frac{\bar{y} - \bar{x}}{|y - x|^n} n(y)[f(y) - f(x)]dS_y + f(x)|$$

$$\le M_2 H_\alpha(f, \Sigma) \int_\Sigma \frac{1}{|y - x|^{n-1-\alpha}} |dS_y| + |f(x)| \tag{2.6}$$

$$\le M_3 H_\alpha(f, \Sigma) + C(f, \Sigma),$$

where M_2, M_3 are constants independent of $f(x)$. Let $x, \tilde{x}$ be two points on Σ and $\delta = |x - \tilde{x}|$. Suppose that $4\delta < d$, where d is a positive number dependent on Σ, and denote by $\Sigma_{2\delta}$ the part of Σ lying inside a sphere of radius 2δ with centre at the point x and $\tilde{\Sigma}_{2\delta} = \Sigma \backslash \Sigma_{2\delta}$, we have

$$|Kf(x) - Kf(\tilde{x})| \le |\frac{2}{\omega_{n-1}} \int_{\Sigma_{2\delta}} \frac{\bar{y} - \bar{x}}{|y - x|^n} n(y)[f(y) - f(x)]dS_y|$$

$$+ |\frac{2}{\omega_{n-1}} \int_{\Sigma_{2\delta}} \frac{\bar{y} - \bar{\tilde{x}}}{|y - \tilde{x}|^n} n(y)[f(y) - f(\tilde{x})]dS_y|$$

$$+ |\frac{2}{\omega_{n-1}} \int_{\tilde{\Sigma}_{2\delta}} \frac{\bar{y} - \bar{x}}{|y - x|^n} n(y)[f(y) - f(x)]dS_y \tag{2.7}$$

$$- \frac{2}{\omega_{n-1}} \int_{\tilde{\Sigma}_{2\delta}} \frac{\bar{y} - \bar{\tilde{x}}}{|y - \tilde{x}|^n} n(y)[f(y) - f(\tilde{x})]dS_y| + |f(x) - f(\tilde{x})|$$

$$= J_1 + J_2 + J_3 + |f(x) - f(\tilde{x})|.$$

It can be derived

$$J_1 \le M_4 H_\alpha(f, \Sigma) \int_{\Sigma_{2\delta}} \frac{1}{|y - x|^{n-1-\alpha}} |d\Sigma_y| = M_5 H_\alpha(f, \Sigma)$$

$$\times \int_0^{2\delta} \frac{\rho^{n-2}}{\rho^{n-1-\alpha}} d\rho = M_6 H_\alpha(f, \Sigma)\delta^\alpha = M_6 H_\alpha(f, \Sigma)|x - \tilde{x}|^\alpha, \tag{2.8}$$

in which $M_j (j = 5, 6, 7)$ are constants independent of $x, \tilde{x}$. Similarly, we may estimate

$$J_2 \le M_7 H_\alpha(f, \Sigma)|x - \tilde{x}|^\alpha, \tag{2.9}$$

where M_7 is a constant independent of $x, \tilde{x}$. Next we estimate J_3.

$$J_3 = |\frac{2}{\omega_{n-1}} \int_{\tilde{\Sigma}_{2\delta}} (\frac{\bar{y} - \bar{x}}{|y - x|^n} - \frac{\bar{y} - \bar{\tilde{x}}}{|y - \tilde{x}|^n})n(y)[f(y) - f(x)]dS_y$$

$$-\frac{2}{\omega_{n-1}} \int_{\tilde{\Sigma}_{2\delta}} \frac{\bar{y} - \bar{\tilde{x}}}{|y - \tilde{x}|^n}n(y)[f(x) - f(\tilde{x})]dS_y|$$

$$\leq |\frac{2}{\omega_{n-1}} \int_{\tilde{\Sigma}_{2\delta}} (\frac{\bar{y} - \bar{x}}{|y - x|^n} - \frac{\bar{y} - \bar{\tilde{x}}}{|y - \tilde{x}|^n})n(y)[f(y) - f(x)]dS_y| \tag{2.10}$$

$$+|\frac{2}{\omega_{n-1}} \int_{\tilde{\Sigma}_{2\delta}} \frac{\bar{y} - \bar{\tilde{x}}}{|y - \tilde{x}|^n}[f(x) - f(\tilde{x})]dS_y|$$

$$\leq M_8 H_\alpha(f, \Sigma) \int_{\tilde{\Sigma}_{2\delta}} |\bar{y} - \bar{x}|^{-n+\alpha}dS_y|x - \tilde{x}|$$

$$+M_9 H_\alpha(f, \Sigma)|x - \tilde{x}|^\alpha \leq M_{10}H_\alpha(f, \Sigma)|x - \tilde{x}|^\alpha,$$

in which from (1.19) we obtain

$$\left|\frac{\bar{y} - \bar{x}}{|y - x|^n} - \frac{\bar{y} - \bar{\tilde{x}}}{|y - \tilde{x}|^n}\right| \leq \sum_{k=1}^{n-1} |y - x|^{-k}|y - \tilde{x}|^{k-n}|x - \tilde{x}|, \tag{2.11}$$

and $|y - \tilde{x}| \geq |y - x|/2$, $M_j(j = 8, 9, 10)$ are constants which do not depend on $x, \tilde{x}$. Thus, for $4|x - \tilde{x}| < d$, it can be derived

$$|Kf(x) - Kf(\tilde{x})| \leq M_{11}H_\alpha(f, \Sigma)|x - \tilde{x}|^\alpha, \tag{2.12}$$

where M_{11} is a constant independent of x and $\tilde{x}$. As for $4|x - \tilde{x}| \geq d$, it is clear that there exists an estimate similar to (2.12). Combining (2.6) and (2.12), the estimate (2.5) is obtained.

Theorem 2.3 *Suppose that the function $G(x)$ in (2.2) satisfies the condition*

$$M_{12} = 2^{n-2}(M_1 + 1)C_\alpha[1 - G(x), \Sigma] < 1, \tag{2.13}$$

where M_1 is the constant in (2.5). Then Problem R for regular functions in Clifford analysis possesses a unique solution.

Proof We shall find a solution of Problem R of the form

$$F(x) = \frac{1}{\omega_{n-1}} \int_\Sigma \frac{\bar{y} - \bar{x}}{|y - x|^n}n(y)f(y)dS_y, \tag{2.14}$$

in which $f(x) \in C_\alpha(\Sigma)$. From Lemma 2.1, Problem R can be reduced to an equivalent singular integral equation for $f(x)$:

$$f(x) = \frac{1 - G(x)}{2}[f(x) - \frac{2}{\omega_{n-1}} \int_\Sigma \frac{\bar{y} - \bar{x}}{|y - x|^n}n(y)f(y)dS_y] + g(x), \quad x \in \Sigma. \tag{2.15}$$

Let A denote the integral operator defined by the right–hand side of (2.15), namely

$$Af(x) = \frac{1 - G(x)}{2}[f(x) - Kf(x)] + g(x). \tag{2.16}$$

From the condition (2.13) and Lemma 2.2, we see that the integral operator A is a contraction operator which maps the Banach space $C_\alpha(\Sigma)$ into itself. Hence there exists a unique fixed point for the operator A. This shows that the integral equation (2.15) has a unique solution $f(x)$, and the function $F(x)$ in (2.14) is just a solution of Problem R.

Next, we discuss the nonlinear Riemann boundary value problem (Problem R^*) for regular functions, the boundary condition of which is as follows:

$$F^+(x) = G(x)f^-(x) + \varepsilon g(x, F^+(x), F^-(x)), \ x \in \Sigma, \tag{2.17}$$

and $F^-(\infty) = 0$, where ε is a positive constant, $G(x) \in C_\alpha(\Sigma), 0 < \alpha < 1$, and $g(x, F^{(1)}, F^{(2)}) \in C_\alpha(\Sigma)$ for any Clifford numbers $F^{(1)}$ and $F^{(2)}$ and satisfies

$$|g(x, F^{(1)}, F^{(2)}) - g(\tilde{x}, \tilde{F}^{(1)}, \tilde{F}^{(2)})| \le d_1|x - \tilde{x}|^\alpha + d_2|F^{(1)} - \tilde{F}^{(1)}| + d_3|F^{(2)} - \tilde{F}^{(2)}| \tag{2.18}$$

for any $x, \tilde{x} \in \Sigma$, where $d_j(j = 1, 2, 3)$ are nonnegative constants.

Theorem 2.4 *If the function $G(x)$ satisfies condition (2.13) and the constant ε is small enough, then Problem R^* for regular functions in Clifford analysis has a solution.*

Proof Similarly to the proof of Theorem 2.3, we shall find a solution of the form (2.14). The nonlinear Riemann Problem R^* can be transformed into an equivalent nonlinear singular integral equation

$$f(x) = \frac{1 - G(x)}{2}[f(x) - \frac{2}{\omega_{n-1}} \int_\Sigma \frac{\bar{y} - \bar{x}}{|y - x|^n} n(y)f(y)dS_y]$$

$$+\varepsilon g[x, \frac{f(x)}{2} + \frac{1}{\omega_{n-1}} \int_\Sigma \frac{\bar{y} - \bar{x}}{|y - x|^n} n(y)f(y)dS_y, -\frac{f(x)}{2} \tag{2.19}$$

$$+\frac{1}{\omega_{n-1}} \int_\Sigma \frac{\bar{y} - \bar{x}}{|y - x|^n} n(y)f(y)dS_y], \ x \in \Sigma.$$

Let A^* denote the integral operator defined by the right–hand side of (2.19), namely

$$A^*f(x) = \frac{1 - G(x)}{2}[f(x) - \frac{2}{\omega_{n-1}} \int_\Sigma \frac{\bar{y} - \bar{x}}{|y - x|^n} n(y)f(y)dS_y]$$

$$+\varepsilon g[x, \frac{f(x)}{2} + \frac{1}{\omega_{n-1}} \int_\Sigma \frac{\bar{y} - \bar{x}}{|y - x|^n} n(y)f(y)dS_y, -\frac{f(x)}{2} \tag{2.20}$$

$$+\frac{1}{\omega_{n-1}} \int_\Sigma \frac{\bar{y} - \bar{x}}{|y - x|^n} n(y)f(y)dS_y].$$

We introduce a closed, convex subset B_M of the continuous function space $C(\bar{D})$, the elements of which are all functions $f(x)$ with values in the Clifford algebra satisfying

$$C_\alpha[f(x), \Sigma] \leq M = M_{13}, \ 0 < \alpha < 1, \tag{2.21}$$

where M_{13} is a nonnegative constant. By the constant (2.13) and (2.18), it is not difficult to obtain the following inequality

$$C_\alpha[Bf(x), \Sigma] \leq M_{12} C_\alpha[f, \Sigma] + \varepsilon[M_{14} C_\alpha(f, \Sigma) + d_1] \tag{2.22}$$

for any function $f(x) \in B_M$, in which M_{12} is the constant in (2.13) and $M_{14} = (1 + M_1)(d_2 + d_3)/2$. If the constant ε is sufficiently small so that

$$\varepsilon < M(1 - M_{12})/(d_1 + MM_{14}) \tag{2.23}$$

then the operator B maps B_M into itself. To prove that A^* is a continuous operator, we choose an arbitrary sequence of functions: $f^{(n)}(x) \in B_M$ $(n = 0, 1, 2, \cdots)$ such that $C[f^{(n)}(x) - f^{(0)}(x), \Sigma] \to 0$ as $n \to \infty$. For any $x \in \Sigma$, as in the proof of Lemma 2.2, denote by $\Sigma_{2\delta}$ the part of Σ lying inside a sphere of radius 2δ with centre at the point x and $\tilde{\Sigma}_{2\delta} = \Sigma \backslash \Sigma_{2\delta}$, we have

$$|Kf^{(n)}(x) - Kf^{(0)}(x)|$$

$$\leq |\frac{2}{\omega_{n-1}} \int_{\Sigma_{2\delta}} \frac{\bar{y} - \bar{x}}{|y - x|^n} n(y)[f^{(n)}(y) - f^{(0)}(y)]dS_y|$$

$$+ |\frac{2}{\omega_{n-1}} \int_{\tilde{\Sigma}_{2\delta}} \frac{\bar{y} - \bar{x}}{|y - x|^n} n(y)[f^{(n)}(y) - f^{(0)}(y)]dS_y| = J_4 + J_5. \tag{2.24}$$

Suppose $2\delta < d$, we can obtain

$$J_4 = |\frac{2}{\omega_{n-1}} \int_{\Sigma_{2\delta}} \frac{\bar{y} - \bar{x}}{|y - x|^n} n(y)[f^{(n)}(y) - f^{(n)}(x)]dS_y$$

$$- \frac{2}{\omega_{n-1}} \int_{\Sigma_{2\delta}} \frac{\bar{y} - \bar{x}}{|y - x|^n} n(y)[f^{(0)}(y) - f^{(0)}(x)]dS_y$$

$$+ \frac{2}{\omega_{n-1}} \int_{\Sigma_{2\delta}} \frac{\bar{y} - \bar{x}}{|y - x|^n} n(y)[f^{(n)}(x) - f^{(0)}(x)]dS_y|$$

$$\leq |\frac{2}{\omega_{n-1}} \int_{\Sigma_{2\delta}} \frac{\bar{y} - \bar{x}}{|y - x|^n} n(y)[f^{(n)}(y) - f^{(n)}(x)]dS_y| \tag{2.25}$$

$$+ |\frac{2}{\omega_{n-1}} \int_{\Sigma_{2\delta}} \frac{\bar{y} - \bar{x}}{|y - x|^n} n(y)[f^{(0)}(y) - f^{(0)}(x)]dS_y|$$

$$+ 2^{n-1} |\frac{2}{\omega_{n-1}} \int_{\Sigma_{2\delta}} \frac{\bar{y} - \bar{x}}{|y - x|^n} n(y)dS_y| \, |f^{(n)}(x) - f^{(0)}(x)|$$

$$\leq M_{15} \int_0^{2\delta} \frac{\rho^{n-2}}{\rho^{n-1-\alpha}} d\rho + M_{16} \int_0^{2\delta} \frac{\rho^{1+\delta}}{\rho^n} \rho^{n-2} d\rho = M_{17}\delta^\alpha + M_{18}\delta^\beta,$$

where M_j $(j = 15, \cdots, 18)$ are constants independent of $f^{(n)}(x)$, $f^{(0)}(x)$, x and δ, and β is a positive constant. For any given positive number η, provided that δ is small enough, $J_4 < \eta/2$ may be obtained. Moreover, $J_5 \leq M_{19}C[f^{(n)}(x) - f^{(0)}(x), \Sigma]$ is evident, where $M_{19} = M_{19}(\delta)$, and then $J_5 < \eta/2$, for $n > N$, N is a sufficiently large number. Thus for $n > N$, $|A^* f^{(n)}(x) - A^* f^0(x)| < \varepsilon$. In addition, it is clear that A^* is a compact operator. On the basis of the Schauder fixed point theorem, there exists a function $f(x) \in B_M$, which is a solution of the nonlinear integral equation (2.18). Therefore we find a solution of Problem R^*, which can be expressed as (2.14).

If we impose some additional conditions on the function $g(x, w^{(1)}, w^{(2)})$, the uniqueness of solutions of Problem R^* can be proved (see [60]).

3 Schwarz Formulae for Halfspace and Ball

In this section, we discuss the 3–dimensional space $I\!\!R^3$ and the corresponding Clifford algebra $A_3(I\!\!R)$ over $I\!\!R^3$. In this case, $x = x_1 + x_2 e_2 + x_3 e_3 + x_{23} e_{23} \in A_3(I\!\!R)$, we define by

$$\operatorname{Re} x = x_1 + x_2 e_2, \quad \operatorname{Im} x = x_3 - x_{23} e_2, \tag{3.1}$$

the real part and imaginary part of x. It is evident that

$$x = \operatorname{Re} x + e_3 \operatorname{Im} x, \quad \operatorname{Re} x = \frac{1}{2}(x + \tilde{x}), \quad \operatorname{Im} x = -e_3 \frac{x - \tilde{x}}{2}, \tag{3.2}$$

where $\tilde{x} = \operatorname{Re} x - e_3 \operatorname{Im} x$. Let $y = \operatorname{Re} y + e_3 \operatorname{Im} y \in A_3(I\!\!R)$, we can verify that $\widetilde{xy} = \tilde{x}\tilde{y}$.

3.1 Schwarz formula for the halfspace

Theorem 3.1 *Suppose that $u(y)$ is a Hölder continuous function on the plane $E = \{x_3 = 0\}$, and $u(y) = 0$ if $|y|$ is sufficiently large. Then there exists a regular function $f(x)$ in the upper halfspace D, such that*

$$\operatorname{Re} f^+(y) = u(y), \, y \in E. \tag{3.3}$$

Then the function $f(x)$ can be expressed as

$$f(x) = \frac{1}{2\pi} \int\!\!\int_E \frac{x_3 + e_3[(x_1 - y_1) + e_2(x_2 - y_2)]}{|y - x|^3} u(y) dS_y + e_3 g, \tag{3.4}$$

where we assume that $\lim_{|x| \to \infty} f(x) = f(\infty) = e_3(a_3 - a_{23} e_2)$, a_3, a_{23} are real constants, and then $e_3 g = f(\infty)$. This is also a representation of the solution of the Dirichlet problem for regular functions in the halfspace D.

Proof We assume that $f(x)$ is a desired regular function, where x is any point in D. Let Σ_R be the upper half sphere with the centre at the origin and radius R, and denote by D_R a domain with the boundary Σ_R and $E_R = \{|x| < R, x \in E\}$. We choose R so large that $x \in D_R$ and $|y - x| \geq R/2$, $y \in E_R$. By the Cauchy formula we have

$$f(x) = \frac{1}{4\pi} \int_{\Sigma_R} \frac{\bar{y} - \bar{x}}{|y - x|^3} n(y) f(y) dS_y +$$

$$+ \frac{1}{4\pi} \int_{E_R} \frac{\bar{y} - \bar{x}}{|y - x|^3} n(y) f(y) dS_y = I_1 + I_2. \tag{3.5}$$

From Lemma 1.2,

$$\left| \frac{\bar{y} - \bar{x}}{|y - x|^3} - \frac{\bar{y}}{|y|^3} \right| \leq |x| \left(\frac{1}{|y - x||y|^2} + \frac{1}{|y - x|^2|y|} \right)$$

$$\leq |x| \left(\frac{2}{R^3} + \frac{4}{R^3} \right) = \frac{6|x|}{R^3}. \tag{3.6}$$

Hence

$$\left| \frac{1}{4\pi} \int_{\Sigma_R} \left[\frac{\bar{y} - \bar{x}}{|y - x|^3} - \frac{\bar{y}}{|y|^3} \right] n(y) f(y) dS_y \right|$$

$$\leq \frac{1}{4\pi} \int_{\Sigma_R} \frac{6|x|}{R^3} |f(y)| dS_y \leq \frac{M|x|}{4\pi R^3} \int_{\Sigma_R} dS_y \tag{3.7}$$

$$= \frac{M|x|}{2R} \to 0 \text{ as } R \to \infty.$$

For arbitrarily given positive constant ε, there exists a large positive constant R_0, such that $|f(y) - f(\infty)| < \varepsilon$ if $|y| \geq R_0$. Thus

$$\left| \int_{\Sigma_R} \frac{\bar{y}}{|y|^3} n(y) [f(y) - f(\infty)] dS_y \right| \leq \frac{4}{\pi} \varepsilon,$$

which implies that

$$\int_{\Sigma_R} \frac{\bar{y}}{|y|^3} n(y) [f(y) - f(\infty)] dS_y \to 0 \text{ as } R \to \infty. \tag{3.8}$$

Noting that

$$I_1 = \frac{1}{4\pi} \int_{\Sigma_R} \frac{\bar{y} - \bar{x}}{|y - x|^3} n(y) f(y) dS_y = \frac{1}{4\pi} \int_{\Sigma_R} \left[\frac{\bar{y} - \bar{x}}{|y - x|^3} - \frac{\bar{y}}{|y|^3} \right] n(y) f(y) dS_y$$

$$+ \frac{1}{4\pi} \int_{\Sigma_R} \frac{\bar{y}}{|y|^3} n(y) [f(y) - f(\infty)] dS_y + \frac{1}{4\pi} \int_{\Sigma_R} \frac{\bar{y}}{|y|^3} n(y) f(\infty) dS_y$$

and

$$\frac{1}{4\pi} \int_{\Sigma_R} \frac{\bar{y}}{|y|^3} n(y) f(\infty) dS_y = \frac{1}{2} f(\infty),$$

and applying (3.7)and (3.8),

$$\lim_{R \to \infty} I_1 = \frac{1}{2}f(\infty) \tag{3.9}$$

follows.

As for the integral I_2, taking $n(y) = -e_3$, $y = y_1 + e_2 y_2$ on E_R into account, we have

$$I_2 = \frac{1}{4\pi} \int_{E_R} \frac{\bar{y} - \bar{x}}{|y - x|^3} n(y) f(y) dS_y$$

$$= \frac{1}{4\pi} \int_{E_R} \frac{(y_1 - x_1) - e_2(x_2 - y_2) + e_3 x_3}{|y - x|^3} (-e_3) f(y) dS_y \tag{3.10}$$

$$= \frac{1}{4\pi} \int_{E_R} \frac{x_3 + e_3[(x_1 - y_1) + e_2(x_2 - y_2)]}{|y - x|^3} n(y) f(y) dS_y.$$

Letting R tend to ∞

$$f(x) = \frac{1}{4\pi} \int_{E_R} \frac{x_3 + e_3[(x_1 - y_1) + e_2(x_2 - y_2)]}{|y - x|^3} f(y) dS_y + \frac{1}{2}f(\infty) \tag{3.11}$$

can be derived. If $\operatorname{Im} x < 0$, then according to the Cauchy theorem for regular functions, we can similarly obtain

$$0 = \frac{1}{4\pi} \int_{E_R} \frac{-x_3 + e_3[(x_1 - y_1) + e_2(x_2 - y_2)]}{|y - x|^3} f(y) dS_y + \frac{1}{2}f(\infty).$$

From the above formula, it is easy to derive

$$0 = \frac{1}{4\pi} \int_{E} \frac{x_3 + e_3[(x_1 - y_1) + e_2(x_2 - y_2)}{|y - x|^3} \tilde{f}(y) dS_y - \frac{1}{2}\tilde{f}(\infty). \tag{3.12}$$

Adding (3.11) and (3.12), and noting that $\tilde{f}(\infty) = -f(\infty)$, gives

$$f(x) = \frac{1}{4\pi} \int_{E} \frac{x_3 + e_3[(x_1 - y_1) + e_2(x_2 - y_2)]}{|y - x|^3} [f(y) + \tilde{f}(y)] dS_y + f(\infty)$$

$$= \frac{1}{2\pi} \int_{E} \frac{x_3 + e_3[(x_1 - y_1) + e_2(x_2 - y_2)]}{|y - x|^3} u(y) dS_y + f(\infty).$$

It remains to verify that $f(x)$ is just the desired function. It is sufficient to prove that

$$\operatorname{Re} f^+(x^0) = u(x^0), \tag{3.13}$$

where x^0 is any point on E. We rewrite (3.4) in the form

$$f(x) = \frac{1}{2\pi} \int_{E_R} \frac{x_3 + e_3[(x_1 - y_1) + e_2(x_2 - y_2)]}{|y - x|^3} u(y) dS_y +$$

$$+ \frac{1}{2\pi} \int_{E \backslash E_R} \frac{x_3 + e_3[(x_1 - y_1) + e_2(x_2 - y_2)]}{|y - x|^3} u(y) dS_y + f(\infty),$$

156   *Heinrich Begehr and Guo Chun Wen*

where $E_R = \{\, |y| < R, y \in E \,\}$ and Σ_R are as stated before. Due to $u(y) = 0$ if $|y|$ is sufficiently large, and for piecewise smooth surface $\Sigma_R \cup E_R$ and sectionally Hölder–continuous function $u(y)$, the Plemelj formula is still true. Hence

$$g^+(x^0) - g^-(x^0) = 2u(x^0), \tag{3.14}$$

where

$$g^+(x) = \frac{1}{2\pi} \int_{E_R \cup \Sigma_R} \frac{x_3 + e_3[(x_1 - y_1) + e_2(x_2 - y_2)]}{|y - x|^3} u(y)dS_y, \ x \in D_R.$$

Setting $\tilde{\zeta} = x, \zeta \notin \bar{D}_R$,

$$\tilde{g}^+(x) = \frac{1}{2\pi} \int_{E_R \cup \Sigma_R} \frac{x_3 - e_3[(x_1 - y_1) + e_2(x_2 - y_2)]}{|y - x|^3} u(y)dS_y$$

$$= \tilde{g}^+(\tilde{\zeta}) = \frac{1}{2\pi} \int_{E_R \cup \Sigma_R} \frac{-\zeta_3 - e_3[(x_1 - y_1) + e_2(x_2 - y_2)]}{|y - \zeta|^3} u(y)dS_y$$

can be obtained. Letting x tend to x^0, we know that $\tilde{g}^+(x^0) = -g^-(x^0)$. From (3.14), it follows that

$$g^+(x^0) - g^-(x^0) = g^+(x^0) + \tilde{g}^+(x^0) = 2u(x^0), \ \text{i.e.}$$

$$\mathrm{Re}\, g^+(x^0) = u(x^0). \tag{3.15}$$

Noting that

$$\lim_{x \to x^0} \frac{1}{2\pi} \iint_{E \setminus E_R} \frac{x_3 + e_3[(x_1 - y_1) + e_2(x_2 - y_2)]}{|y - x|^3} u(y)dS_y + f(\infty)$$

$$= \frac{1}{2\pi} \iint_{E \setminus E_R} \frac{e_3[(x_1^0 - y_1) + e_2(x_2^0 - y_2)]}{|y - x^0|^3} u(y)dS_y + f(\infty),$$

it is easily seen that

$$\mathrm{Re}\left[\frac{1}{2\pi} \int_{E \setminus E_R} \frac{e_3[(x_1^0 - y_1) + e_2(x_2^0 - y_2)]}{|y - x^0|^3} u(y)dS_y + f(\infty) \right] = 0.$$

Thus we obtain (3.13).

Finally, on account of $u(y) = 0$ if $|y|$ is large enough,

$$\lim_{|x| \to \infty} f(x) = f(\infty)$$

can be derived. This shows that the above regular function $f(x)$ is unique.

Remark   By using a similar method, we can also get the expression for the solution of the Dirichlet problem for regular functions in the halfspace D with weaker conditions, namely we suppose $u(y)$ is a bounded continuous function in the plane $E = \{x_3 = 0\}$, and the solution $f(x)$ of the Dirichlet problem satisfies the condition $|\,\mathrm{Re} f(x)\,| = O(1/|x|)$ as $|x|$ tends to ∞. Then $f(x)$ may be expressed in the form (3.4), where $g(x_1, x_2) = g_1(x_1, x_2) + e_2 g_2(x_1, x_2)$ is an arbitrary function satisfying $g_{x_1} = e_2 g_{x_2}$.

3.2 Schwarz formula for a ball

Next, we shall give the Schwarz formula for regular functions in a ball $G = \{|x| < R,\ 0 < R < \infty\}$ in $I\!\!R^3$. We need the following lemmas.

Lemma 3.2 *A function $f(x)$ with values in the Clifford algebra $A_3(I\!\!R)$ is regular in D if and if only $\operatorname{Re} f(x)$ and $\operatorname{Im} f(x)$ satisfy the system of first order equations*

$$2\frac{\partial}{\partial x_*}\operatorname{Re} f = \frac{\partial}{\partial x_3}\operatorname{Im} f,\ x \in D, \tag{3.16}$$

$$\frac{\partial}{\partial x_3}\operatorname{Re} f = -2\frac{\partial}{\partial x^*}\operatorname{Im} f,\ x \in D, \tag{3.17}$$

where $x^* = x_1 + x_2 e_2$, $x_* = x_1 - x_2 e_2$, and

$$\frac{\partial}{\partial x^*} = \frac{1}{2}\left(\frac{\partial}{\partial x_1} - e_2\frac{\partial}{\partial x_2}\right),\ \frac{\partial}{\partial x_*} = \frac{1}{2}\left(\frac{\partial}{\partial x_1} + e_2\frac{\partial}{\partial x_2}\right).$$

Proof It is clear that

$$2\frac{\partial}{\partial x^*}(e_3\operatorname{Im} f) = 2e_3\frac{\partial}{\partial x_*}\operatorname{Im} f.$$

Hence

$$\bar\partial f = \left(2\frac{\partial}{\partial x_*} + e_3\frac{\partial}{\partial x_3}\right)(\operatorname{Re} f + e_3\operatorname{Im} f) =$$

$$= \left(2\frac{\partial}{\partial x_*}\operatorname{Re} f - \frac{\partial}{\partial x_3}\operatorname{Im} f\right) + e_3\left(\frac{\partial}{\partial x_3}\operatorname{Re} f + 2\frac{\partial}{\partial x^*}\operatorname{Im} f\right). \tag{3.18}$$

If $f(x)$ is regular in D, i.e. $\bar\partial f = 0$, from (3.18), it follows (3.16) and (3.17). The inverse statement is also true.

Suppose that $f(x)$ is a regular function and $f(x) \in C^2(D)$. Then

$$\bar\partial\partial\operatorname{Re} f = \partial\bar\partial\operatorname{Re} f = \Delta\operatorname{Re} f = 0,\ \Delta\operatorname{Im} f = 0.$$

So $\operatorname{Re} f$ and $\operatorname{Im} f$ are called harmonic functions in Clifford analysis, and $\operatorname{Im} f$ is called the conjugate harmonic function of $\operatorname{Re} f$.

Lemma 3.3 *Let $f(x) = u_1(x) + e_2 u_2(x)$, $u_1(x)$ and $u_2(x)$ be real harmonic functions in D and $u_1(x), u_2(x) \in C^1(\bar D)$. Then the conjugate harmonic function $v(x)$ of $u(x)$ is given by the formula*

$$v(x) = \int_0^{x_3} 2\frac{\partial}{\partial x_*}u\,dx_3 - \frac{1}{2}\tilde T\left(\frac{\partial}{\partial x_3}u(x_1, x_2, 0)\right) + g(x_1, x_2), \tag{3.19}$$

where $g(x_1, x_2) = g_1(x_1, x_2) + e_2 g_2(x_1, x_2)$ is arbitrary function satisfying $\partial/\partial x^* g = 0$, and

$$\tilde T u(x) = -\frac{1}{\pi}\iint_{E_R}\frac{u(y^*)}{y_* - x_*}\,d\sigma_{y^*}, \tag{3.20}$$

where $E_R = \{(x_1, x_2, 0) \,|\, x_1^2 + x_2^2 \,| < R^2\}$.

Proof   From (3.16), it follows that

$$v(x) = \int_0^{x_3} 2\frac{\partial}{\partial x_*} u\, dx_3 + w(x_1, x_2).$$

Substituting the above expression into (3.15) and letting $x_3 = 0$, we obtain

$$\frac{\partial}{\partial x_3} u(x_1, x_2, x_3)|_{x_3=0} + 2\frac{\partial}{\partial x^*} w(x_1, x_2) = 0.$$

According the result in [50],

$$w(x_1, x_2) = -\frac{1}{2}\tilde{T}(\frac{\partial}{\partial x_3} u(x_1, x_2, x_3)|_{x_3=0}) + g(x_1, x_2)$$

can be obtained, in which $g(x_1, x_2)$ satisfies $\partial g/\partial x^* = 0$. Thus formula (3.19) holds.

Moreover, if $v(x)$ is given by (3.19), it can be seen that $u(x)$ and $v(x)$ satisfy (3.16). Since $u(x)$ is a harmonic function and

$$4\frac{\partial}{\partial x^*}\frac{\partial}{\partial x_*}u = (\frac{\partial^2}{\partial x_1^2} + \frac{\partial^2}{\partial x_2^2})u = -\frac{\partial^2}{\partial x_3^2}u,$$

it is easy to verify that $u(x)$ and $v(x)$ satisfy (3.17).

Lemma 3.4   *Suppose that $u(x) = u_1(x) + e_2 u_2(x)$ is a harmonic function in $D = \{|x| < R\}$ in Clifford analysis and $u(x)$ is continuous on $\bar{D}$. Then $u(x)$ can be expressed as*

$$u(x) = -\int_{|y|=R}\frac{\partial}{\partial n}\Big(\frac{1}{4\pi}\frac{1}{|x-y|} - \frac{1}{4\pi}\frac{R}{|x||\tilde{x}-y|}\Big)u(y)dS_y, \qquad (3.21)$$

where $x = x_1 + x_2 e_2 + x_3 e_3$, $y = y_1 + y_2 e_2 + y_3 e_3$, $\tilde{x} = R^2 x/|x|^2$, $\partial/\partial n$ denotes the exterior normal derivative with respect to y on the sphere $|y| = R$, and dS_y is the area element of $|y| = R$.

Proof   Due to $\Delta u = 0$ in D and because $u(x)$ is continuous on $\bar{D}$, from the Poisson formula for harmonic functions in a ball, formula (3.21) is true.

Now, we find the conjugate harmonic functions of $1/|x-y|$ and $R/|x||\tilde{x}-y|$ with respect to x. Noting that

$$\int_0^{x_3} 2\frac{\partial}{\partial x_*}\frac{1}{|x-y|}dx_3 = \int_0^{x_3}\Big(-\frac{x^*-y^*}{|x-y|^3}\Big)dx_3$$

$$= -\frac{(x^*-y^*)(x_3-y_3)}{|x^*-y^*|^2(x-y)} - \frac{(x^*-y^*)y_3}{|x^*-y^*|^2[(x_1-y_1)^2 + (x_2-y_2)^2 + y_3^2]^{1/2}}$$

and

$$-\frac{1}{2}\frac{\partial}{\partial x_3}\frac{1}{|x-y|}\Big|_{x_3=0} = -\frac{1}{2}\frac{y_3}{[(x_1-y_1)^2+(x_2-y_2)^2+y_3^2]^{3/2}}$$
$$= \frac{\partial}{\partial x^*}\frac{(x^*-y^*)y_2}{|x^*-y^*|^2[(x_1-y_1)^2+(x_2-y_2)^2+y_3^2]^{3/2}},$$

we obtain

$$-\frac{1}{2}\tilde{T}\Big(\frac{\partial}{\partial x_3}\frac{1}{|x-y|}\Big|_{x_3=0}\Big) = \frac{(x^*-y^*)y_2}{|x^*-y^*|^2[(x_1-y_1)^2+(x_2-y_2)^2+y_3^2]^{1/2}} + w(x_1,x_2),$$

where $w(x_1,x_2)$ is a function satisfying $\frac{\partial}{\partial x^*}w = 0$. On the basis of Lemma 3.3, we know that the conjugate harmonic function of $1/|x-y|(x\neq y)$ with respect to x possesses the form

$$-\frac{(x^*-y^*)(x_3-y_3)}{|x^*-y^*|^2|x-y|} + c(x_1,x_2) + w(x_1,x_2).$$

In particular, choosing $c(x_1,x_2) = -w(x_1,x_2)$, it is easy to see that $-(x^*-y^*)(x_3-y_3)/|x^*-y^*|^2|x-y|$ is a conjugate harmonic function of $1/|x-y|$. Similarly, we can find that a conjugate function of $R/|x||\tilde{x}-y|$ with respect to y possess the form

$$-\frac{R(y^*-\tilde{x}^*)(y_3-\tilde{x}_3)}{|x||y^*-\tilde{x}^*|^2|y-\tilde{x}|} = -\frac{R}{|x|}\frac{(y^*-R^2x^*/|x|^2)(y_3-R^2|x_3|/|x|^2)}{|y-R^2x^*/|x|^2|^2|y-R^2x/|x|^2|}$$
$$= -\frac{R|x|(|x|^2y^*-R^2x^*)(|x|^2y_3-R^2x_3)}{|x_1^2y^*-R^2x^*|^2||x|^2y-R^2x|},$$

and a conjugate function of $R/|x||\tilde{x}-y|$ with respect to x possesses the form

$$-\frac{R|y|(|y|^2x^*-R^2y^*)(|y|^2x_3-R^2y_3)}{||y|^2x^*-R^2y^*|^2||y|^2x-R^2y|}.$$

Denoting

$$g(x,y) = \frac{1}{4\pi}\Big(\frac{1}{|x-y|} - \frac{R}{|x|}\frac{1}{|\tilde{x}-y|}\Big),$$
$$h(x,y) = -\frac{1}{4\pi}\Big(\frac{(x^*-y^*)(x_2-y_2)}{|x^*-y^*|^2|x-y|}$$
$$-\frac{R|y|(|y|^2x^*-R^2y^*)(|y|^2x_3-R^2y_3)}{||y|^2x^*-R^2y^*|^2||y|^2x-R^2y|}\Big),$$

then for $x\neq y$, the function $s(x,y) = g(x,y) + e_3 h(x,y)$ satisfies $\bar{\partial}s = 0$ with respect to x. Since

$$\frac{\partial}{\partial n} = \frac{\partial}{\partial y_1}\frac{y_1}{|y|} + \frac{\partial}{\partial y_2}\frac{y_2}{|y|} + \frac{\partial}{\partial y_3}\frac{y_3}{|y|},$$

by calculation, we obtain

$$-\frac{\partial}{\partial n}s(x,y)\Big|_{|y|=R} = \frac{1}{4\pi R}\Big\{\frac{R^2-|x|^2}{|x-y|^3} + e_3\Big[-\frac{2(x^*(x_3-y_3)+x_3(x^*-y^*))}{|x^*-y^*|^2|x-y|}$$

$$+ \frac{4(x^* - y^*)(x_3 - y_3)(|x^*|^2 - (x_1 y_1 + x_2 y_2))}{|x^* - y^*|^4 |x - y|}$$

$$+ \frac{(x^* - y^*)(x_3 - y_3)(|x|^2 - R^2)}{|x^* - y^*|^2 |x - y|^3} \Bigg]\Bigg\} . \tag{3.22}$$

Theorem 3.5   *Let $u(x) = u_1(x) + e_2 u_2(x)$ be a continuous function on the sphere $|x| = R$. Then there exists a continuous function $f(x)$ on $\bar{G}$ which satisfies the system of first order equations*

$$\bar{\partial} f = 0 \ \text{in} \ D = \{|x| < R\} \tag{3.23}$$

and the boundary condition

$$\operatorname{Re} f = u(x) \ \text{on} \ \partial D = \{|x| = R\}, \tag{3.24}$$

and $f(x)$ can be expressed as

$$f(x) = \frac{1}{4\pi R} \int_{\partial D} \Bigg\{ \frac{R^2 - |x|^2}{|x - y|^3} + e_3 \Bigg[-\frac{2(x^*(x_3 - y_3) + x_3(x^* - y^*))}{|x^* - y^*||x - y|}$$

$$+ \frac{4(x^* - y^*)(x_3 - y_3)(|x^*|^2 - (x_1 y_1 + x_2 y_2))}{|x^* - y^*|^4 |x - y|}$$

$$+ \frac{(x^* - y^*)(x_3 - y_3)(|x|^2 - R^2)}{|x^* - y^*|^2 |x - y|^3} \Bigg]\Bigg\} u(y) dS_y + e_3 c(x_1, y_2), \tag{3.25}$$

where $c(x_1, x_2) = c_1(x_1, x_2) + e_2 c_2(x_1, x_2)$ is an arbitrary function satisfying

$$\frac{\partial}{\partial x^*} c(x_1, x_2) = 0.$$

Proof   From the above discussion, we see that the function expressed by (3.25) is a solution of the Dirichlet boundary value problem (3.23) and (3.24). Conversely, if the Dirichlet problem (3.23) and (3.24) has a solution $f(x)$, we denote the integral on the right–hand side of (3.25) by $F(x)$, i.e.

$$F(x) = \frac{2}{4\pi R} \int_{\partial G} \Bigg\{ \frac{R^2 - |x|^2}{|x - y|^3} + e_3 \Bigg[-\frac{2x^*(x_3 - y_3) + x_3(x^* - y^*)}{|x^* - y^*||x - y|}$$

$$+ \frac{4(x^* - y^*)(x_3 - y_3)(|x^*|^2 - (x_1 y_1 x_2 y_2))}{|x^* - y^*|^4 |x - y|}$$

$$+ \frac{(x^* - y^*)(x_3 - y_3)(|x|^2 - R^2)}{|x^* - y^*|^2 |x - y|^3} \Bigg]\Bigg\} u(y) dS_y.$$

Then $\bar{\partial} F(x) = 0$ in D and $\operatorname{Re} F|_{|x|=R} = u(x)$. Hence $\Delta(\operatorname{Re} f - \operatorname{Re} F) = 0$ in D, and $(\operatorname{Re} f - \operatorname{Re} F)|_{|x|=R} = 0$. This implies that $\operatorname{Re} f = \operatorname{Re} F$ on $\bar{D}$. According to Lemma 3.2, we obtain

$$\frac{\partial}{\partial x^*}(\operatorname{Im} f - \operatorname{Im} F) = 0, \ \frac{\partial}{\partial x_3}(\operatorname{Im} f - \operatorname{Im} F) = 0.$$

Thus $\operatorname{Im} f - \operatorname{Im} F = c(x_1, x_2)$ and $\partial c(x_1, x_2)/\partial x^* = 0$. The theorem is proved (see [163]3)).

4 Riemann–Hilbert Problem for Generalized Regular Functions in Clifford Analysis

In this section, we introduce the generalized Cauchy–Riemann systems in a bounded space domain and give a criterion for the Fredholm property of the Riemann–Hilbert boundary value problem, and then according to this criterion, the question of existence and non–existence of the above boundary value problem is discussed (see [123]1),2)).

4.1 Generalized Cauchy–Riemann systems and representations of the Clifford algebra

For a system of hypercomplex numbers $e_0 = 1, e_1, \cdots, e_n$ with the following properties

$$e_j^2 = -1, e_i e_j + e_j e_i = 0, i \neq j, (e_i e_j)e_k = e_i(e_j e_k), i, j, k = 1, \cdots, n. \qquad (4.1)$$

every product of which can be transformed into one of the following expressions apart from the sign:

$$e_0 := 1, e_1, \cdots, e_n, e_1 e_2, e_1 e_3, \cdots, e_1 e_2 e_3, \cdots, e_1 e_2 \cdots e_n. \qquad (4.2)$$

They are all different from each other and linearly independent. Assuming the commutativity with respect to the multiplication by complex numbers, the basis elements (4.2) span a complex vector space of dimension 2^n. This vector space is called the Clifford algebra C_n generated by $e_0, e_1, e_2, \cdots, e_n$. The Clifford algebra C_n is a ring with respect to multiplication, the basis elements (4.2) together with their inverse elements from a multiplicative group.

Let G be a group and R_m a vector space. We are speaking of a representation of G, if every element $s \in G$ corresponds to a linear mapping in R_m, i.e. a matrix $D(s)$ after fixing a base in R_m, in such a way that the multiplication of the group matches the multiplication of the matrices

$$D(s \cdot t) = D(s) \cdot D(t). \qquad (4.3)$$

If we demand $D(1) = E$, the identity, then we have

$$D(s^{-1}) = D^{-1}(s) \qquad (4.4)$$

and all matrices are non–singular. The dimension m of the space of the representation R_m is called the degree of the representation. The representation is true, if the mapping $s \to D(s)$ is one–to–one. Two representations $A(s)$ and $B(s)$ of the same degree are called equivalent, if there is a non–singular matrix P with $A(s) = PB(s)P^{-1}$, for all $s \in G$. When performing a coordinate transformation P in R_m the system of matrices $D(s)$ becomes $PD(s)P^{-1}$, which is a representation too because of (4.3). That means an equivalent system of matrices is nothing else than the same system of linear transformations expressed in another coordinate system. The system $D(s)$ is said to be reducible, if there is an invariant subspace of R_m, i.e. every vector of this subspace is mapped onto a vector of the same subspace by every linear transformation $D(s)$, $s \in G$. Otherwise, $D(s)$ is called irreducible.

Choosing the coordinate system suitably a reducible system takes the form

$$D(s) = \begin{pmatrix} D_1(s) & K(s) \\ 0 & D_2(s) \end{pmatrix}.$$

If R_m is the direct sum of two invariant subspaces, then assuming an appropriate basis, we have

$$D(s) = \begin{pmatrix} D_1(s) & 0 \\ 0 & D_2(s) \end{pmatrix} = D_1(s) + D_2(s).$$

where $D_1(s)$ and $D_2(s)$ describe transformations in the subspaces. We say that the system $D(s)$ is split up. A system of representation matrices is called completely reducible, if it is irreducible or it splits up into several irreducible systems.

By defining
$$D(\alpha \cdot s + \beta \cdot t) = \alpha D(s) + \beta D(t), \ \alpha, \beta \in \mathbb{C},$$

we get from a representation of this group a representation of the algebra too. Because of (4.3), it is sufficient to specify only the representations of the generating system to obtain from them the representation of the algebra.

When representing the algebra C_n, we have to distinguish two cases, whether n is even or odd.

Case 1. $n = 2k$. We define the following matrices

$$\rho = \begin{pmatrix} 0 & 1 \\ 1 & 0 \end{pmatrix}, \sigma = \begin{pmatrix} 0 & i \\ -i & 0 \end{pmatrix}, \tau = i\rho\sigma = \begin{pmatrix} 1 & 0 \\ 0 & -1 \end{pmatrix}, \varepsilon = \begin{pmatrix} 1 & 0 \\ 0 & 1 \end{pmatrix}.$$

The matrices ρ, σ and τ are the so–called Pauli matrices. For $j = 1, \cdots, k$, we put

$$\rho_j = i \cdot \tau \times \cdots \times \tau \times \rho \times \varepsilon \times \cdots \times \varepsilon, \sigma_j = i \cdot \tau \times \cdots \times \tau \times \sigma \times \varepsilon \times \cdots \times \varepsilon, \quad (4.5)$$

where the factors ρ and σ respectively occur at the jth position and the Kronecker product contains k factors. One can prove that the $2k$ matrices (4.5) have the properties (4.1). Therefore

$$\sigma_{2j-1} \to \rho_j, \ e_{2j} \to \sigma_j \quad (4.6)$$

is a representation of C_{2k} of degree 2^k. Thus we have

Theorem 4.1 *Every representation of the Clifford algebra C_{2k} is completely reducible. C_{2k} has the only irreducible representation (4.6) and is isomorphic to the complete matrix ring M_2^k. Every representation is true and its degree a multiple of 2^k.*

Case 2. $n = 2k + 1$. Here we have with

$$\tau_0 = i \cdot \tau \times \tau \times \cdots \times \tau (k \text{ factors}),$$

the two representations

$$e_{2j-1} \to \rho_j,\ e_{2j} \to \sigma_j,\ e_n \to \tau_0,\ j = 1, \cdots, k, \tag{4.7}$$

and

$$e_{2j-1} \to -\rho_j,\ e_{2j} \to -\sigma_j,\ e_n \to -\tau_0. \tag{4.8}$$

Theorem 4.2 *Every representation of the Clifford algebra C_{2k+1} is completely reducible. There are two irreducible representation (4.7) and (4.8) of degree 2^k, and C_{2k+1} is isomorphic to the direct sum of two matrix rings M_{2^k}. Any representation is true, if and only if (4.7) and (4.8) are contained in it at least once. Every degree of a representation is a multiple of 2^k.*

We shall show how the matrices ρ_j, σ_j and τ_0 may be achieved recursively. Let ρ_j', σ_j' and τ_0' be the representation matrices of degree 2^{k-1}, then we have for the matrices ρ_j, σ_j and τ_0 of degree 2^k :

$$\rho_1 = \rho_1' \times \varepsilon = \begin{pmatrix} 0 & iE \\ iE & 0 \end{pmatrix},\ \sigma_1 = \sigma' \times \varepsilon = \begin{pmatrix} 0 & -E \\ E & 0 \end{pmatrix},$$

$$\tau_0 = \tau \times \tau_0' = \begin{pmatrix} \tau_0' & 0 \\ 0 & -\tau_0' \end{pmatrix},\ \rho_{j+1} = \tau \times \rho_j' = \begin{pmatrix} \rho_j' & 0 \\ 0 & -\rho_j' \end{pmatrix}, \tag{4.9}$$

$$\sigma_{j+1} = \tau \times \sigma_j' = \begin{pmatrix} \sigma_j' & 0 \\ 0 & -\sigma_j' \end{pmatrix},\ j = 1, \cdots, k - 1.$$

Secondly, we introduce the system of partial differential equations of the form

$$E\frac{\partial w}{\partial x_0} + A_1 \frac{\partial w}{\partial x_1} + \cdots + A_n \frac{\partial w}{\partial x_n} = 0, \tag{4.10}$$

where w is a vector–function with values in $\mathbb{C}^m$, $A_j\ (j = 1, \cdots, n)$ are $(m \times m)$ complex constant matrices, $A_0 = E$ is the identity matrix, and $A_j\ (j = 1, \cdots, n)$ satisfy the relations

$$A_j^2 = -E,\ A_i A_j + A_j A_i = 0,\ i \neq j,\ i, j = 1, \cdots, n. \tag{4.11}$$

The system (4.10) is called a generalized Cauchy–Riemann system. Let w be a solution of system (4.10) from the Sobolev space $W_2^1(D)$, where D is a bounded domain in $\mathbb{R}^{n+1}$. For a twice continuously differentiable solution w we have

$$(E\frac{\partial}{\partial x_0} - \sum_{j=1}^{n} A_j \frac{\partial}{\partial x_j})(E\frac{\partial}{\partial x_0} + \sum_{j=1}^{n} A_j \frac{\partial}{\partial x_j})w(x) = 0.$$

A formal calculation using (4.11) shows that the second–order differential operator is just the Laplacian and so $w(x) \in C^2(D)$ is proved to be harmonic. For functions $w(x) \in W_2^1(D)$, the Weyl lemma can be used to complete the proof. We omit this for reasons of space.

In particular, if A_j is identified with $e_j, j = 1, \cdots, n$, then the generalized Cauchy–Riemann operator is defined as

$$\partial = e_0 \frac{\partial}{\partial x_0} + e_1 \frac{\partial}{\partial x_1} + \cdots + e_n \frac{\partial}{\partial x_n}. \tag{4.12}$$

A hypercomplex function gets the form

$$w(x) = w_0(x)e_0 + w_1(x)e_1 + \cdots + w_n(x)e_n + w_{12}(x)e_1e_2$$

$$+ \cdots + w_{n-1,n}(x)e_{n-1}e_n + \cdots + w_{12\cdots n}(x)e_1e_2 \cdots e_n = \sum_A w_A(x)e_A, \tag{4.13}$$

$$e_A = e_{j_1} \cdots e_{j_k}, \quad A = \{j_1, \cdots, j_k\} \subset \{1, \cdots, n\}, \quad e_\phi = e_0 = 1 \text{ for } A = \phi,$$

with complex functions $w_l(x)$, and the generalized Cauchy–Riemann system is stated as

$$\partial w = 0. \tag{4.14}$$

We can multiply out this hypercomplex equation and write a single equation for every component of the 2^n–dimensional vector space. So we obtain a complex system with coefficient matrices A_j of dimension 2^n. The coefficient matrices $A_j\,(j = 1, \cdots, n)$ form the so–called regular representation of the generating system of the Clifford algebra C_n.

4.2   A criterion for the Fredholm property of the Riemann–Hilbert problem

We consider the generalized Cauchy–Riemann system in the form

$$E\frac{\partial w}{\partial x_0} + A_1 \frac{\partial w}{\partial x_1} + \cdots + A_n \frac{\partial w}{\partial x_n} + Aw = f(x) \tag{4.15}$$

in a bounded domain $D \subset \mathbb{R}^{n+1}$ of class $\mathbb{C}^k\,(k \geq 1)$, $A_j\,(j = 0, 1, \cdots, n)$ are $(m \times m)$-complex constant coefficient matrices satisfying condition (4.11), Aw is a linear lower order term with $A \in C^{k-1}(\bar{D})$, $f(x) \in W_2^{l-1}(D), 1 \leq l \leq k, l$ is an integer.

Problem R-H   The so–called Riemann–Hilbert problem is to find a solution $w(x) \in W_2^l(D)(1 \le l \le k)$ of (4.15) satisfying the boundary condition

$$(B_1 \cdot B_2) \cdot w = g(x) \text{ on } \partial D, \tag{4.16}$$

where B_1 and B_2 are complex matrices of dimension $m/2$ depending on the boundary point, the rows of the $(m/2 \times m)$–matrix (B_1, B_2) are orthonormal, $B_1, B_2 \in C^k(\partial D)$ (and so extendable to functions of $C^k(\bar{D})$), and $g(x) \in W_2^{l-1/2}(\partial D)$ is a vector function with values in $\mathbb{C}^{m/2}$.

Lemma 4.3   *For the $(m \times m)$–matrices $A_j, j = 1, \cdots, n$, let the condition (4.11) hold. Then the matrix*

$$i \sum_{j=1}^{n} \xi_j A_j \text{ with } \xi = (\xi_1, \cdots, \xi_n) \in \mathbb{R}^n \backslash \{0\}$$

has got exactly the eigenvalues $\lambda = \pm|\xi|$, and all eigenvectors of the eigenvalues $\pm|\xi|$ have the form

$$\nu = (\pm|\xi| + i \sum_{j=1}^{n} \xi_j A_j) \cdot c \text{ with } c \in \mathbb{C}^m.$$

If c runs through all vector in $\mathbb{C}^m$, then ν is either an eigenvector or it equals the zero vector. For every eigenvalue there are exactly $m/2$ linearly independent eigenvectors.

Proof   Let

$$\det(i \sum_{j=1}^{n} \xi_j A_j - \lambda E) = 0,$$

then we have to show $\lambda = \pm|\xi|$. Indeed, it is

$$(i \sum_{j=1}^{n} \xi_j A_j + \lambda E) \cdot (i \sum_{j=1}^{n} \xi_j A_j - \lambda E) = (|\xi|^2 - \lambda^2)E$$

and the preposition is obvious.

We demonstrate the matrices $(i \sum \xi_j A_j \pm |\xi| E)$ to have rank $\ge m/2$. By an equivalence transformation, the matrices $A_j \, (j = 1, \cdots, n)$ take the form, i.e. the matrices $(i \sum \xi_j A_j \pm |\xi| E)$ are equivalent to matrices with block structure, where every block matrix is of the form

$$\begin{pmatrix} \pm|\xi|E + i \sum_{j=3}^{n} \xi_j A_{j-2} & -(\xi_1 + i\xi_2)E \\ (-\xi_1 + i\xi_2)E & \pm|\xi|E - i \sum_{j=3}^{n} \xi_j A_{j-2} \end{pmatrix}$$

and the dimension $2^k (m = 2^k l, \; n = 2k$ or $2k + 1)$. Obviously the 2^{k-1}–dimensional submatrices $(-\xi_1 \pm i\xi_2)E$ are of maximal rank, if $\xi_1^2 + \xi_2^2 \ne 0$. Otherwise, i.e. if $\xi_1 = \xi_2 = 0$, we have to demonstrate that the other two submatrices both have a rank $\geq 2^{k-2}$. We restrict ourselves to the left upper submatrix, because the other is constructed quite analogously. Again using the recusive representation (4.9) gives us

$$\pm|\xi|E + i\sum_{j=3}^{n}\xi_j A'_{j-2} = \begin{pmatrix} \pm|\xi|E + i\sum_{j=5}^{n}\xi_j A''_{j-4} & -(\xi_3 + i\xi_4)E \\ (-\xi_3 + i\xi_4)E & \pm|\xi|E - i\sum_{j=5}^{n}\xi_j A''_{j-4} \end{pmatrix}.$$

If $\xi_3^2 + \xi_4^2 \neq 0$, then we see that the rank of this matrix is $\geq \dim/2 = 2^{k-2}$. In the case of $\xi_3 = \xi_4 = 0$, we split up the submatrices recusively again and prove that they have a rank$\geq \dim/2$.

This procedure either leads us to a pair of numbers (ξ_{2j-1}, ξ_{2j}) with $\xi_{2j-1}^2 + \xi_{2j}^2 \neq 0$, where the proposition is obvious, or finally we obtain the (2×2)–matrices

$$\begin{pmatrix} \pm|\xi| & -\xi_{n-1} - i\xi_n \\ -\xi_{n-1} + i\xi_n & \pm|\xi| \end{pmatrix} \text{ for } n \text{ even,}$$

$$\begin{pmatrix} \pm|\xi| - \xi_n & -\xi_{n-2} - i\xi_{n-1} \\ -\xi_{n-2} + i\xi_{n-1} & \pm|\xi| + \xi_n \end{pmatrix} \text{ for } n \text{ odd,}$$

whose rank equals 1.

By this recursive procedure, we have shown that the block matrices of dimension 2^k have a rank$\geq 2^{k-1}$. So the whole matrix has got a rank $\geq 2^{k-1}\, l = m/2$. Because of

$$(i\sum\xi_j A_j \mp |\xi|E) \cdot (i\sum\xi_j A_j \pm |\xi|E) \cdot c = 0, \text{ for all } c \in \mathbb{C}^m,$$

the vectors $(i\sum\xi_j A_j \pm |\xi|E)\cdot c \neq 0$ are eigenvectors of the eigenvalues $\lambda = \pm|\xi|$. So the rank of the matrices $(i\sum\xi_j A_j \pm |\xi|E)$ must be equal to $m/2$, because the number of linearly independent eigenvectors can not exceed m. Furthermore, there can not exist any other eigenvectors besides the ones given.

Theorem 4.4 *Let $D \in \mathbb{C}^k(k \geq 1)$ be a bounded domain in $\mathbb{R}^{n+1}(n \geq 2)$ and the $(m \times m)$–matrices $A_j(j = 1, \cdots, n)$ in (4.11) are unitary, i.e. $A_j^* = -A_j, j = 1, \cdots, n$. Then Problem $R - H$ for (4.15) has the Fredholm property if and only if the following relation is valid for all $x \in \partial D$*

$$\det[(B_1\ B_2) \cdot (\sum_{j=0}^{n}\alpha_j A_j^*) \cdot (\sum_{j=0}^{n}t_j A_j) \cdot \begin{pmatrix} B_1^* \\ B_2^* \end{pmatrix} - iE] \neq 0, \tag{4.17}$$

where $\alpha = (\alpha_0, \cdots, \alpha_n)$ denotes the vector of the inner normal on $x \in \partial D$ and $t = (t_0, \cdots, t_n)$ runs through all tangential unit vectors belonging to this point.

Proof. The Fredholm property is equivalent to the ellipticity of Problem R-H (see [159]). So there remains the proof that (4.17) is equivalent to the condition of Lopatinski–Shapiro.

Let $x^0 \in \partial D$ be any boundary point. Choose a local coordinate system $X_0, X_1, \cdots, X_n$ in the following way: (1) The origin lies in x^0. (2) The X_0–axis coincides with the direction of the inner normal. (3) The $X_1, X_2, \cdots, X_n$–axes lie in the tangential plane. (4) The local $X_0, X_1, \cdots, X_n$–system is obtained from the global $x_0, x_1, \cdots, x_n$–system be a translation and a rotation, i.e.

$$x_i = \sum_{j=0}^{n} d_{ij} X_j + x_i^0, \quad X_j = \sum_{j=0}^{n} d_{ij}(x_i - x_i^0)$$

with a proper orthogonal matrix $(d_{ij})_{i,j=0}^n$. Therefore we have

$$\frac{\partial}{\partial x_i} = \sum_{j=0}^{n} d_{ij} \frac{\partial}{\partial X_j} \text{ and } \sum_{j=0}^{n} A_j \frac{\partial w}{\partial x_i} = \sum_{i,j} A_i d_{ij} \frac{\partial w}{\partial X_j} = \sum_{j=0}^{n} (\sum_{j=0}^{n} d_{ij} A_i) \frac{\partial w}{\partial X_j}.$$

Because of the rotational invariance of the system of differential equations, we obtain a new generalized Cauchy–Riemann system

$$\sum_{j=0}^{n} \tilde{A}_j \frac{\partial w}{\partial X_j} + (\sum_i d_{i0} A_j^*) A w = (\sum_i d_{i0} A_i^*) f,$$

where $\tilde{A}_j = (\sum_i d_{i0} A_i^*)(\sum_i d_{ij} A_i)$ are unitary namely $\tilde{A}_j^* = -\tilde{A}_j, j = 1, \cdots, n$.

The main part of the differential operator is $\sum_{j=0}^{n} \tilde{A}_j \partial w / \partial X_j$. We apply to it the Fourier transformation with respect to the tangential coordinates $X_1, \cdots, X_n$. Putting $X_0 = t$, we have the corresponding homogeneous system of ordinary differential equations for $t \geq 0$:

$$\frac{d\nu(t)}{dt} + i \sum_{j=1}^{n} \xi_j \tilde{A}_j \nu(t) = 0.$$

Freezing the coefficients of the boundary condition with their values in the origin of the local coordinate system, the homogeneous boundary condition is transferred by the above Fourier transformation into the homogeneous initial value condition

$$(B_1 \; B_2)\nu(0) = 0, \; B_i = B_i(0), \; i = 1, 2.$$

The condition of Lopatinski–Shapiro at the boundary point $x^0 \in \partial D$ is formulated as follows: The initial value problem has only the trivial solution in the space of stable solutions for all $\xi = (\xi_1, \cdots, \xi_n) \neq 0$.

Making the ansatz $\nu(t) = e^{\lambda t} c$, we have

$$(\lambda E + i \sum_{j=1}^{n} \xi_j \tilde{A}_j) \cdot c = 0.$$

Because of Lemma 4.3, there are exactly the two eigenvalues $\lambda = \pm|\xi|$. Only for $\lambda = -|\xi|$ do we get stable solutions, the corresponding $m/2$ linearly independent eigenvectors are of the form

$$c = (|\xi|E + i\sum_{j=1}^{n} \xi_j \tilde{A}_j) \cdot c', \ c' \in \mathbb{C}^m.$$

So the space of stable solutions of the system of ordinary differential equations is constituted by all vector–function

$$\nu(t) = e^{-|\xi|t}(|\xi|E + i\sum_{j=1}^{n} \xi_j \tilde{A}_j) \cdot c', \ c' \in \mathbb{C}^m.$$

The matrix $i\sum \xi_j \tilde{A}_j$ is Hermitean, because the $\tilde{A}_j(j = 1, \cdots, n)$ are unitary. Consequently we can perform a main diagonal transformation

$$i\sum_{j=1}^{n} \xi_j \tilde{A}_j = |\xi|P \cdot \left(\begin{array}{cc} E & 0 \\ 0 & E \end{array} \right) \cdot P^*, \ P = P(\xi, (d_{ij}))$$

with a unitary matrix P. We obtain

$$\nu(t) = 2|\xi|e^{-|\xi|t}P \cdot \left(\begin{array}{cc} E & 0 \\ 0 & E \end{array} \right) \cdot P^* \cdot c',$$

and by substituting

$$P = \left(\begin{array}{cc} P_1 & P_2 \\ P_3 & P_4 \end{array} \right), \ \dim P_j = m/2 \text{ and } P^* \cdot c' = \left(\begin{array}{c} x \\ y \end{array} \right), \ x, y \in \mathbb{C}^{m/2},$$

we have

$$\nu(t) = 2|\xi|e^{-|\xi|t} \left(\begin{array}{c} P_1 \\ P_2 \end{array} \right) x, \ x \in \mathbb{C}^{m/2} \text{ arbitrary},$$

for the stable solutions. The initial value condition is stated as

$$(B_1 \ B_2)\nu(0) \equiv 2|\xi|(B_1 \ B_2) \left(\begin{array}{c} P_1 \\ P_2 \end{array} \right) x = 0.$$

Obviously there is $\nu(t) \equiv 0$, if and only if $x = 0$. So demanding that the initial value problem has got only the trivial solution is equivalent to the condition

$$\det \left[(B_1 \ B_2) \left(\begin{array}{c} P_1 \\ P_2 \end{array} \right) \right] \neq 0. \tag{4.18}$$

From short we write

$$(Q_1 \ Q_2) = (B_1 \ B_2)P, \ \dim Q_j = m/2,$$

and then (4.18) means $\det Q_1 \neq 0$. This is equivalent to $\det Q_1 Q_1^* \neq 0$. Using the equation

$$Q_1 Q_1^* + Q_2 Q_2^* = E,$$

we have

$$2Q_1 Q_1^* = (Q_1 \ Q_2) \begin{pmatrix} E & 0 \\ 0 & -E \end{pmatrix} \begin{pmatrix} Q_1^* \\ Q_2^* \end{pmatrix} + E$$

$$= (B_1 \ B_2) \, P \begin{pmatrix} E & 0 \\ 0 & -E \end{pmatrix} P^* \begin{pmatrix} B_1^* \\ B_2^* \end{pmatrix} + E$$

$$= (B_1 \ B_2) \frac{i}{|\xi|} \sum_{j=1}^{n} \xi_j \tilde{A}_j \begin{pmatrix} B_1^* \\ B_2^* \end{pmatrix} + E.$$

Moreover,

$$\sum_{j=1}^{n} \frac{\xi_j}{|\xi|} \tilde{A}_j = (\sum_{i=0}^{n} d_{i0} A_i^*) \cdot (\sum_{i,j} \frac{\xi_j}{|\xi|} d_{ij} A_i).$$

The vectors

$$(d_{i0})_{i=0}^{n} \text{ and } (\sum_{j=1}^{n} \frac{\xi_j}{|\xi|} d_{ij})_{i=0}^{n}$$

describe the inner normal $(\alpha_i)_{i=0}^{n}$ and the tangential vectors $(t_i)_{i=0}^{n}$ of unit length respectively. Hence, we write

$$\sum_{j=1}^{n} \frac{\xi_j}{|\xi|} \tilde{A}_j = (\sum_{j=1}^{n} \alpha_j A_j^*) \cdot (\sum_{j=0}^{n} t_j A_j)$$

and obtain the relation equivalent to (4.18):

$$\det [(B_1 \ B_2) \cdot (\sum_{j=0}^{n} \alpha_j A_j^*) \cdot (\sum_{j=0}^{n} t_j A_j) \cdot \begin{pmatrix} B_1^* \\ B_2^* \end{pmatrix} - iE] \neq 0,$$

which was to be shown.

A. Avantaggiati [5] investigated the Riemann–Hilbert problem for real elliptic systems of partial differential equations of first order with quadratic coefficient matrices. The problem was transferred to singular integral equations and a condition for the Fredholm property was given. The condition stated is Theorem 4.4 for generalized Cauchy–Riemann systems exceeds this result and is practically more usable, because the criterion of Avantaggiati contains an unknown matrix, which must be suitably chosen, while nothing is known about its existence. The integral equations considered are not equivalent to the boundary value problem, only propositions concerning the Fredholm property were transferred.

Futhermore, according to Theorem 4.4, I. Stern [123]2) investigates the question about existence and non–existence of the Riemann–Hilbert boundary value problem, the main result is as follows.

Theorem 4.5 (1) *There does not exist any Fredholm Problem $R-H$ for* (4.15), *if the space dimension is even, i.e. $n = 2k + 1$ odd, and the representation $A_1, \cdots, A_n$ of the Clifford algebra C_n contains the irreducible components in an odd number (i.e. degree $m = 2^k l$, l odd).*

(2) *Problem $R-H$ for* (4.15) *is Fredholm, if*

1) *n is odd and both irreducible representations appear in the same number, or*

2) *n is even and so is the number of irreducible components.*

In the above theorem, we assume the function spaces and smoothness properties to be as in Theorem 4.4 and $G \in \mathbb{C}^{k+1}$.

5 Oblique Derivative Problems for Generalized Regular Functions in Clifford Analysis

First of all, we consider the case of the Clifford algebra A_3 over the space $\mathbb{R}^3$. Let Ω be a ball, i.e. $\Omega = \{\sum_{j=1}^3 |x_j|^2 < R^2(< \infty)\}$, and for convenience, $e_2 e_3$ be denoted by e_4. Any point x in Ω may be written as $x = \sum_{j=1}^3 x_j e_j$, and any function $f(x)$ in Ω with values in the Clifford algebra may be denoted by $w(x) = \sum_{j=1}^4 w_j(x) e_j$.

5.1 Oblique derivative problems for generalized regular functions

A generalized regular functions $w(x)$ in Ω is defined as a solution $w(x) = \sum_{j=1}^4 w_j(x) e_j$ ($\in C^2(\Omega)$) for the elliptic system of first order equations in the form

$$\bar{\partial} w = aw + b\bar{w} + C \text{ in } \Omega, \tag{5.1}$$

where

$$a(x) = \sum_{j=1}^4 a_j(x) e_j, \ b(x) = \sum_{j=1}^4 b_j(x) e_j, \ C(x) = \sum_{j=1}^4 C_j(x) e_j \in C_\alpha^1(\Omega), 0 < \alpha < 1.$$

Problem P The oblique derivative problem for system (5.1) is to find a solution $w(x) \in C_\alpha^1(\bar{\Omega}) \cap C^2(\Omega)$ of (5.1) satisfying the boundary condition

$$\frac{\partial w_j}{\partial \nu_j} + \sigma_j(x) w_j(x) = \tau_j(x) + h_j, \ x \in \partial\Omega, w_j(R) = u_j, j = 1, 2, \tag{5.2}$$

where $\sigma_j(x), \tau_j(x) \in C^1_\alpha(\partial\Omega)$, $\sigma_j(x) \geq 0$ on $\partial\Omega, j = 1, 2$, $h_j(j = 1, 2)$ are unknown real constants to be determined appropriately, $u_j(j = 1, 2)$ are real constants, and $\nu_j(j = 1, 2)$ are vectors at the point $x \in \partial\Omega$, $\cos(\nu_j, n) \geq 0, j = 1, 2$, n is the outward normal at $x \in \partial\Omega$, and $\cos(\nu_j, n) \in C^1_\alpha(\partial\Omega)$.

If $w(x)$ is a solution of Problem P for system (5.1), then we can verify that $[w_1(x), w_2(x)]$ is a solution of the system of second order

$$\begin{cases} \Delta w_1 = \sum_{j=1}^{3}(A_j + B_j)w_{1x_j} + B_5 w_1 + B_6, \\[2mm] \Delta w_2 = 2\sum_{j=1}^{3} B_j w_{2x_j} + B_7 w_2 + (A_2 w_1)_{x_1} - (A_1 w_1)_{x_2} \\[2mm] \quad -(A_4 w_1)_{x_3} - B_2 w_{1x_1} + B_1 w_{1x_2} + B_4 w_{1x_3} + B_8 w_1 + B_9 \end{cases} \qquad (5.3)$$

satisfying the boundary condition (5.2), where $A_j = a_j + b_j$, $B_j = a_j - b_j$, $j = 1, \cdots, 4$, B_j is only a function of x_j, $j = 1, 2, 3$, B_4 is a real constant, and

$$B_5 = -\sum_{j=1}^{4} A_j B_j + \sum_{j=1}^{3} A_{jx_j}, \; B_6 = -\sum_{j=1}^{4} B_j C_j + \sum_{j=1}^{3} C_{jx_j},$$

$$B_7 = \sum_{j=1}^{3} B_{jx_j} - \sum_{j=1}^{4} B_j^2, \; B_8 = A_1 B_2 - A_2 B_1 - A_3 B_4 + A_4 B_3,$$

$$B_9 = -C_{1x_2} + C_{2x_1} - C_{4x_3} - B_1 C_2 + B_2 C_1 + B_3 C_4 - B_4 C_3.$$

In fact, it follows from (5.1) that

$$\begin{cases} w_{1x_1} - w_{2x_2} - w_{3x_3} = A_1 w_1 - B_2 w_2 - B_3 w_3 - B_4 w_4 + C_1, \\[2mm] w_{2x_1} + w_{1x_2} + w_{4x_3} = A_2 w_1 + B_1 w_2 - B_4 w_3 + B_3 w_4 + C_2, \\[2mm] w_{3x_1} - w_{4x_2} + w_{1x_3} = A_3 w_1 + B_4 w_2 + B_1 w_3 - B_2 w_4 + C_3, \\[2mm] w_{4x_1} + w_{3x_2} - w_{2x_3} = A_4 w_1 - B_3 w_2 + B_2 w_3 + B_1 w_4 + C_4. \end{cases} \qquad (5.4)$$

When $B_j = B_j(x_j), j = 1, 2, 3$, and B_4 is a real constant, from (5.4) we can derive the first equation in (5.3), and then the second equation in (5.3) can be obtained.

Conversely, if the following conditions hold:

$$B_5 \geq 0, \; B_7 \geq 0 \text{ on } \bar{\Omega}, \qquad (5.5)$$

then according to the result in [50], the boundary value problem (5.3),(5.2) has a solution $[w_1(x), w_2(x)]$. Afterwards, if

$$B_3 = B_4 = 0 \text{ on } \bar{\Omega}, \qquad (5.6)$$

then we can find $w_3(x), w_4(x)$ by the following integrals

$$\begin{cases} w_3(x) = \int_0^{x_3}[w_{1x_1} - w_{2x_2} - A_1 w_1 + B_2 w_2 - C_1]dx_3 + \phi_3(x_1, x_2), \\[2mm] w_4(x) = \int_0^{x_3}[-w_{2x_1} - w_{1x_2} + A_2 w_1 + B_1 w_2 + C_2]dx_3 + \phi_4(x_1, x_2), \end{cases} \qquad (5.7)$$

where $\phi_3(x_1, x_2), \phi_4(x_1, x_2)$ satisfy the conditions

$$\begin{cases} \phi_{3x_1} - \phi_{4x_2} - B_1\phi_3 + B_2\phi_4 - C_3 = \psi_3(x_1, x_2), \\[2mm] \phi_{3x_2} + \phi_{4x_1} - B_2\phi_3 - B_1\phi_4 - C_4 = \psi_4(x_1, x_2), \end{cases} \tag{5.8}$$

in which

$$\psi_3(x_1, x_2) = [-w_{1x_3} + A_3 w_1]|_{x_3=0}, \ \psi_4(x_1, x_2) = [w_{2x_3} + A_4 w_1]|_{x_3=0},$$

and A_j, B_j, C_j satisfy some conditions such that the integrals in (5.7) are single–valued. Suppose that $\phi = \phi_3 + \phi_4 e_2 = \phi_3 + \phi_4 i$ satisfies the Riemann–Hilbert boundary conditions:

$$\begin{cases} \operatorname{Re}[\overline{\lambda(t)}\phi(t)] = r(t) + h(t), \ t = t_1 + it_2 \in \Gamma = \partial\Omega \cap \{x_3 = 0\}, \\[2mm] \operatorname{Im}[\overline{\lambda(d_j)}\phi(d_j)] = g_j, \ j = 1, \cdots, 2K + 1 \text{ for } K \geq 0, \end{cases} \tag{5.9}$$

where $|\lambda(t)| = 1$, $\lambda(t), r(t) \in C_\alpha^2(\Gamma)$, $d_j \ (j = 1, \cdots, 2K + 1 \text{ for } K \geq 0)$ are distinct points on Γ, $g_j \ (j = 1, \cdots, 2K + 1 \text{ for } K \geq 0)$ are known real constants, and

$$h(t) = \begin{cases} 0 \text{ on } \Gamma \text{ for } K = \frac{1}{2\pi}\Delta_\Gamma \arg\lambda(t) \geq 0, \\[2mm] h_0 + \operatorname{Re}\sum_{m=1}^{-K-1}(h_m^+ + ih_m^-)t^m \text{ on } \Gamma \text{ for } K < 0, \end{cases}$$

herein $h_0, h_m^\pm (m = 1, \cdots, -K - 1)$ are undetermined real constants. According to Theorem 4.1 and Theorem 4.6 of Chapter 2 in [140], the functions $\phi_3(x_3, x_4), \phi_4(x_1, x_2)$ may be uniquely determined, and then $w_3(x), w_4(x)$ may be also uniquely determined. Then we obtain

Theorem 5.1   *If the coefficients of the system* (5.1) *satisfy the conditions* (5.5), (5.6), *and* $W(t) = w_3(t) + iw_4(t)$ *satisfies the boundary conditions*

$$\operatorname{Re}[\overline{\lambda(t)}W(t)] = r(t) + h(t), \ t \in \Gamma = \partial\Omega \cap \{x_3 = 0\}, \tag{5.10}$$

$$\operatorname{Im}[\overline{\lambda(d_j)}W(d_j)] = g_j, \ j = 1, \cdots, 2K + 1 \text{ for } K \geq 0, \tag{5.11}$$

where $\lambda(t), r(t), h(t), d_j, g_j$ *and* K *are as stated in* (5.9), *then Problem P has a unique solution* $w(x) = \sum_{j=1}^{4} w_j(x)e_j \in C_\alpha^1(\bar{\Omega}) \cap C^2(\Omega), \ 0 < \alpha < 1$ (see [139]15)).

Next, we consider the case of arbitrary dimension n. Let $A_n(\mathbb{R})$ be a real Clifford algebra and Ω be a polycylinder $D_1 \times \cdots \times D_n$ in the space $\mathbb{R}^n$, $D_j = \{|x_j| \leq R_j\}, 0 < R_j < \infty, j = 1, \cdots, n$. We only give a simple introduction (see [70]).

Problem P'   The oblique derivative problem for generalized regular functions in $\Omega \subset \mathbb{R}^n$ is defined to be the problem of finding a generalized regular function

$$w(x) = \sum_{j=1}^{2^{n-1}} w_j(x)e_j = \sum_A w_A(x)e_A \text{ in } \Omega,$$

$$A = \{j_1, \cdots, j_k\} \subset \{1, \cdots, n\}, \ e_A = e_{j_1} \cdots e_{j_k}, \ 1 \leq j_1 < \cdots < j_k \leq n,$$

satisfying the boundary condition

$$\frac{\partial w_j}{\partial \nu_j} + \sigma_j(x)w_j(x) = \tau_j + h_j,\ x \in \partial\Omega,\ w_j(R) = u_j, j = 1, \cdots, 2^{n-2}. \tag{5.12}$$

Here $j = A$ if $j \leq n$ and if A includes at least two integers greater than 1 then $j(n < j \leq 2^{n-1})$ denotes one of integers system A such that j and A possess a one to one relation, ν_j is a vector at the point $x \in \partial\Omega$, $\cos(\nu_j, n) \geq 0$, $\sigma_j(x) \geq 0$, $x \in \partial\Omega$, $h_j\,(j = 1, \cdots, 2^{n-1})$ are undetermined constants.

Theorem 5.2 *A function*

$$w(x) = \sum_A w_A(x)e_A \text{ in } \Omega$$

is a generalized regular function if and only if the $w_A(x)$ satisfy the following real elliptic system of first order equations

$$\sum_{k=1}^{n} \delta_{\overline{kB}} w_{Bx_k} = \sum_{d=1}^{n}(a_d + b_d)w_M \delta_{\overline{dM}} + \sum_{d=1}^{n}(a_d - b_d)w_N \delta_{\overline{dN}} + c_A, \tag{5.13}$$

where $\overline{kB} = A$, $\overline{dM} = A$, $\overline{dN} = A$ and $\delta_{\overline{bB}}$ etc. are proper signs(see [70]).

Proof Since $w(x)$ is a generalized regular function, it is clear that $w(x)$ satisfies the elliptic system of first order equations

$$\bar{\partial}w = aw + b\bar{w} + c. \tag{5.14}$$

By the formula in Section 4, we have

$$\bar{\partial}w = \sum_{A}\sum_{k=1}^{n} \delta_{\overline{kB}} w_{Bx_k} e_A = \sum_{k,A} \delta_{\overline{kB}} w_{Bx_k} e_A.$$

Thus

$$aw + b\bar{w} + c = \sum_{d,A}(a_d + b_d)w_M \delta_{\overline{dM}} e_A +$$

$$+ \sum_{d,A}(a_d - b_d)w_N \delta_{\overline{dN}} e_A + \sum_A c_A e_A,$$

where $\overline{kB} = A$, $\overline{dM} = A$ and $\overline{dN} = A$, it shows that $w_A(x)$ satisfies (5.13).

Under certain conditions, by introducing various quasipermutations, from (5.13) we can obtain

$$\Delta w_k = \sum_{m=1}^{n} d_{mk} w_{kx_m} + f_k w_k + g_k,\ k = 1, \cdots, 2^{n-2}. \tag{5.15}$$

On the basis of the result in [112], a solution $[w_1(x), \cdots, w_{2^{n-2}}(x)]$ of (5.15) satisfying the boundary condition (5.12) can be found. Moreover, we rewrite (5.13) in the form

$$W_{k\overline{x_k}} = F_{kl}(z_1, \cdots, z_m, W_{2^{n-3}+1}, \cdots, W_{2^{n-2}}),$$
$$k = 2^{n-3} + 1, \cdots, 2^{n-2}, \; l = 1, \cdots, m, \tag{5.16}$$

where

$$n = 2m, \; x_{2k-1} + ix_{2k} = z_k, \; w_{2k-1} + iw_{2k} = W_k, \; k = 1, \cdots, 2^{n-2}.$$

If Ω is a polycylinder, under some conditions, we can obtain $[W_{2^{n-3}+1}(x), \cdots, W_{2^{n-2}}(x)]$ satisfying

$$\text{Re}[\overline{z_1}^{K_1} \cdots \overline{z_m}^{K_m} W_j(z_1, \cdots, z_m)] = r_j(z_1, \cdots, z_m) + h_j,$$
$$z(z_1, \cdots, z_m) \in \partial D_1 \times \cdots \times \partial D_m, \; j = 2^{n-3} + 1, \cdots, 2^{n-2}. \tag{5.17}$$

The above result can be written as

Theorem 5.3   *Under some conditions, Problem P' for (5.14) has a solution $[w_1(x), \cdots, w_{2^{n-1}}(x)]$, where Ω is a polycylinder* (see [70]).

5.2   Oblique derivative problem for degenerate elliptic system of first order.

Now, we discuss the degenerate elliptic system of first order equations

$$\begin{cases} w_{1x_1} - w_{2x_2} - x_3^\mu w_{3x_3} = A_1 w_1 - B_2 w_2 - B_3 x_3^{1+\mu} w_3 - B_4 w_4 + C_1, \\[1mm] w_{2x_1} + w_{1x_2} + w_{2x_3} = A_2 w_1 + B_1 w_2 - B_4 x_3^\mu w_3 + B_3 w_4 + C_2, \\[1mm] x_3^\mu w_{3x_1} - w_{4x_2} + w_{1x_3} = A_3 w_1 + B_2 w_2 + B_1 x_3^\mu w_3 - B_2 w_4 + C_3, \\[1mm] w_{4x_1} + x_3^\mu w_{3x_2} - w_{2x_3} = A_4 w_1 - B_3 w_2 + B_2 x_3^\mu w_3 + B_1 w_4 + C_4, \end{cases} \tag{5.18}$$

where $A_j(x)$, $B_j(x)$, $C_j(x)(j = 1, \cdots, 4)$ are known functions as stated in (5.4), and $C_1(x) = x_3^{1+\mu} c_1(x)$, $C_j(x) = x_3^\mu c_j(x)$, $j = 2,3,4$, $c_j(x) \in C_\alpha^1(\bar{\Omega})$, $j = 1, \cdots, 4$, μ is a nonnegative constant, and Ω is a domain in the upper halfspace $x_3 > 0$ with the boundary $\partial\Omega = G \cup D \in C_\alpha^2$, where D is in $x_3 = 0$ and G in $x_3 > 0$, $\Gamma = \bar{G} \cap \{x_3 = 0\}$. The oblique derivative problem (Problem O) for (5.18) is to find a bounded solution $w(x) = \sum_{j=1}^4 w_j(x)e_j \in C_\alpha^1(\bar{\Omega}) \cap C^2(\Omega)$ of (5.18) satisfying the boundary condition

$$\begin{cases} \dfrac{\partial w_3}{\partial \nu_3} + \sigma_3(x)w_3(x) = \tau_3(x), \; x \in G, \\[4mm] \dfrac{\partial w_4}{\partial \nu_4} + \sigma_4(x)w_4(x) = \tau_4(x), \; x \in \partial\Omega, \end{cases} \tag{5.19}$$

$$\begin{cases} \operatorname{Re}[\overline{\lambda(t)}(w_1(t) + iw_2(t))] = r(t) + h[\zeta(t)], \ t = x_1 + ix_2 \in \Gamma \\ \operatorname{Im}[\overline{\lambda(d_j)}(w_1(d_j) + iw_2(d_j))] = g_j, \ j = 1, \cdots, 2K + 1 \text{ for } K \geq 0, \end{cases} \tag{5.20}$$

where $\cos(\nu_j, n) > 0$, $j = 3, 4$, n is the outward normal at $x \in \partial\Omega$, and $\sigma_j(x) > 0$, $\tau_j(x) \in C_\alpha^1(\partial D)$, $j = 3, 4$, $\lambda(t), r(t), h(t), d_j, g_j$ and K are similar as in (5.9), $\zeta(t)$ is a conformal mapping from the unit disk $\{t| < 1\}$ onto D. Similarly to before, if B_j is only a function of $x_j, j = 1, 2, 3$, B_4 is a real constant, and $A_j = B_j, j = 1, \cdots, 4$, then the solution $w(x) = \sum_{j=1}^4 w_j(x)e_j$ of Problem O for (5.18) is also a solution of the following boundary value problem for the degenarate system of second order equations

$$\begin{cases} \Delta w_3 = 2B_1 w_{3x_1} + 2B_2 w_{3x_2} + A_5 x_3^{-1} w_{3x_3} + A_6 w_3 + A_7, \\ \Delta w_4 = \sum_{j=1}^3 (A_j + B_j) w_{4x_j} + A_8 w_4 + A_9 w_3 + A_{10} \end{cases} \tag{5.21}$$

with the boundary condition (5.19), in which

$$A_5 = -\mu + B_3 x_3(1 + x_3), \ C_5 \in C^2(\bar{\Omega}),$$
$$A_6 = (1 + \mu)B_3 - \sum_{j=1}^4 B_j^2 + \sum_{j=1}^3 B_{jx_j} + (B_3^2 - B_{3x_2})(1 - x_3),$$
$$A_7 = (-B_1 C_3 - B_2 C_4 + B_3 C_1 + B_4 C_2 - C_{1x_3} + C_{3x_1} + C_{4x_2})x_3^{-\mu},$$
$$A_8 = -\sum_{j=1}^4 B_j^2 + \sum_{j=1}^3 B_{jx_j}, \ A_9 = A_4 x_3^{\mu-1}(B_4\mu + A_3 x_3 - B_3 x_3^2),$$
$$A_{10} = -A_1 C_4 + A_2 C_3 - A_3 C_2 + A_4 C_1 + C_{2x_3} - C_{3x_2} + C_{4x_1}.$$

Conversely, if the following conditions hold:

$$A_6 \geq 0, \ A_8 \geq 0 \text{ on } \bar{\Omega}, \tag{5.22}$$

then on the basis of the result in [90], the boundary value problem (5.21),(5.19) is solvable. Moreover, if

$$A_1 = A_2 = 0 \text{ on } \bar{\Omega}, \tag{5.23}$$

then $[w_1(x), w_2(x)]$ can be found by the following integrals

$$\begin{cases} w_1(x) = \int_0^{x_3}[-x_3^\mu w_{3x_1} + w_{4x_2} + B_1 x_3^\mu w_3 - B_2 w_4 + C_3]dx_3 + \phi_1(x_1, x_2), \\ w_2(x) = \int_0^{x_3}[x_3^\mu w_{3x_2} + w_{4x_1} - B_2 x_3^\mu w_3 - B_1 w_4 - C_4]dx_3 + \phi_2(x_1, x_2), \end{cases} \tag{5.24}$$

in which $\phi_1(x_1, x_2), \phi_2(x_1, x_2)$ satisfy the conditions

$$\begin{cases} \phi_{1x_1} - \phi_{2x_2} - A_1\phi_1 + B_2\phi_2 = \psi_1(x_1, x_2), \\ \phi_{2x_1} + \phi_{1x_2} - A_2\phi_1 - B_1\phi_2 = \psi_2(x_1, x_2), \end{cases} \tag{5.25}$$

where

$$\psi_1(x_1, x_2) = [x_3^\mu \mu w_{3x_3} + C_1]|_{x_3=0}, \ \psi_2(x_1, x_2) = [-w_{4x_3} + C_2]|_{x_3=0},$$

and A_j, B_j, C_j satisfy some conditions.

Besides, we require that $\phi = \phi_1 + \phi_2 e_2 = \phi_1 + \phi_2 i$ satisfies the boundary condition (5.20), i.e.

$$\begin{cases} \mathrm{Re}[\overline{\lambda(t)}\phi(t)] = r(t) + h(t), \ t \in \Gamma, \\ \mathrm{Im}[\overline{\lambda(d_j)}\phi(d_j)] = g_j, \ j = 1, \cdots, 2K + 1 \text{ for } K \geq 0. \end{cases} \tag{5.26}$$

Similarly to (5.10),(5.11), the function $[\phi_1(x_1, x_2), \phi_2(x_1, x_2)]$ is uniquely determined, and hence $[w_1(x), w_2(x)]$ is also uniquely determined. Thus we have

Theorem 5.4 *Suppose that the coefficients of system* (5.18) *satisfy conditions* (5.22), (5.23) *etc. stated as before. Then Problem O for degenerate elliptic system* (5.18) *has a solution* $w(x) = \sum_{j=1}^{4} w_j(x)e_j \in C_\alpha^1(\Omega \cup G) \cap C^2(\Omega), \ 0 < \alpha < 1$ *(see* [139]15)).

Finally, we mention that when the boundary condition (5.19) is replaced by

$$w_3(x) = \tau_3(x), \ w_4(x) = \tau_4(x), \ x \in \partial\Omega, \tag{5.27}$$

we can similarly discuss the solvability of the boundary value problem (5.18) and (5.27).

VI   Boundary Value Problems for Elliptic Equations and Systems of Higher Order

In this chapter, we first transform the uniformly elliptic equations and systems of fourth order into complex forms, afterwards we discuss the representations, existence theorems of solutions and some boundary value problems for elliptic equations and systems of fourth order. Finally, we simply introduce some properties and boundary value problems for elliptic equations and systems of $2n$th order.

1   Reduction of Elliptic Equations and Systems of Fourth Order

1.1   Complex form of linear and nonlinear elliptic equations of fourth order

Let $\Phi(x, y, u_{00}, u_{10}, u_{01}, \cdots, u_{40}, u_{31}, u_{22}, u_{13}, u_{04})$ be a continuous real function in the $(x, y) \in D$ and the real variables $u_{00}, u_{10}, u_{01}, \cdots, u_{40}, u_{31}, u_{22}, u_{13}, u_{04}$, and possess continuous partial derivatives with respect to $u_{40}, u_{31}, u_{22}, u_{13}$ and u_{04}, where D is a bounded domain in the (x, y)–plane. The nonlinear equation of fourth order

$$\Phi(x, y, u, u_x, u_y, \cdots, u_{x^4}, u_{x^3y}, u_{x^2y^2}, u_{xy^3}, u_{y^4}) = 0 \tag{1.1}$$

is called elliptic, if the following inequalities hold:

$$\Phi_{u_{x^4}} \lambda^4 + \Phi_{u_{x^3y}} \lambda^3 + \Phi_{u_{x^2y^2}} \lambda^2 + \Phi_{u_{xy^3}} \lambda + \Phi_{u_{y^4}} > 0,$$
$$\Phi_{u_{x^4}} > 0 \text{ in } D, \tag{1.2}$$

where λ is any real number. Introduce the notations $z = x + iy$, $(\ )_{\bar{z}} = [(\ )_x + i(\ )_y]/2$, $(\ )_z = [(\ )_x - i(\ )_y]/2$, we obtain

$$\begin{cases} (\ )_{z^2\bar{z}^2} = \frac{1}{16}[(\ )_{x^4} + 2(\ )_{x^2y^2} + (\ )_{y^4}], \\[2mm] (\ )_{z\bar{z}^3} = \frac{1}{16}[(\ )_{x^4} + 2i(\ )_{x^3y} + 2i(\ )_{xy^3} - (\ )_{y^4}], \\[2mm] (\ )_{\bar{z}^4} = \frac{1}{16}[(\ )_{x^4} + 4i(\ )_{x^3y} - 6(\ )_{x^2y^2} - 4i(\ )_{xy^3} + (\ )_{y^4}], \end{cases} \tag{1.3}$$

and

$$\begin{cases} (\ )_{x^4} = (\ )_{z^4} + 4(\ )_{z^3\bar{z}} + 6(\ )_{z^2\bar{z}^2} + 4(\ )_{z\bar{z}^3} + (\ )_{\bar{z}^4}, \\[2mm] (\ )_{x^3y} = i[(\ )_{z^4} + 2(\ )_{z^3\bar{z}} - (\ )_{z\bar{z}^3} - (\ )_{\bar{z}^4}], \\[2mm] (\ )_{x^2y^2} = -(\ )_{z^4} + 2(\ )_{z^2\bar{z}^2} - (\ )_{\bar{z}^4}, \\[2mm] (\ )_{xy^3} = i[-(\ )_{z^4} + 2(\ )_{z^3\bar{z}} - 2(\ )_{z\bar{z}^3} + (\ )_{\bar{z}^4}], \\[2mm] (\ )_{y^4} = (\ )_{z^4} - 4(\ )_{z^3\bar{z}} + 6(\ )_{z^2\bar{z}^2} - 4(\ )_{z\bar{z}^3} + (\ )_{\bar{z}^4}. \end{cases} \tag{1.4}$$

Moreover, $(\ )_{x^j y^k}(0 \le j, k \le 3, j + k \le 3)$ can be expressed by $(\ )_{z^j \bar{z}^k}(0 \le j, k \le 3, j + k \le 3)$. Substitute the above expressions of $u_{x^j y^k}(0 \le j, k \le 4, j + k \le 4)$ into the equation (1.1), it can be seen that Φ is a function with respect to $z, u,$ $u_z, \cdots, u_{z^4}, u_{z^3 \bar{z}}, u_{z^2 \bar{z}^2}, u_{z \bar{z}^3}$ and $u_{\bar{z}^4}$. From $\Phi = 0$ we find out the partial derivative with respect to $u_{z^2 \bar{z}^2}$, namely

$$\Phi_{u_{z^2 \bar{z}^2}} = 6\Phi_{u_{x^4}} + 2\Phi_{u_{x^2 y^2}} + 6\Phi_{u_{y^4}}. \tag{1.5}$$

Theorem 1.1   *Let the equation (1.1) satisfy the above ellipticity condition (1.2) and the conditions from the existence theorem for implicit functions. Then (1.1) is solvable with respect to $u_{z^2 \bar{z}^2}$, and the corresponding elliptic complex equation of fourth order*

$$u_{z^2 \bar{z}^2} = F(z, u, u_z, u_{z^2}, u_{z\bar{z}}, u_{z^3}, u_{z^2 \bar{z}}, u_{z^4}, u_{z^3 \bar{z}}) \tag{1.6}$$

can be obtained.

Proof   By the ellipticity condition (1.2) of equation (1.1), we know that for any point in D, the algebraic equation of fourth order

$$\Phi_{u_{x^4}}\left[\lambda^4 + \lambda^3 \frac{\Phi_{u_{x^3 y}}}{\Phi_{u_{x^4}}} + \lambda^2 \frac{\Phi_{u_{x^2 y^2}}}{\Phi_{u_{x^4}}} + \lambda \frac{\Phi_{u_{xy^3}}}{\Phi_{u_{x^4}}} + \frac{\Phi_{u_{y^4}}}{\Phi_{u_{x^4}}}\right] = 0 \tag{1.7}$$

has two pair of complex conjugate roots, which are denoted by $\lambda_1, \overline{\lambda_1}, \lambda_2, \overline{\lambda_2}$. According to the relations between roots and coefficients of the equation (1.7)

$$\begin{cases} \Phi_{u_{x^3 y}}/\Phi_{u_{x^4}} = -(\lambda_1 + \overline{\lambda_1} + \lambda_2 + \overline{\lambda_2}) = -2\mathrm{Re}(\lambda_1 + \lambda_2), \\[2mm] \Phi_{u_{x^2 y^2}}/\Phi_{u_{x^4}} = |\lambda_1|^2 + |\lambda_2|^2 + (\lambda_1 + \overline{\lambda_1})(\lambda_2 + \overline{\lambda_2}) = \\[2mm] \qquad\qquad = |\lambda_1|^2 + |\lambda_2|^2 + 4\mathrm{Re}\lambda_1\mathrm{Re}\lambda_2, \\[2mm] \Phi_{u_{xy^3}}/\Phi_{u_{x^4}} = -|\lambda_1|^2(\lambda_2 + \overline{\lambda_2}) - |\lambda_2|^2(\lambda_1 + \overline{\lambda_1}) \\[2mm] \qquad\qquad = -2(|\lambda_1|^2\mathrm{Re}\lambda_2 + |\lambda_2|^2\mathrm{Re}\lambda_1), \\[2mm] \Phi_{u_{y^4}}/\Phi_{u_{x^4}} = \lambda_1\overline{\lambda_1}\lambda_2\overline{\lambda_2} = |\lambda_1\lambda_2|^2, \end{cases} \tag{1.8}$$

and noting that $|\lambda_1| > \mathrm{Re}\lambda_1, |\lambda_2| > \mathrm{Re}\lambda_2$, from (1.5) and (1.8), we can obtain

$$\Phi_{u_{z^2 \bar{z}^2}} = 2\Phi_{u_{x^4}}[3 + 3|\lambda_1\lambda_2|^2 + |\lambda_1|^2 + |\lambda_2|^2 + 4\mathrm{Re}\lambda_1\mathrm{Re}\lambda_2]$$

$$> 2\Phi_{u_{x^4}}[3 + 3|\lambda_1\lambda_2|^2 + (\mathrm{Re}\lambda_1)^2 + (\mathrm{Re}\lambda_2)^2 +$$

$$+ 2\mathrm{Re}\lambda_1\mathrm{Re}\lambda_2 + (\lambda_1 + \overline{\lambda_1})(\lambda_2 + \overline{\lambda_2})/2] = \tag{1.9}$$

$$= 2\Phi_{u_{x^4}}[2 + 2|\lambda_1\lambda_2|^2 + (\mathrm{Re}\lambda_1 + \mathrm{Re}\lambda_2)^2 +$$

$$+ |1 + \lambda_1\lambda_2|^2/2 + |1 + \lambda_1\overline{\lambda_2}|^2/2] > 0.$$

Hence we can solve (1.1) for $u_{z^2\bar{z}^2}$ and obtain the complex form of (1.1), i.e. (1.6).

In particular, we discuss the linear elliptic equation of fourth order

$$a_1 u_{x^4} + a_2 u_{x^3 y} + a_3 u_{x^2 y^2} + a_4 u_{xy^3} + a_5 u_{y^4} + \sum_{j+k=0}^{3} a_{jk} u_{x^j y^k} = a_0, \qquad (1.10)$$

where the coefficients $a_j = a_j(x,y)\,(j = 0,1,\cdots,5)$, $a_{jk} = a_{jk}(x,y)\,(j,k \geq 0, 0 \leq j + k \leq 3)$ are known functions in D. Using the relation (1.4), equation (1.10) can be written in the complex form

$$b_1 u_{z^4} + b_2 u_{z^3 \bar{z}} + b_3 u_{z^2 \bar{z}^2} + b_4 u_{z \bar{z}^3} + b_5 u_{\bar{z}^4} + \sum_{j+k=0}^{3} b_{jk} u_{z^j \bar{z}^k} = b_0, \qquad (1.11)$$

in which $b_{jk} = b_{jk}(z)$ are functions of $a_{jk}\,(j,k \geq 0, 0 \leq j + k \leq 3)$, $b_0 = b_0(z) = a_0(z)$, $b_j = b_j(z)$ are functions of $a_j\,(j = 1,\cdots,5)$. More precisely

$$\begin{cases} b_1 = a_1 + ia_2 - a_3 - ia_4 + a_5, \quad b_2 = 4a_1 + 2ia_2 + 2ia_4 - 4a_5, \\[2mm] b_3 = 6a_1 + 2a_3 + 6a_5, \quad b_4 = 4a_1 - 2ia_2 - 2ia_3 - 4a_5, \\[2mm] b_5 = a_1 - ia_2 - a_3 + ia_4 + a_5. \end{cases} \qquad (1.12)$$

Now, as in the proof of Theorem 1.1, we can solve (1.11) for $u_{z^2 \bar{z}^2}$ and obtain the complex form of (1.11)

$$\begin{cases} u_{z^2 \bar{z}^2} = F(z, u, u_z, u_{z^2}, u_{z\bar{z}}, u_{z^2 \bar{z}}, u_{z^4}, u_{z^3 \bar{z}}), \\[2mm] F = \mathrm{Re}[Q_1(z) u_{z^4} + Q_2(z) u_{z^3 \bar{z}}] + \sum_{j+k=0}^{3} A_{jk}(z) u_{z^j \bar{z}^k} + A_0(z), \end{cases} \qquad (1.13)$$

where

$$Q_1 = Q_1(z) = -2b_1/b_3 = -[a_1 - a_3 + a_5 + i(a_2 - a_4)]/(3a_1 + a_3 + 3a_5),$$

$$Q_2 = Q_2(z) = -b_2/b_3 = -[4a_1 - 4a_5 + 2i(a_2 + a_4)]/(3a_1 + a_3 + 3a_5),$$

$$A_{jk} = A_{jk}(z) = -b_{jk}/b_3\,(j,k \geq 0, 0 \leq j + k \leq 3), \quad A_0 = A_0(z) = b_0/b_3.$$

1.2 Complex form of linear and nonlinear elliptic systems of fourth order

We consider the nonlinear system of fourth order equations

$$\begin{aligned} \Phi_j(x, y, u, v, u_x, \cdots, u_{x^4}, u_{x^3 y}, u_{x^2 y^2}, u_{xy^3}, \\[2mm] u_{y^4}, v_{x^4}, v_{x^3 y}, v_{x^2 y^2}, v_{xy^3}, v_{y^4}) = 0, \quad j = 1, 2, \end{aligned} \qquad (1.14)$$

180   *Heinrich Begehr and Guo Chun Wen*

where the function $\Phi_j(x, y, u_{00}, v_{00}, u_{10}, \cdots, u_{40}, u_{31}, u_{22}, u_{13}, u_{04}, v_{40}, v_{31}, v_{22}, v_{13}, v_{04})$
$(j = 1, 2)$ are continuous real functions in $(x, y) \in D$ and the real variables u_{jk}, v_{jk}
$(j, k \geq 0, 0 \leq j + k \leq 4)$, and possess continuous partial derivatives with respect to
u_{jk}, v_{jk} $(j, k \geq 0, j + k = 4)$. System (1.14) is called elliptic, if the inequalities

$$\begin{cases} |A_{40}\lambda^4 + A_{31}\lambda^3 + A_{22}\lambda^2 + A_{13}\lambda + A_{04}| > 0, \ |A_{40}| > 0, \\ A_{jk} = \begin{pmatrix} \Phi_{1u_{jk}} & \Phi_{1v_{jk}} \\ \Phi_{2u_{jk}} & \Phi_{2v_{jk}} \end{pmatrix}, \ u_{jk} = u_{x^j y^k}, \ v_{jk} = v_{x^j y^k}, \ j + k = 4, \end{cases} \tag{1.15}$$

hold in D for any real number λ. Applying (1.4) to u, v, and substituting the resulting
expressions into (1.14), it can be seen that $\Phi_j(j = 1, 2)$ on the left–hand side of (1.14)
are function of $u_{z^4}, u_{z^3\bar{z}}, u_{z^2\bar{z}^2}, v_{z^4}, v_{z^3\bar{z}}, v_{z^2\bar{z}^2}, \cdots$. Thus we have

$$\begin{cases} \Phi_{j\sigma_{z^4}} = \Phi_{j\sigma_{x^4}} + i\Phi_{j\sigma_{x^3 y}} - \Phi_{j\sigma_{x^2 y^2}} - i\Phi_{j\sigma_{xy^3}} + \Phi_{j\sigma_{y^4}}, \\ \Phi_{j\sigma_{z^3\bar{z}}} = 4\Phi_{j\sigma_{x^4}} + 2i\Phi_{j\sigma_{x^3 y}} + 2i\Phi_{j\sigma_{xy^3}} - 4\Phi_{ju_{y^4}}, \\ \Phi_{j\sigma_{z^2\bar{z}^2}} = 6\Phi_{j\sigma_{x^4}} + 2\Phi_{j\sigma_{x^2 y^2}} + 6\Phi_{j\sigma y^4}, \ \sigma = u, v, \ j = 1, 2. \end{cases} \tag{1.16}$$

Theorem 1.2 *Let the system (1.14) satisfy the conditions*

$$|A_{04}| \geq \delta > 0, \ \sup_{m=1,2,\, j+k=4} (|\Phi_{mu_{x^j y^k}}|, |\Phi_{mv_{x^j y^k}}|) \leq \delta^{-1}, \tag{1.17}$$

where δ is a positive constant. Then through a linear transformation

$$\xi = tx + y, \ \eta = -tx + y, \tag{1.18}$$

*in which t is an undetermined positive constant, system (1.14) is solvable with respect
to $w_{\zeta^2\bar{\zeta}^2}(w = u + iv, \zeta = \xi + i\eta)$, and a complex equation of new variables*

$$w_{\zeta^2\bar{\zeta}^2} = F(\zeta, w, w_\zeta, \cdots, w_{\zeta^4}, w_{\zeta^3\bar{\zeta}}, \bar{w}_{\zeta^3\bar{\zeta}}, \bar{w}_{\zeta^4}) \tag{1.19}$$

can be obtained.

Proof From (1.18), it follows that

$$\sigma_{x^2} = (-2\sigma_{\xi\eta} + \sigma_{\xi^2} + \sigma_{\eta^2})t^2, \ \sigma_{xy} = (\sigma_{\xi^2} - \sigma_{\eta^2})t,$$

$$\sigma_{y^2} = 2\sigma_{\xi\eta} + \sigma_{\xi^2} + \sigma_{\eta^2}, \ \sigma = u, v,$$

and

$$\begin{cases} \sigma_{x^4} = (\sigma_{\xi^4} - 4\sigma_{\xi^3\eta} + 6\sigma_{\xi^2\eta^2} - 4\sigma_{\xi\eta^3} + \sigma_{\eta^4})t^4, \\ \sigma_{x^3 y} = (\sigma_{\xi^4} - 2\sigma_{\xi^3\eta} + 2\sigma_{\xi\eta^3} - \sigma_{\eta^4})t^3, \\ \sigma_{x^2 y^2} = (\sigma_{\xi^4} - 2\sigma_{\xi^2\eta^2} + \sigma_{\eta^4})t^2, \ \sigma = u, v, \end{cases} \tag{1.20}$$

where $\sigma_{xy^3}, \sigma_{y^4}$ can similarly be written. Hence

$$\Phi_{j\sigma_{\xi^2\eta^2}} = 6\Phi_{j\sigma_{x^4}}t^4 - 2\Phi_{j\sigma_{x^2y^2}}t^2 + 6\Phi_{j\sigma_{y^4}}, \quad \sigma = u, v, \ j = 1, 2, \tag{1.21}$$

moreover $\Phi_{j\sigma_{\xi^4}}$, $\Phi_{j\sigma_{\eta^4}}$ can also be given. Taking the third formula in (1.16) into account, we obtain

$$\Phi_{j\sigma_{\zeta^2\bar{\zeta}^2}} = 24\Phi_{j\sigma_{x^4}}t^4 + 8\Phi_{j\sigma_{x^2y^2}}t^2 + 24\Phi_{j\sigma_{y^4}}, \quad \sigma = u, v, \ j = 1, 2.$$

Due to

$$u_{\zeta^2\bar{\zeta}^2} = \frac{1}{2}(w_{\zeta^2\bar{\zeta}^2} + \bar{w}_{\zeta^2\bar{\zeta}^2}), \quad v_{\zeta^2\bar{\zeta}^2} = \frac{1}{2i}(w_{\zeta^2\bar{\zeta}^2} - \bar{w}_{\zeta^2\bar{\zeta}^2}),$$

it can be derived

$$\frac{D(\Phi_1, \Phi_2)}{D(W, \overline{W})} = \begin{vmatrix} \Phi_{1W} & \Phi_{1\overline{W}} \\ \Phi_{2W} & \Phi_{2\overline{W}} \end{vmatrix} = \frac{i}{2}\begin{vmatrix} \Phi_{1u_{\zeta^2\bar{\zeta}^2}} & \Phi_{1v_{\zeta^2\bar{\zeta}^2}} \\ \Phi_{2u_{\zeta^2\bar{\zeta}^2}} & \Phi_{2v_{\zeta^2\bar{\zeta}^2}} \end{vmatrix}$$

$$= 32i \begin{vmatrix} 3\Phi_{1u_{x^4}}t^4 + \Phi_{1u_{x^2y^2}}t^2 + 3\Phi_{1u_{y^4}} & 3\Phi_{1v_{x^4}}t^4 + \Phi_{1v_{x^2y^2}}t^2 + 3\Phi_{1v_{y^4}} \\ 3\Phi_{2u_{x^4}}t^4 + \Phi_{2u_{x^2y^2}}t^2 + 3\Phi_{2u_{y^4}} & 3\Phi_{2v_{x^4}}t^4 + \Phi_{2v_{x^2y^2}}t^2 + 3\Phi_{2v_{y^4}} \end{vmatrix} \tag{1.22}$$

$$= 288i(t^2 A^* + |A_{04}|),$$

where $W = w_{\zeta^2\bar{\zeta}^2}$. We choose $t(> 0)$ small enough, such that $t^2 A_* \le \delta/2$, thus $t^2 A_* + |A_{04}| \ge \delta/2 > 0$. Then (1.22) implies that

$$\frac{D(\Phi_1, \Phi_2)}{D(W, \overline{W})} \ne 0 \text{ in } D.$$

Therefore from the system (1.14), we can obtain the complex equation (1.19).

In particular, for the linear elliptic system of fourth order equations

$$\sum_{m=1}^{2} \sum_{j+k=0}^{4} a_{mjk}^{(n)}(x, y)u_{mx^jy^k} = a^{(n)}(x, y), \ n = 1, 2, \tag{1.23}$$

where the coefficients $a_{mjk}^{(n)}(x, y)(j, k \ge 0, 0 < j + k \le 4, m, n = 1, 2)$ are continuous in D, according to Theorem 1.2, provided that

$$\begin{vmatrix} a_{104}^{(1)} & a_{204}^{(1)} \\ a_{104}^{(2)} & a_{204}^{(2)} \end{vmatrix} \ge \delta > 0, \quad \sup_{m,n=1,2,\, j+k=4}(|a_{mjk}^{(n)}|) \le \delta^{-1} \text{ in } D, \tag{1.24}$$

then through the linear transformation (1.18), system (1.23) can be reduced to the complex form (1.19). However, by using (1.4) we can rewrite (1.23) in the following complex form

$$\sum_{m=1}^{2} \sum_{j+k=0}^{4} b_{mjk}^{(n)}(z)u_{mx^jy^k} = b^{(n)}(z), \ n = 1, 2, \tag{1.25}$$

in which $b_{mjk}^{(n)}(z)(j, k \geq 0, 0 \leq j+k \leq 4, m, n = 1, 2)$ are known functions of $a_{mjk}^{(n)}(j, k \geq 0, 0 \leq j + k \leq 4, m, n = 1, 2)$. If

$$\left| \begin{matrix} b_{122}^{(1)}(z) & b_{222}^{(1)}(z) \\ b_{122}^{(2)}(z) & b_{222}^{(2)}(z) \end{matrix} \right| \neq 0, \tag{1.26}$$

then we can solve (1.25) for $u_{1z^2\bar{z}^2}, u_{2z^2\bar{z}^2}$ and obtain the complex system

$$u_{mz^2\bar{z}^2} = F_m(z, u_1, u_2, u_{1z}, \cdots, u_{1z^4}, u_{1z^3\bar{z}}, u_{2z^4}, u_{2z^3\bar{z}}), \ m = 1, 2. \tag{1.27}$$

Denote $w = u_1 + iu_2$, $u_1 = (w + \bar{w})/2$, $u_2 = (w - \bar{w})/(2i)$, the foregoing complex system can be written as the following complex equation

$$w_{z^2\bar{z}^2} = F(z, w, w_z, \cdots, w_{z^4}, w_{z^3\bar{z}}, \bar{w}_{z^3\bar{z}}, \bar{w}_{z^4}). \tag{1.28}$$

1.3   Conditions for uniformly elliptic complex equations

Let D be a bounded simply connected domain with boundary $\Gamma \in C_\mu^4 (0 < \mu < 1)$. Without loss of generality, we may assume that D is the unit disk. We first discuss the elliptic equation of fourth order in the complex form (1.6), and consider the equation of the following form

$$\begin{cases} u_{z^2\bar{z}^2} = F(z, u, u_z, u_{z^2}, u_{z\bar{z}}, u_{z^3}, u_{z^2\bar{z}}, u_{z^4}, u_{z^3\bar{z}}), \\ F = \mathrm{Re}[Q_1 u_{z^4} + Q_2 u_{z^3\bar{z}}] + \displaystyle\sum_{j+k=0}^{3} A_{jk} u_{z^j\bar{z}^k} + A_0, \end{cases} \tag{1.29}$$

where $Q_j(j = 1, 2)$ are functions of $z, u, u_z, u_{z^2}, u_{z\bar{z}}, u_{z^3}, u_{z^2\bar{z}}, u_{z^4}, u_{z^3\bar{z}}$, and $A_{jk}(j, k \geq 0, j + k \leq 3)$ are functions of $z, u, u_z, u_{z\bar{z}}, u_{z^3}, u_{z^2\bar{z}}$. We assume that (1.29) satisfies the following conditions.

Condition C   1) The functions $Q_j(z, u, \cdots, u_{z^2\bar{z}}, U, V)$, $A_j(z, u, \cdots, u_{z^2\bar{z}})$ are measurable in $z \in D$ for all real functions $u(z) \in C_\mu^3(\bar{D})$ and complex measurable functions $U(z), V(z) \in L_{p_0}(\bar{D})$, and satisfy the conditions

$$L_p[A_0, \bar{D}] \leq k_0, \ L_p[A_{jk}, \bar{D}] \leq k_1, \ j, k \geq 0, \ 0 \leq j + k \leq 3, \tag{1.30}$$

in which $p, p_0(2 < p_0 < p), k_0, k_1, \mu(0 < \mu < 1)$ are nonnegative constants.

2) Q_j, A_{jk} are continuous in $u, u_z, u_{z^2}, u_{z\bar{z}}, u_{z^3}, u_{z^2\bar{z}}$ for almost every point $z \in D$ and $U, V \in \mathbb{C}$.

3) Equation (1.29) satisfies the following uniform ellipticity condition

$$|F(z, u, u_z, u_{z^2}, u_{z\bar{z}}, u_{z^3}, u_{z^2\bar{z}}, U_1, V_1)$$

$$-F(z, u, u_z, u_{z^2}, u_{z\bar{z}}, u_{z^3}, u_{z^2\bar{z}}, U_2, V_2)| \tag{1.31}$$

$$\leq q_0|U_1 - U_2| + q_0'|V_1 - V_2|, \ q_0 + q_0' < 1,$$

for almost every point $z \in D$, $u(z) \in C_\mu^3(\bar{D})$ and $U_j, V_j \in \mathbb{C}(j = 1, 2)$, where q_0, q_0' are nonnegative constants.

Next we discuss the elliptic complex equation (1.19) or (1.28) in the following form

$$
\begin{cases}
w_{z^2 \bar{z}^2} = F(z, w, w_z, \bar{w}_z, \cdots, w_{z^4}, w_{z^3 \bar{z}}, \bar{w}_{z^3 \bar{z}}, \bar{w}_{z^4}), \\[2mm]
F = \displaystyle\sum_{j+k=0,(j,k)\neq(2,2)}^{4} (Q_{jk} w_{z^j \bar{z}^k} + \tilde{Q}_{jk} \bar{w}_{z^j \bar{z}^k}) \\[2mm]
\quad + \displaystyle\sum_{j+k=0}^{3} (A_{jk} w_{z^j \bar{z}^k} + \tilde{A}_{jk} \bar{w}_{z^j \bar{z}^k}) + A_0,
\end{cases}
\tag{1.32}
$$

where Q_{jk} and $\tilde{Q}_{jk}$ are functions of $z, w, w_z, \bar{w}_z, \cdots, w_{z^4}, w_{z^3 \bar{z}}, \bar{w}_{z^3 \bar{z}}, \bar{w}_{z^4}$, $A_{jk}, \tilde{A}_{jk}$ and A_0 are functions of $z, w, w_z, \cdots, w_{z^3}, w_{z^2 \bar{z}}, \bar{w}_{z^2 \bar{z}}, \bar{w}_{z^3}$. We suppose the complex equation (1.32) satisfies

Condition C' 1) The functions $Q_{jk}(= Q_{jk}(z, w, \cdots, \bar{w}_{z^3}, U, V, X, Y))$, $\tilde{Q}_{jk}$, $A_{jk}(= A_{jk}(z, w, \cdots, \bar{w}_{z^3}))$, $\tilde{A}_{jk}, A_0$ are measurable in $z \in D$ for any complex function $w(z) \in C_\mu^3(\bar{D})$ and $U(z), V(z), X(z), Y(z) \in L_{p_0}(\bar{D})$, and satisfy the conditions

$$
L_p[A_0, \bar{D}] \le k_0, \quad L_p[A_{ij}, \bar{D}] \le k_1, \quad L_p[\tilde{A}_{ij}, \bar{D}] \le k_1,
\tag{1.33}
$$

where $p, p_0 (2 < p_0 < p), k_0, k_1, \mu(0 < \mu < 1)$ are positive constants.

2) The above functions are continuous in $w, w_z, \bar{w}_z, \cdots, \bar{w}_{z^3}$ for almost every point $z \in D$ and $U, V, X, Y \in \mathbb{C}$.

3) The complex equation (1.32) satisfies the uniform ellipticity condition

$$
\begin{aligned}
&|F(z, w, w_z, \cdots, \bar{w}_{z^3}, U_1, V_1, X_1, Y_1) \\
&\quad -F(z, w, w_z, \cdots, \bar{w}_{z^3}, U_2, V_2, X_2, Y_2)| \\
&\le q_0|U_1 - U_2| + q_0'|V_1 - V_2| + q_0''|X_1 - X_2| + q_0'''|Y_1 - Y_2|
\end{aligned}
\tag{1.34}
$$

for almost every point $z \in D$ and $U_j, V_j, X_j, Y_j \in \mathbb{C}(j = 1, 2)$, where q_0, q_0', q_0'', q_0''' are nonnegative constants satisfying $q_0 + q_0' + q_0'' + q_0''' < 1$.

Under the foregoing conditions, a function $u(z)$ (or $w(z)$) being continuous function in D satisfying (1.29)(or (1.32)) for almost every point $z \in D$ and the condition $u(z) \in W_{p_0}^4(D_*)$ (or $w(z) \in W_{p_0}^4(D_*)$) will be called a solution of (1.29) (or (1.32)), where D_* is any closed subset in D(see [139]13)).

2   Existence Theorem and Boundary Value Problems for Elliptic Equations of Fourth Order

In this section, we first discuss the representation and an existence theorem of solutions for the equation (1.29) and then mainly consider the Dirichlet boundary value problem for (1.29).

2.1   A double integral and its properties

We introduce a double integral

$$f(z) = \frac{1}{\pi} \int\!\!\int_D Y(z,\zeta)\rho(\zeta)d\sigma_\zeta, \tag{2.1}$$

where $\rho(z)$ is a real measurable function, $\rho(z) \in L_p(\bar{D}), p > 1$, and $Y(z,\zeta)$ is a biharmonic function in D except for a point $z = \zeta \in D$, i.e.

$$\begin{cases} Y(z,\zeta) = 2|z-\zeta|^2 \ln|z-\zeta|, \; Y_z = (\bar{z}-\bar{\zeta})(2\ln|z-\zeta|+1), \\[2mm] Y_{z^2} = (\bar{z}-\bar{\zeta})/(z-\zeta), \; Y_{z\bar{z}} = 2(\ln|z-\zeta|+1), \\[2mm] Y_{z^3} = -(\bar{z}-\bar{\zeta})/(z-\zeta)^2, \; Y_{z^2\bar{z}} = 1/(z-\zeta), \; Y_{z^3\bar{z}} = -1/(z-\zeta)^2, \\[2mm] Y_{z^4} = 2(\bar{z}-\bar{\zeta})/(z-\zeta)^3, \; Y_{z^2\bar{z}^2} = 0, \; z(\neq \zeta) \in D. \end{cases} \tag{2.2}$$

From (2.1) and (2.2),

$$\begin{cases} f_z = -\frac{1}{\pi} \int\!\!\int_D (\bar{\zeta}-\bar{z})(2\ln|z-\zeta|+1)\rho(\zeta)d\sigma_\zeta, \; f_{z^2} = \frac{1}{\pi} \int\!\!\int_D \frac{\bar{\zeta}-\bar{z}}{\zeta-z}\rho(\zeta)d\sigma_\zeta, \\[3mm] f_{z\bar{z}} = \frac{2}{\pi} \int\!\!\int_D (\ln|\zeta-z|+1)\rho(\zeta)d\sigma_\zeta, \; f_{z^3} = \frac{1}{\pi} \int\!\!\int_D \frac{\bar{\zeta}-\bar{z}}{(\zeta-z)^2}\rho(\zeta)d\sigma_\zeta = \tilde{T}\rho, \\[3mm] f_{z^2\bar{z}} = -\frac{1}{\pi} \int\!\!\int_D \frac{\rho(\zeta)}{\zeta-z}d\sigma_\zeta = T\rho, \; f_{z^4} = \frac{2}{\pi} \int\!\!\int_D \frac{\bar{\zeta}-\bar{z}}{(\zeta-z)^3}\rho(\zeta)d\sigma_\zeta = \tilde{\Pi}\rho, \\[3mm] f_{z^3\bar{z}} = -\frac{1}{\pi} \int\!\!\int_D \frac{\rho(\zeta)}{(\zeta-z)^2}d\sigma_\zeta = \Pi\rho, \; f_{z^2\bar{z}^2} = \rho(z), \end{cases} \tag{2.3}$$

can be obtained. Using a similar way as in Chapter 1 of [131]1) and Theorem 4.5, Chapter 1 in [140], we can derive the above formulae.

Lemma 2.1   (1) $\Pi\rho$ *and* $\tilde{\Pi}\rho$ *are linear bounded operators on* $L_p(\bar{D}), p > 1$ *and possess the following properties:*

$$L_p[\Pi\rho, \bar{D}] \leq \Lambda_p L_p[\rho, \bar{D}], \; L_p[\tilde{\Pi}\rho, \bar{D}] \leq \tilde{\Lambda}_p L_p[\rho, \bar{D}], \tag{2.4}$$

where Λ_p, $\tilde{\Lambda}_p$ are least positive constants only dependent on $p > 1$, such that (2.4) hold, and

$$\Lambda_2 = \tilde{\Lambda}_2 = 1. \tag{2.5}$$

Moreover, for two nonnegative constants q_0 and $q_0'(q_0 + q_0' < 1)$, there exist constants p_0 and $p_1(2 < p_0 < p_1)$, so that

$$q_0\Lambda_{p_j} + q_0'\tilde{\Lambda}_{p_j} < 1, \; j = 0, 1. \tag{2.6}$$

(2) If $\rho(z) \in L_p(\bar{D}), p > 2$, then

$$C_\beta^3[f, \bar{D}] \le M_1 L_p[\rho, \bar{D}], \tag{2.7}$$

where $\beta = 1 - 2/p$, $M_1 = M_1(p)$.

Proof (1) From Chapter 1 in [131]1), it can be seen that Π is a linear bounded operator on $L_p(\bar{D})$, $p > 1$, the first formula in (2.4) holds and $\Lambda_2 = 1$. $\tilde{\Pi}$ is a linear bounded operator too and the second formula in (2.4) can be proved by a similar method. In the following, we give the proof for $\tilde{\Lambda}_2 = 1$. Let $\rho(z) \in D_\infty^0$ and $\rho(z) = 0$, for $z \notin D$. Denote $E_R = \{|z| \le R\}(1 \le R < \infty)$ and $\Gamma_R = \{|z| = R\}$. Applying Green's formula, we have

$$\parallel \tilde{\Pi}\rho \parallel_{L_2(C)}^2 = [L_2(\tilde{\Pi}\rho, \mathcal{C})]^2 = \lim_{R \to \infty} \iint_{E_R} |\tilde{\Pi}\rho|^2 d\sigma_z = \lim_{R \to \infty} \iint_{E_R} |(\tilde{T}\rho)_z|^2 d\sigma_z$$

$$= \lim_{R \to \infty} \iint_{E_R} [(\tilde{\Pi}\rho\,\overline{\tilde{T}\rho})_{\bar{z}} - (\tilde{T}\rho)_{z\bar{z}}\,\overline{\tilde{T}\rho}] d\sigma_z$$

$$= \lim_{R \to \infty} [\frac{1}{2i} \int_{\Gamma_R} \tilde{\Pi}\rho\,\overline{\tilde{T}\rho}\,dz - \iint_{E_R} (\Pi\rho)_z\,\overline{\tilde{T}\rho}\,d\sigma_z] \tag{2.8}$$

$$= \lim_{R \to \infty} \{\frac{1}{2i} \int_{\Gamma_R} \tilde{\Pi}\rho\,\overline{\tilde{T}\rho}\,dz - \iint_{E_R} [(\Pi\rho\,\overline{\tilde{T}\rho})_z - |\Pi\rho|^2] d\sigma_z\}$$

$$= \lim_{R \to \infty} \{\frac{1}{2i} \int_{\Gamma_R} [\tilde{\Pi}\rho\,\overline{\tilde{T}\rho}\,dz + \Pi\rho\,\overline{\tilde{T}\rho}\,d\bar{z}] + \iint_{E_R} |\Pi\rho|^2 d\sigma_z\}.$$

Noting that

$$\lim_{R \to \infty} \iint_{E_R} |\Pi\rho|^2 d\sigma_z = \parallel \Pi\rho \parallel_{L_2(\mathcal{C})}^2 = \parallel \rho \parallel_{L_2(\mathcal{C})}^2 = \parallel \rho \parallel_{L_2(\bar{D})}^2,$$

and $\rho(z) = 0$ for $z \notin D$, we can obtain

$$\tilde{T}\rho = O(\frac{1}{R}), \; \Pi\rho = O(\frac{1}{R^2}), \; \tilde{\Pi}\rho = O(\frac{1}{R^2}).$$

Hence

$$\begin{cases} \lim_{R\to\infty} \dfrac{1}{2i}\displaystyle\int_{\Gamma_R} \tilde{\Pi}\rho\,\overline{\tilde{T}\rho}\,dz = \lim_{R\to\infty}\dfrac{1}{2i}\int_0^{2\pi} O(\dfrac{1}{R^2})O(\dfrac{1}{R})Rd\theta = 0, \\[3mm] \lim_{R\to\infty} \dfrac{1}{2i}\displaystyle\int_{\Gamma_R} \Pi\rho\,\overline{\tilde{T}\rho}\,d\bar{z} = \lim_{R\to\infty}\dfrac{1}{2i}\int_0^{2\pi} O(\dfrac{1}{R^2})O(\dfrac{1}{R})Rd\theta = 0. \end{cases} \tag{2.9}$$

From (2.8), (2.9), we have

$$\| \tilde{\Pi}\rho \|^2_{L_2(\bar{D})} \leq \| \tilde{\Pi}\rho \|^2_{L_2(\mathcal{C})} = \| \rho \|^2_{L_2(\bar{D})}\,.$$

By the density–property of $D^0_\infty(D)$ in $L_2(\bar{D})$ we see that for any function $\rho(z) \in L_2(\bar{D})$, there is $\| \tilde{\Pi}\rho \|_{L_2(\bar{D})} \leq \| \rho \|_{L_2(\bar{D})}$, this shows that $\tilde{\Lambda}_2 = 1$. On the basis of the Riesz theorem (see [131]1)), the inequality (2.6) is derived.

(2) If $\rho(z) \in L_p(\bar{D}), p > 2$, according to Theorem 1.19 in [131]1), $f_{z^2\bar{z}} = T\rho$ satisfies

$$C_\beta[f_{z^2\bar{z}}, \bar{D}] \leq M_2 L_p[\rho, \bar{D}], \ M_2 = M_2(p). \tag{2.10}$$

To prove that

$$C_\beta[f_{z^3}, \bar{D}] \leq M_3 L_p[\rho, \bar{D}], \ M_3 = M_3(p), \tag{2.11}$$

we choose any two points $z_1, z_2 \in D$ and consider

$$\begin{aligned} f_{z^3}(z_1) - f_{z^3}(z_2) &= \frac{1}{\pi}\iint_D \left[\frac{\bar{\zeta}-\bar{z}_1}{(\zeta-z_1)^2} - \frac{\bar{\zeta}-\bar{z}}{(\zeta-z_2)^2}\right]\rho(\zeta)d\sigma_\zeta \\ &= \frac{z_1-z_2}{\pi}\iint_D \frac{(\bar{\zeta}-\bar{z}_1)\rho(\zeta)}{(\zeta-z_1)^2(\zeta-z_2)}d\sigma_\zeta \\ &\quad + \frac{\bar{z}_2-\bar{z}_1}{\pi}\iint_D \frac{(\zeta-z_1)\rho(\zeta)}{(\zeta-z_1)^2(\zeta-z_2)}d\sigma_\zeta \\ &\quad + \frac{z_1-z_2}{\pi}\iint_D \frac{(\bar{\zeta}-\bar{z}_2)\rho(\zeta)}{(\zeta-z_1)(\zeta-z_2)^2}d\sigma_\zeta. \end{aligned}$$

By using the Hölder inequality, we have

$$|f_{z^3}(z_1) - f_{z^3}(z_2)| \leq \frac{3|z_1-z_2|}{\pi}\left[\iint_D (|\zeta-z_1||\zeta-z_2|)^{-q}d\sigma_\zeta\right]^{1/q}$$

$$\times L_p[\rho, \bar{D}] \leq M_4|z_1-z_2|^\beta L_p[\rho, \bar{D}],$$

where $q = p/(p-1), \beta = 1 - 2/p, M_4 = M_4(p, D)$, and

$$|f_{z^3}(z)| \leq \frac{1}{\pi}\left[\iint_D |\zeta-z|^{-q}d\sigma_\zeta\right]^{1/q} L_p[\rho, \bar{D}] \leq M_5 L_p[\rho, \bar{D}],$$

in which $M_5 = M_5(p, D)$. Hence, (2.11) is obtained. In addition, we can prove the inequality of f_{z^2} similarly to (2.10) and (2.11). Furthermore from

$$
\begin{cases}
f_{z\bar{z}}(z) - f_{z\bar{z}}(0) = \displaystyle\int_0^z f_{z^2\bar{z}}dz + \int_0^{\bar{z}} f_{\bar{z}^2 z}d\bar{z}, \\[2mm]
f_z(z) - f_z(0) = \displaystyle\int_0^z f_{z^2}dz + \int_0^{\bar{z}} f_{z\bar{z}}d\bar{z}, \\[2mm]
f(z) - f(0) = \displaystyle\int_0^z f_z dz + \int_0^{\bar{z}} f_{\bar{z}}d\bar{z},
\end{cases}
\tag{2.12}
$$

similar properties of $f_{z\bar{z}}, f_z$ and $f(z)$ can be derived. Thus formula (2.7) is concluded.

2.2　Representation and existence theorem of solutions for elliptic equations of fourth order

We first give the following representation of solutions for the equation (1.29).

Lemma 2.2　*Let equation* (1.29) *satisfy Condition* C *and* $u(z)$ *be a solution of* (1.29), $u(z) \in W_{p_0}^4(D)(2 < p_0 < p)$. *Then* $u(z)$ *may be expressed as*

$$
u(z) = U(z) + f(z) = U(z) + \frac{1}{\pi}\iint_D Y(z,\zeta)\rho(\zeta)d\sigma_\zeta,
\tag{2.13}
$$

where $U(z)$ *is a biharmonic function, and* $\rho(z) \in L_{p_0}(\bar{D})$ *is a solution of the following integral equation*

$$
\rho(\zeta) = F(z, U + f, U_z + f_z, \cdots, U_{z^4} + f_{z^4}, U_{z^3\bar{z}} + f_{z^3\bar{z}}).
\tag{2.14}
$$

Proof　Setting $\rho(z) = u_{z^2\bar{z}^2}$, and by the condition in this theorem, it can be seen that $\rho(z) \in L_{p_0}(\bar{D})$. Let $f(z)$ be the function as stated in (2.1), from (2.3) we know that $[u(z) - f(z)]_{z^2\bar{z}^2} = 0$, i.e. $U(z) = u(z) - f(z)$ is a biharmonic function in D. So $u(z)$ can be expressed as (2.13), where $\rho(z)$ is a solution of the integral equation (2.14).

Next, using Lemma 2.2, we prove the following existence theorem of solutions for (1.29).

Theorem 2.3　*Let equation* (1.29) *satisfy Condition* C *and* $U(z)$ *be a biharmonic function,* $U(z) \in C^3(\bar{D}) \cap W_{p_0}^4(D), 2 < p_0 < p$. *If the constant* k_1 *in* (1.30) *is small enough, then* (1.29) *has a solution possessing the form* (2.13), *where* $\rho(z)$ *is a solution of the integral equation* (2.14):

Proof　Let us substitute $U, U_z, \cdots, U_{z^4}, U_{z^3\bar{z}}$ into the corresponding positions of equation (1.29). If we can find the solution $\rho(z)$ of the integral equation (2.14), then $u(z)$ in the form (2.13) is just a solution of (1.29) in D. By (2.7),

$$
C[f_{z^j\bar{z}^k}, \bar{D}] \le M_{jk}L_{p_0}[\rho, \bar{D}], \quad j, k \ge 0, \quad 0 < j + k \le 3,
\tag{2.15}
$$

can be obtained, where $M_{jk} \leq M_1$. Provided that the constant k_1 in (1.30) is so small that

$$M_6 = q_0 \Lambda_{p_0} + q_0' \tilde{\Lambda}_{p_0} + k_1 \sum_{j+k=0}^{3} M_{jk} < 1, \tag{2.16}$$

then we can give an estimate of the solution $u(z) = U(z) + f(z)$ for the equation (1.29). In fact, since $\rho(z) \in L_{p_0}(\bar{D})$ is a solution of the integral equation (2.14) in D, i.e.

$$\rho(z) = \mathrm{Re}[Q_1 \tilde{\Pi} \rho] + \mathrm{Re}[Q_2 \Pi \rho] + \sum_{j+k=0}^{3} A_{jk} f_{z^j \bar{z}^k} + A, \tag{2.17}$$

where

$$A = \mathrm{Re}[Q_1 U_{z^4}] + \mathrm{Re}[Q_2 U_{z^3 \bar{z}}] + \sum_{j+k=0}^{3} A_{jk} U_{z^j \bar{z}^k} + A_0.$$

It is easy to see that

$$L_{p_0}[A, \bar{D}] \leq q_0 L_{p_0}[U_{z^4}, \bar{D}] + q_0' L_{p_0}[U_{z^3 \bar{z}}, \bar{D}]$$

$$+ k_1 \sum_{j+k=1}^{3} C[U_{z^j \bar{z}^k}, \bar{D}] + k_0 = M_7. \tag{2.18}$$

From (2.17),

$$L_{p_0}[\rho, \bar{D}] \leq M_7/(1 - M_6) = M_8 = M_8(q_0, q_0', p_0, k_0, D, U), \tag{2.19}$$

can be obtained, in which the constants M_6, M_7 are as stated in (2.16) and (2.18).

In the following, we prove the existence of solutions for the integral equation (2.14) by using the Schauder fixed–point theorem. Denote by B_M a bounded, closed and convex set in the Banach space $L_{p_0}(\bar{D})$, the elements of which are all measurable functions $\rho(z)$ satisfying condition (2.19). Choosing an arbitrary function $\rho(z) \in B_M$ and substituting it into (2.13), we obtain a function $u(z) = U(z) + f(z)$ and substitute it in the coefficients of equation (2.14) except for $u_{z^4}, u_{z^3 \bar{z}}$. Afterwards we consider the integral equation

$$\rho^*(z) = \mathrm{Re}[Q_1(U_{z^4} + \tilde{\Pi} \rho^*) + Q_2(U_{z^3 \bar{z}} + \Pi \rho^*)]$$

$$+ \sum_{j+k=0}^{3} A_{jk}(U_{z^j \bar{z}^k} + f^*_{z^j \bar{z}^k}) + A_0, \tag{2.20}$$

where

$$f^*(z) \;=\; \frac{1}{\pi} \iint_D Y(z, \zeta) \rho^*(\zeta) d\sigma_\zeta = \frac{1}{\pi} \iint_D 2|z - \zeta|^2 \ln|z - \zeta| \rho^*(\zeta) d\sigma_\zeta,$$

$$Q_j \;=\; Q_j(z, u, u_z, \cdots, u_{z^2 \bar{z}}, U_{z^4} + f^*_{z^4}, U_{z^3 \bar{z}} + f^*_{z^3 \bar{z}}), \; j = 1, 2,$$

$$A_{jk} \;=\; A_{jk}(z, u, u_z, \cdots, u_{z^2 \bar{z}}), \; j, k \geq 0, \; 0 \leq j + k \leq 3.$$

According to the principle of contracting mappings, from (2.20) we can find a unique solution $\rho^*(z) \in L_{p_0}(\bar{D})$, which satisfies the estimate (2.19). Denote by $\rho^* = S(\rho)$ the mapping from $\rho(z) \in B_M$ to $\rho^*(z)$. The above discussion shows that $\rho^* = S(\rho)$ maps B_M into itself. By Condition C, we can prove that $\rho^* = S(\rho)$ continuously maps B_M onto a compact set in B_M. On the basis of the Schauder theorem (see [117]), the integral equation (2.14) has a unique solution $\rho(z) \in B_M$. Hence the partial differential equation (1.29) possesses a solution $u(z)$ as the form (2.13).

2.3 Another double integral and its properties

In order to discuss the solvability of the Dirichlet boundary value problem for equation (1.29) in the unit disk D, it is needful to construct some integral operator satisfying the homogeneous boundary conditions. Hence we introduce the double integral

$$g(z) = \frac{1}{\pi} \iint_D Z(z,\zeta)\rho(\zeta)d\sigma_\zeta, \tag{2.21}$$

where $\rho(z)$ is a measurable function in $L_p(\bar{D}), p > 1$ and $Z(z,\zeta)$ is the Green's function, it is a biharmonic function of z in D except for the point $\zeta \in D$, given by

$$\left\{ \begin{aligned} &Z(z,\zeta) = 2|z - \zeta|^2 \ln\left|\frac{z-\zeta}{1-\bar{\zeta}z}\right| + (1 - |z|^2)(1 - |\zeta|^2), \\[2mm] &Z_z = 2(\bar{z} - \bar{\zeta})\ln\left|\frac{z-\zeta}{1-\bar{\zeta}z}\right| - \frac{(1 - |z|^2)(1 - |\zeta|^2)\bar{\zeta}}{1-\bar{\zeta}z}, \\[2mm] &Z_{z^2} = (\bar{z} - \bar{\zeta})\left[\frac{1}{z-\zeta} + \frac{\bar{\zeta}}{1-\bar{\zeta}z} + \frac{(1 - |\zeta|^2)\bar{\zeta}}{(1-\bar{\zeta}z)^2}\right], \\[2mm] &Z_{z\bar{z}} = 2\ln\left|\frac{z-\zeta}{1-\bar{\zeta}z}\right| + \frac{(\bar{z} - \bar{\zeta})\zeta}{1-\bar{\zeta}z} + \frac{1 - |\zeta|^2}{1-\bar{\zeta}z} + |\zeta|^2, \\[2mm] &Z_{z^3} = (\bar{z} - \bar{\zeta})\left[\frac{-1}{(z-\zeta)^2} + \frac{\bar{\zeta}^2}{(1-\bar{\zeta}z)^2} + \frac{2(1 - |\zeta|^2)\bar{\zeta}^2}{(1-\bar{\zeta}z)^3}\right], \\[2mm] &Z_{z^2\bar{z}} = \frac{1}{z-\zeta} + \frac{\bar{\zeta}}{1-\bar{\zeta}z} + \frac{(1 - |\zeta|^2)\bar{\zeta}}{(1-\bar{\zeta}z)^2}, \\[2mm] &Z_{z^4} = 2(\bar{z} - \bar{\zeta})\left[\frac{1}{(z-\zeta)^3} + \frac{\bar{\zeta}^3}{(1-\bar{\zeta}z)^3} + \frac{3(1 - |\zeta|^2\bar{\zeta}^3}{(1-\bar{\zeta}z)^4}\right], \\[2mm] &Z_{z^3\bar{z}} = \frac{-1}{(\zeta - z)^2} + \frac{\bar{\zeta}^2}{(1-\bar{\zeta}z)^2} + \frac{2(1 - |\zeta|^2)\bar{\zeta}^2}{(1-\bar{\zeta}z)^3}. \end{aligned} \right. \tag{2.22}$$

From (2.21) and (2.22),

$$
\begin{cases}
g_z = \dfrac{1}{\pi}\iint_D Z_z\rho(\zeta)d\sigma_\zeta, \quad g_{z^2} = \dfrac{1}{\pi}\iint_D Z_{z^2}\rho(\zeta)d\sigma_\zeta, \\[2mm]
g_{z\bar z} = \dfrac{1}{\pi}\iint_D Z_{z\bar z}\rho(\zeta)d\sigma_\zeta, \quad g_{z^3} = \dfrac{1}{\pi}\iint_D Z_{z^3}\rho(\zeta)d\sigma_\zeta, \\[2mm]
g_{z^2\bar z} = \dfrac{1}{\pi}\iint_D Z_{z^2\bar z}\rho(\zeta)d\sigma_\zeta, \quad g_{z^4} = \dfrac{1}{\pi}\iint_D Z_{z^4}\rho(\zeta)d\sigma_\zeta, \\[2mm]
g_{z^3\bar z} = \dfrac{1}{\pi}\iint_D Z_{z^3\bar z}\rho(\zeta)d\sigma_\zeta
\end{cases}
\tag{2.23}
$$

can be obtained. As in (2.3) and Theorem 3.5, Chapter 1 in [140], we can prove the above formulae.

Lemma 2.4   (1) $g_{z^4}, g_{z^3\bar z}$ *are linear bounded operators on* $L_p(\bar D), p > 1,$ *and*

$$
L_p[g_{z^4},\bar D] \le M_9 L_p[\rho,\bar D], \quad L_p[g_{z^3\bar z},\bar D] \le M_{10} L_p[\rho,\bar D],
\tag{2.24}
$$

where $M_j = M_j(p), j = 9,10.$

    (2) *If* $\rho(z) \in L_p(\bar D), p > 2,$ *then*

$$
C_\beta^3[g,\bar D] \le M_{11} L_p[\rho,\bar D],
\tag{2.25}
$$

where $\beta = 1 - 2/p,\ M_{11} = M_{11}(p).$ *Moreover,* $g(z)$ *satisfies the homogeneous boundary conditions*

$$
g(z) = 0, \quad \frac{\partial g}{\partial n} = 0, \quad z \in \Gamma = \partial D,
\tag{2.26}
$$

in which n *is the outward normal of the boundary* $\Gamma.$

Proof   (1) In order to prove that (2.24) holds, we rewrite $g_{z^3\bar z}$ in the form

$$
g_{z^3\bar z} = \frac{1}{\pi}\iint_D Z_{z^3\bar z}\rho(\zeta)d\sigma_\zeta = -\frac{1}{\pi}\iint_D \frac{\rho(\zeta)}{(\zeta - z)^2}d\sigma_\zeta
$$

$$
+\frac{1}{\pi}\iint_D \frac{\bar\zeta^2\rho(\zeta)}{(1-\bar\zeta z)^2}d\sigma_\zeta + \frac{2}{\pi}\iint_D \frac{(1-|\zeta|^2)\bar\zeta^2\rho(\zeta)}{(1-\bar\zeta z)^3}d\sigma_\zeta
$$

$$
= \Pi\rho + \Pi_1\rho + \Pi_2\rho,
\tag{2.27}
$$

and by $(\bar z - \bar\zeta)\zeta = 1 - |\zeta|^2 - (1 - \bar\zeta z),$ g_{z^4} can be rewritten in the form

$$
g_{z^4} = \frac{1}{\pi}\iint_D Z_{z^4}\rho(\zeta)d\sigma_\zeta = \frac{2}{\pi}\iint_D \frac{(\bar\zeta - \bar z)\rho(\zeta)}{(\zeta - z)^3}d\sigma_\zeta
$$

$$
-\frac{2}{\pi}\iint_D \frac{(1-\bar\zeta z)\bar\zeta^3\zeta^{-1}\rho(\zeta)}{(1-\bar\zeta z)^3}d\sigma_\zeta + \frac{2}{\pi}\iint_D \frac{(1-|\zeta|^2)\bar\zeta^3\zeta^{-1}\rho(\zeta)}{(1-\bar\zeta z)^3}d\sigma_\zeta
$$

$$-\frac{6}{\pi}\iint_D \frac{(1-\zeta\bar z)(1-|\zeta|^2)\bar\zeta^3\zeta^{-1}\rho(\zeta)}{(1-\bar\zeta z)^4}d\sigma_\zeta + \frac{6}{\pi}\iint_D \frac{(1-|\zeta|^2)^2\bar\zeta^3\zeta^{-1}\rho(\zeta)}{(1-\bar\zeta z)^4}d\sigma_\zeta$$

$$= \tilde\Pi\rho + \tilde\Pi_1\rho + \tilde\Pi_2\rho + \tilde\Pi_3\rho + \tilde\Pi_4\rho \tag{2.28}$$

in (2.27) and (2.28), $\Pi\rho, \Pi_j\rho\,(j=1,2), \tilde\Pi\rho, \tilde\Pi_j\rho\,(j=1,\cdots,4)$ are the integrals which arrange in sequence from first to last. The properties of $\Pi\rho, \tilde\Pi\rho$ are given in Lemma 2.1. $\Pi_j\rho, \tilde\Pi_j\rho$ are analytic in D and there appears a singularity at the point $z=\zeta=\exp(i\theta)$. The property of $\Pi_1\rho$ can be derived as in Lemma 1.2 of Chapter 2 in [139]13). For $\tilde\Pi_1\rho$ setting $\zeta=1/\bar\zeta_1$ and $\rho(z)=0$ in $|z|>1$, we have

$$\tilde\Pi_1\rho = -\frac{2}{\pi}\iint_D \frac{(1-\zeta\bar z)\bar\zeta^3\zeta^{-1}\rho(\zeta)}{(1-\bar\zeta z)^3}d\sigma_\zeta = -\frac{2}{\pi}\iint_{\tilde D} \frac{(\bar\zeta_1-\bar z)|\zeta_1|^{-4}\rho(1/\bar\zeta_1)}{(\zeta_1-z)^3}d\sigma_{\zeta_1},$$

where $\tilde D = \{|\zeta_1|\geq 1\}$. Noting that

$$\iint_{\tilde D} \||\zeta_1|^{-4}\rho(\frac{1}{\bar\zeta_1})|^p d\sigma_{\zeta_1} = \iint_D |\zeta|^{4p-4}|\rho(\zeta)|^p d\sigma_\zeta \leq \iint_D |\rho(\zeta)|^p d\sigma_\zeta$$

for $\rho(z)\in L_p(\bar D), p>1$. Hence as in (2.4), we get

$$L_p[\tilde\Pi_1\rho,\bar D] \leq L_p[\tilde\Pi_1\rho,C] \leq M_{12}L_p[\rho,C] \leq M_{12}L_p[\rho,\bar D], \tag{2.29}$$

where $M_{12}=M_{12}(p)$. Moreover,

$$\frac{1-|\zeta|^2}{|1-\bar\zeta z|} \leq \frac{1-|\zeta|^2}{1-|\bar\zeta z|} \leq 2 \text{ for } |\zeta|<1, |z|<1,$$

we can prove that the remained $\Pi_2\rho, \tilde\Pi_j\rho(j=2,3,4)$ possess properties similar to (2.29). Thus, (2.24) is derived.

(2) If $\rho(z)\in L_p(\bar D), p>2$, we can verify

$$C_\beta[g_{z^3},\bar D] \leq M_{13}L_p[\rho,\bar D],\ C_\beta[g_{z^2\bar z},\bar D] \leq M_{14}L_p[\rho,\bar D], \tag{2.30}$$

where $\beta=1-2/p$, $M_j=M_j(p), j=13,14$. In fact, $g_{z^2\bar z}, g_{z^3}$ can be written in the forms

$$g_{z^2\bar z} = \frac{1}{\pi}\iint_D Z_{z^2\bar z}\rho(\zeta)d\sigma_\zeta = -\frac{1}{\pi}\iint_D \frac{\rho(\zeta)}{\zeta-z}d\sigma_\zeta$$

$$+\frac{1}{\pi}\iint_D \frac{\bar\zeta\rho(\zeta)}{1-\bar\zeta z}d\sigma_\zeta + \frac{1}{\pi}\iint_D \frac{(1-|\zeta|^2)\bar\zeta\rho(\zeta)}{(1-\bar\zeta z)^2}d\sigma_\zeta$$

$$= T\rho + g_1(z) + g_2(z), \tag{2.31}$$

$$g_{z^3} = \frac{1}{\pi}\iint_D Z_{z^3}\rho(\zeta)d\sigma_\zeta = \frac{1}{\pi}\iint_D \frac{(\bar\zeta-\bar z)\rho(\zeta)}{(\zeta-z)^2}d\sigma_\zeta$$

$$+\frac{1}{\pi}\iint_D \frac{(\bar{z}-\bar{\zeta})\bar{\zeta}^2\rho(\zeta)}{(1-\bar{\zeta}z)^2}d\sigma_\zeta + \frac{2}{\pi}\iint_D \frac{(\bar{z}-\bar{\zeta})(1-|\zeta|^2)\bar{\zeta}^2\rho(\zeta)}{(1-\bar{\zeta}z)^3}d\sigma_\zeta$$

$$= \tilde{T}\rho + \tilde{g}_1(z) + \tilde{g}_2(z). \tag{2.32}$$

The above mentioned $T\rho, g_j(z), \tilde{T}\rho, \tilde{g}_j(z)(j=1,2)$ are the integrals arranged in sequence. The Hölder continuity of $T\rho, g_1(z), \tilde{T}\rho$ is given in Lemma 2.1 (see also Chapter 1 in [140]). We have

$$C_\beta[T\rho,\bar{D}] \le M_{15}L_p[\rho,\bar{D}], \ C_\beta[g_1,\bar{D}] \le M_{16}L_p[\rho,\bar{D}], \ C_\beta[\tilde{T}\rho,\bar{D}] \le M_{17}L_p[\rho,\bar{D}],$$
$$\tag{2.33}$$

where $\beta = 1 - 2/p$, $M_j = M_j(p), j = 15, 16, 17$. Furthermore,

$$g_2(z_1) - g_2(z_2) = \frac{1}{\pi}\iint_D \left[\frac{(z_1-z_2)(1-|\zeta|^2)\bar{\zeta}}{(1-\bar{\zeta}z_1)^2(1-\bar{\zeta}z_2)} + \frac{(z_1-z_2)(1-|\zeta|^2)\bar{\zeta}}{(1-\bar{\zeta}z_1)(1-\bar{\zeta}z_2)^2}\right]\rho(\zeta)d\sigma_\zeta,$$

where $|z_k| < 1, k = 1, 2$. Noting that $|\zeta| \le 1, |z| \le 1$ and $1 - \bar{\zeta}z_k \ne 0$ except for $\zeta = z_k \in \Gamma = \{|z| = 1\}, k = 1, 2$, similarly to the proof of (2.10), we can obtain

$$|g_2(z_1) - g_2(z_2)| \le |z_1 - z_2|M_{18}L_p[\rho,\bar{D}]$$

$$\times \left[\iint_D(|1-\bar{\zeta}z_1||1-\bar{\zeta}z_2|)^{-q}d\sigma_\zeta\right]^{1/q} \le M_{19}L_p[\rho,\bar{D}]|z_1-z_2|^\beta,$$

and

$$|g_2(z)| \le \frac{1}{\pi}L_p[\rho,\bar{D}]\left[\iint_D\left|\frac{1-|\zeta|^2}{(1-\bar{\zeta}z)^2}\right|^q d\sigma_\zeta\right]^{1/q} \le M_{20}L_p[\rho,\bar{D}],$$

where $M_j = M_j(p), j = 18, 19, 20$. This shows that $g_2(z)$ satisfies

$$C_\beta[g_2,\bar{D}] \le M_{21}L_p[\rho,\bar{D}], \ M_{21} = M_{21}(p). \tag{2.34}$$

Next,

$$\tilde{g}_1(z_1) - \tilde{g}_1(z_2) = \frac{z_1 - z_2}{\pi}\iint_D \frac{\bar{\zeta}^3(\bar{z}_1-\bar{\zeta})\rho(\zeta)}{(1-\bar{\zeta}z_1)^2(1-\bar{\zeta}z_2)}d\sigma_\zeta$$

$$+\frac{\bar{z}_1 - \bar{z}_2}{\pi}\iint_D \frac{\bar{\zeta}^2\rho(\zeta)}{(1-\bar{\zeta}z_1)(1-\bar{\zeta}z_2)}d\sigma_\zeta + \frac{z_1-z_2}{\pi}\iint_D \frac{\bar{\zeta}^3(\bar{z}_2-\bar{\zeta})\rho(\zeta)}{(1-\bar{\zeta}z_1)(1-\bar{\zeta}z_2)^2}d\sigma_\zeta.$$

Thus

$$|\tilde{g}_1(z_1) - \tilde{g}(z_2)| \le |z_1 - z_2|M_{22}L_p[\rho,\bar{D}]$$

$$\times \left[\iint_D |(1-\bar{\zeta}z_1)(1-\bar{\zeta}z_2)|^{-q}d\sigma_\zeta\right]^{1/q} \le M_{23}L_p[\rho,\bar{D}]|z_1-z_2|^\beta,$$

can be derived and

$$|\tilde{g}(z)| \le \frac{1}{\pi}L_p[\rho,\bar{D}]\left[\iint_D\left|\frac{\bar{\zeta}-\bar{z}}{(1-\bar{\zeta}z)^2}\right|^q d\sigma_\zeta\right]^{1/q} \le M_{24}L_p[\rho,\bar{D}],$$

where $M_j = M_j(p), j = 22, 23, 24$. Finally, we prove the Hölder continuity of $\tilde{g}_2(z)$ on $\bar{D}$. Taking into account

$$\tilde{g}_2(z_1) - \tilde{g}_2(z_2) = \frac{2}{\pi} \iint_D \left[\frac{\bar{\zeta} - \bar{z}_2}{(1 - \bar{\zeta} z_2)^2} - \frac{\bar{\zeta} - \bar{z}_1}{(1 - \bar{\zeta} z_1)} \right] \frac{(1 - |\zeta|^2)\bar{\zeta}^2}{1 - \bar{\zeta} z_2} \rho(\zeta) d\sigma_\zeta$$

$$+ \frac{2}{\pi} \iint_D \left[\frac{1}{1 - \bar{\zeta} z_2} - \frac{1}{1 - \bar{\zeta} z_1} \right] \frac{(\bar{\zeta} - \bar{z}_1)(1 - |\zeta|^2)\bar{\zeta}^2}{(1 - \bar{\zeta} z_1)^2} \rho(\zeta) d\sigma_\zeta.$$

and

$$\frac{1 - |\zeta|^2}{|1 - \bar{\zeta} z_2|} \le 2, \quad \frac{|\zeta(\zeta - z_1)(1 - |\zeta|^2)|}{|1 - \bar{\zeta} z_1|^2} \le 2,$$

for $|\zeta| < 1, |z_1| \le 1, |z_2| \le 1$,

$$|\tilde{g}_2(z_1) - \tilde{g}_2(z_2)| \le M_{25} L_p[\rho, \bar{D}] |z_1 - z_2|^\beta$$

and

$$|\tilde{g}_2(z)| \le M_{26} L_p[\rho, \bar{D}]$$

can be derived, where $M_j = M_j(p), j = 25, 26$. Combining the above estimates, (2.30) is proved. Afterwards using the method stated in (2.12), we can obtain the estimate (2.25). The boundary condition (2.26) can be verified by the property of $Z(z, \zeta)$ satisfying the boundary condition (2.26).

2.4 Dirichlet problem for elliptic equations of fourth order.

Problem D Let the linear and nonlinear elliptic equations (1.13) and (1.29) satisfy Condition C. Find a solution $u(z)$ of (1.13) and (1.29) which possesses continuous partial differentiable derivatives of order ≤ 3 and satisfies the boundary conditions:

$$u(z) = r_1(z), \quad \frac{\partial u}{\partial n} = r_2(z) \text{ on } \Gamma = \{|z| = 1\}, \tag{2.35}$$

where $r_1(z) \in C_\mu^3(\Gamma), r_2(z) \in C_\mu^2(\Gamma), 1/2 < \mu < 1$.

To prove the solvability of Problem D for the equation (1.29), we first find a solution $U(z)$ of Problem D for biharmonic functions by using the method stated in Sections 39-44 of [131]3), and prove that $U(z) \in C^3(\bar{D}) \cap W_{p_0}^4(D)(2 < p_0 < \min(p, 1/(1 - \mu)))$. Moreover, we find a solution $u(z)$ of Problem D for equation (1.29) in the following form

$$u(z) = U(z) + g(z) = U(z) + \frac{1}{\pi} \iint_D Z(z, \zeta) \rho(\zeta) d\sigma_\zeta, \tag{2.36}$$

where $\rho(\zeta) \in L_{p_0}(\bar{D})(2 < p_0 < \min(p, 1/(1 - \mu)))$. It is sufficient to solve the following integral equation

$$\rho(z) = F(z, u, u_z, \cdots, U_{z^4} + g_{z^4}, U_{z^3 \bar{z}} + g_{z^3 \bar{z}}). \tag{2.37}$$

Let the solution $\rho(z) \in L_{p_0}(\bar{D})$ of (2.37) be substituted in (2.36), we know that $u(z)$ is just a solution of Problem D for (1.29). For this purpose, we assume that (1.29) satisfies Condition C and the inequality

$$q_0 M_9 + q_0' M_{10} + k_1 \sum_{j+k=0}^{3} M_{jk} < 1 \qquad (2.38)$$

holds, in which q_0, q_0', k_1 are the constants in (1.30) and (1.31), and M_{jk} is the least constant such that the following formulae remain valid.

$$C[g_{z^j\bar{z}^k}, \bar{D}] \leq M_{jk} L_p[\rho, \bar{D}], \; j, k \geq 0, 0 \leq j + k \leq 3, \qquad (2.39)$$

where $\rho(z)$ is any measurable function in $L_{p_0}(\bar{D})$, herein $p_0 \, (2 < p_0 < p)$ is a constant.

Theorem 2.5   *Suppose that the nonlinear elliptic equation (1.29) satisfies Condition C, and the constants q_0, q_0', k_1 are small enough such that (2.38) holds, then Problem D for (1.29) is solvable.*

Proof   Similarly to the proof of Theorem 2.3, we can prove this theorem by using the principle of contracting mappings and the Schauder fixed–point theorem.

Secondly we discuss the linear elliptic equation (1.13) which we rewrite in the form

$$\begin{cases} U_{z^2\bar{z}^2} - \mathrm{Re}[Q_1(z)u_{z^4} + Q_2(z)u_{z^3\bar{z}}] = \lambda f(z, u, \cdots, u_{z^2\bar{z}}) + A_0(z), \\[2mm] f = \displaystyle\sum_{j+k=1}^{3} A_{jk}(z)u_{z^j\bar{z}^k}, \; |Q_1(z)| \leq q_0, \; |Q_2(z)| \leq q_0', \\[2mm] L_{p_0}[A_{jk}, \bar{D}] \leq k_0, \; L_{p_0}[A_0, \bar{D}] \leq k_0, \end{cases} \qquad (2.40)$$

where q_0, q_0', k_0 are nonnegative constants, and λ is the real parameter. If we choose q_0, q_0' so small that

$$q_0 M_9 + q_0' M_{10} < 1, \qquad (2.41)$$

in which M_9, M_{10} are constants in (2.24), and denote by

$$\lambda_j \, (0 < |\lambda_j| \leq |\lambda_{j+1}|, \; j = 1, 2, \cdots) \qquad (2.42)$$

the isolated eigenvalues of the homogeneous equation

$$\rho(z) - \mathrm{Re}[Q_1(z)g_{z^4} + Q_2(z)g_{z^3\bar{z}}] = \lambda f(z, g, \cdots, g_{z^2\bar{z}}), \qquad (2.43)$$

with a similar method used in the proof of Theorem 3.6, Chapter 5 in [140], we may prove

Theorem 2.6   *Suppose that the linear elliptic equation (2.40) satisfies condition (2.41). If $\lambda \neq \lambda_j(j = 1, 2, \cdots)$, where $\lambda_j(j = 1, 2, \cdots)$ are as stated in (2.42), then*

(2.40) *has the solution of the form* (2.36). *If* $\lambda = \lambda_j (1 \leq j < \infty)$, *we can also give the solvability condition of Problem D for* (2.40).

Besides, we can construct the integral operator satisfying the boundary condition

$$u(z) = r_1(z), \ \Delta u = r_2(z), \ z \in \Gamma = \{|z| = 1\}, \tag{2.44}$$

and discuss the solvability of the boundary value problem (1.29),(2.44) or (2.40),(2.44) satisfying some conditions, the solution of which can be expressed as the form (2.36), where $U(z)$ is a biharminic function satisfying the boundary condition (2.44), and $Z(z, \zeta)$ is the biharmonic Green's function satisfying the homogeneous boundary condition of (2.44), i.e.

$$Z(z, \zeta) = 2|z - \zeta|^2 \ln \left| \frac{z - \zeta}{1 - \bar{\zeta} z} \right| - 2(1 - |z|^2)(1 - |\zeta|^2) \mathrm{Re} \frac{\ln(1 - \bar{\zeta} z)}{\bar{\zeta} z} \tag{2.45}$$

(see [139]13)). As for the solvability of Problem D for equation (1.29) or (2.40) in a multiply connected domain, the problem is well worth continuously investigating.

3 Existence Theorem and Boundary Value Problems for Elliptic Systems of Fourth Order Equations

3.1 Existence of solutions for an elliptic system of fourth order

We introduce a double integral

$$f(z) = \frac{1}{\pi} \iint_D Y(z, \zeta) \rho(\zeta) d\sigma_\zeta, \tag{3.1}$$

where $\rho(z)$ is a complex variable measurable function, $\rho(z) \in L_p(\bar{D})$, $p > 1$, D is a bounded domain, and $Y(z, \zeta)$ as stated in (2.2). Similarly to (2.3), we have

$$\begin{cases} f_{z^3} = \tilde{T}\rho, \ \bar{f}_{z^3} = \tilde{T}\bar{\rho}, \ f_{z^2\bar{z}} = T\rho, \ \bar{f}_{z^2\bar{z}} = T\bar{\rho}, \\ f_{z^4} = \tilde{\Pi}\rho, \ \bar{f}_{z^4} = \tilde{\Pi}\bar{\rho}, \ f_{z^3\bar{z}} = \Pi\rho, \ \bar{f}_{z^3\bar{z}} = \Pi\bar{\rho}, \ f_{z^2\bar{z}^2} = \rho. \end{cases} \tag{3.2}$$

In addition, we can derive the following lemma.

Lemma 3.1   (1) *If* $\rho(z) \in L_p(\bar{D}), p > 1$, *then*

$$L_p[\Pi\rho, \bar{D}] \leq \Lambda_p L_p[\rho, \bar{D}], \ L_p[\tilde{\Pi}\rho, \bar{D}] \leq \tilde{\Lambda}_p L_p[\rho, \bar{D}], \ \Lambda_2 = \tilde{\Lambda}_2 = 1. \tag{3.3}$$

Moreover, for the nonnegative constants q_0, q_0', q_0'', q_0''' *satisfying the condition* $q_0 + q_0' + q_0'' + q_0''' < 1$, *there exist positive constants* $p_0, p_1 (2 < p_0 < p_1)$, *such that*

$$(q_0 + q_0')\tilde{\Lambda}_{p_j} + (q_0'' + q_0''')\Lambda_{p_j} < 1, \ j = 0, 1. \tag{3.4}$$

(2) *If $\rho(z) \in L_p(\bar{D}), p > 2$, then*

$$C_\beta^3[f, \bar{D}] \le M_1 L_p[\rho, \bar{D}], \tag{3.5}$$

where $\beta = 1 - 2/p$, $M_1 = M_1(p, D)$.

By using the same method used in the proof of Lemma 2.2 and Theorem 2.3, we can obtain the representation and the existence theorem of solutions for the uniformly elliptic complex equation (1.32).

Theorem 3.2   *Let* (1.32) *satisfy Condition C.*

(1) *If $w(z)$ is a solution of* (1.32), $w(z) \in W_{p_0}^4(D) \, (p_0 > 2)$, *then $w(z)$ can be expressed as*

$$w(z) = W(z) + f(z) = W(z) + \frac{1}{\pi} \int\!\!\int_D Y(z, \zeta)\rho(\zeta)d\sigma_\zeta, \tag{3.6}$$

in which $W(z)$ is a biharmonic function of a complex variable in D, i.e. $W_{z^2\bar{z}^2} = 0$ in D, and $\rho(z) \in L_{p_0}(\bar{D})$.

(2) *If $W(z)$ is a biharmonic function of a complex variable in D, $W(z) \in C^3(\bar{D}) \cap W_{p_0}^4(D)$, and the constant k_1 in Condition C is small enough, then* (1.32) *possesses a solution in the form* (3.6), *where $\rho(z) \in L_{p_0}(\bar{D})$ is a solution of the following integral equation*

$$\rho(z) = F(z, w, w_z, \cdots, W_{z^4} + \tilde{\Pi}\rho, W_{z^3\bar{z}} + \Pi\rho, \overline{W}_{z^3\bar{z}} + \Pi\bar{\rho}, \overline{W}_{z^4} + \tilde{\Pi}\bar{\rho}) \tag{3.7}$$

in D.

3.2   Riemann–Hilbert problem for an elliptic system of fourth order

First of all, we assume D is the unit disk, and discuss the following Riemann–Hilbert boundary value problem.

Problem A   Find a solution $w(z)$ of the complex equation (1.32), which possesses continuous partial derivatives up to third order and satisfies the boundary condition

$$\begin{cases} \mathrm{Re}[\bar{z}^{K_1}\overline{w(z)}] = r_1(z), \ \mathrm{Re}[\bar{z}^{K_2}w_z] = r_2(z), \\ \mathrm{Re}[\bar{z}^{K_3}\overline{w_{z\bar{z}}}] = r_2(z), \ \mathrm{Re}[\bar{z}^{K_4}w_{z^2\bar{z}}] = r_4(z), \end{cases} z \in \Gamma = \partial D, \tag{3.8}$$

where $r_j(z) \in C_\mu^{4-j}(\Gamma)$, $K_j \, (j = 1, \cdots, 4)$ are integers, $1/2 < \mu < 1$. If $K_j \ge 0$, we suppose that

$$\int_\Gamma [s_j(z) - \Phi_j(z)]e^{-im\theta}d\theta = 0, \ m = 1, \cdots, 2K_j, \ 1 \le j \le 4, \tag{3.9}$$

where $s_1(z) = \overline{w(z)}$, $s_2(z) = w_z$, $s_3(z) = \overline{w_{z\bar z}}$, $s_4(z) = w_{z^2\bar z}$, and $\Phi_j(z)\,(j = 1, \cdots, 4)$ are analytic functions as follows:

$$\Phi_j(z) = \begin{cases} \dfrac{z^{K_j}}{2\pi i} \displaystyle\int_\Gamma r_j(t)\dfrac{t+z}{t-z}\dfrac{dt}{t} + \sum_{m=0}^{2K_j} c_m^{(j)} z^m & \text{for } K_j \geq 0, \\[3ex] \dfrac{1}{\pi i} \displaystyle\int_\Gamma \dfrac{r_j(t)}{t^{-K_j}(t-z)}dt & \text{for } K_j < 0,\ j = 1, \cdots, 4. \end{cases}$$

Here $c_{2K_j-m}^{(j)} = -\overline{c_m^{(j)}}$, $m = 0, 1, \cdots, K_j$ for $K_j \geq 0$, $j = 1, \cdots, 4$ are arbitrary complex constants (see Chapter 5 in [140]).

Problem B   When one of the $K_j\,(j = 1, \cdots, 4)$ is negative integers, Problem A may not be solvable. Hence we introduce the modified Riemann-Hilbert boundary value problem as follows:

$$\begin{cases} \mathrm{Re}[\bar z^{K_1}\overline{w(z)}] = r_1(z) + h_1(z), \ \mathrm{Re}[\bar z^{K_2}w_z] = r_2(z) + h_2(z), \\[1ex] \mathrm{Re}[\bar z^{K_3}\overline{w_{z\bar z}}] = r_3(z) + h_3(z), \ \mathrm{Re}[\bar z^{K_4}w_{z^2\bar z}] = r_4(z) + h_4(z), \end{cases} \quad z \in \Gamma, \qquad (3.10)$$

and

$$h_j(z) = \begin{cases} h_0^{(j)} + \mathrm{Re}\sum_{m=1}^{-K_j-1}(h_m^{(j)} + ih_{-m}^{(j)})z^m & \text{for } K_j < 0, \\[1ex] 0 & \text{for } K_j \geq 0, \ z \in \Gamma, \ 1 \leq j \leq 4, \end{cases} \qquad (3.11)$$

where $h_0^{(j)}, h_m^{(j)}, h_{-m}^{(j)}(m = 1, \cdots, -K_j - 1,\ k = 1, \cdots, 4)$ are undetermined real constants, $r_j(z)\,(j = 1, \cdots, 4)$ are the same as those in (3.8). If $K_j \geq 0$, we assume that the solution $w(z)$ satisfies (3.9).

Let $w(z)$ be a solution of Problem B for (1.32) and $w_{z^2\bar z^2} = \rho(z) \in L_{p_0}(\bar D)\,(2 < p_0 < p)$. According to Theorem 4.1, Chapter 5 in [140], $w_{\bar z z}$ can be expressed as

$$w_{\bar z z} = \overline{\Phi_3(z)} + \overline{P_3}\Phi_4 + H_4\rho + \overline{R_3}H_4\rho, \qquad (3.12)$$

Consequently

$$w(z) = W_1(z) + G(z) = W_1(z) + W(z) + f(z), \qquad (3.13)$$

where

$$\begin{cases} W_1(z) = \overline{\Phi_1(z)} + \overline{P_1}\Phi_2 + H_2(\overline{\Phi_3} + \overline{P_3}\Phi_4) + \overline{R_1}H_2(\overline{\Phi_3} + \overline{P_3}\Phi_4), \\[1.5ex] G(z) = H_2(H_4\rho + \overline{R_3}H_4\rho) + \overline{R_1}H_2(H_4\rho + \overline{R_3}H_4\rho), \\[1.5ex] f(z) = \dfrac{1}{\pi}\displaystyle\iint_D Y(z,\zeta)\rho(\zeta)d\sigma_\zeta, \end{cases} \qquad (3.14)$$

in which

$$H_j\rho = \frac{2}{\pi}\iint_D [\ln|1 - \frac{z}{\zeta}|\rho(\zeta) + g_j(z,\zeta)\overline{\rho(\zeta)}]d\sigma_\zeta, \quad P_j\rho = (H_j\rho)_z,$$

$$g_j(z,\zeta) = \begin{cases} \mathrm{Re}\bar\zeta^{-2K_j-2}[\ln(1-\bar\zeta z) + \sum_{m=1}^{2K_j+1}(\bar\zeta z)^m/m] \text{ for } K_j \geq 0, \\[2mm] \mathrm{Re}\bar\zeta^{-2K_j-2}\ln(1-\bar\zeta z) \text{ for } K_j < 0,\ j = 1,\cdots,4, \end{cases}$$

$$R_j\bar\omega = \begin{cases} \dfrac{-1}{2\pi i}\displaystyle\int_\Gamma \dfrac{\overline{\omega(t)} + \bar t z^{2K_j+1}\omega(t)}{t - z}dt \text{ for } K_j \geq 0, \\[3mm] \dfrac{-1}{2\pi i}\displaystyle\int_\Gamma \dfrac{\overline{\omega(t)} + \bar t^{2|K_j|}\omega(t)}{t - z}dt \text{ for } K_j < 0,\ j = 1,\cdots,4, \end{cases}$$

(see Theorem 4.1, Chapter 5 in [140]). It is not difficult to see that $W_1(z)$ is a solution of the complex biharmonic equation $W_{z^2\bar z^2} = 0$, and $W(z) = G(z) - f(z)$ is a biharmonic complex variable function in D. We arbitrarily choose a function $\rho(z) \in L_{p_0}(\bar D), 2 < p_0 < p$, and construct the two integrals $f(z), G(z)$ from (3.14). Afterwards we find a biharmonic complex function $W(z) = G(z) - f(z)$ satisfying the homogeneous boundary condition

$$\begin{cases} \mathrm{Re}\{\bar z^{K_1}[\overline{W(z)} + \overline{f(z)}]\} = h_1(z), \\[2mm] \mathrm{Re}\{\bar z^{K_2}[W_z + f_z]\} = h_2(z), \\[2mm] \mathrm{Re}\{\bar z^{K_3}[\overline{W_{z\bar z}} + \overline{f_{z\bar z}}]\} = h_3(z), \\[2mm] \mathrm{Re}\{\bar z^{K_4}[W_{z^2\bar z} + f_{z^2\bar z}]\} = h_4(z), \end{cases} \qquad z \in \Gamma, \qquad (3.15)$$

where $h_j(z)\,(j = 1,\cdots,4)$ are as stated in (3.11). Since $\overline{W} + \overline{f}, W_z + f_z, \overline{W_{z\bar z}} + \overline{f_{z\bar z}}, W_{z^2\bar z} + f_{z^2\bar z}$ satisfy the condition (3.9) when $K_j \geq 0(1 \leq j \leq 4)$. It can be seen that the above boundary value problem (3.15) for biharmonic functions has a unique solution $W(z)$. Applying Lemma 3.1 and the method in the proof of Lemma 4.2, Chapter 5 in [140], we can prove that $W(z) + f(z)$ satisfies

$$\begin{cases} C_\alpha^3[W + f, \bar D] \leq M_2 L_{p_0}[\rho, \bar D], \\[2mm] L_{p_0}[|(W + f)_{z^3\bar z}| + |(\overline{W} + \overline{f})_{z^3\bar z}|, \bar D] \leq M_3 L_{p_0}[\rho, \bar D], \\[2mm] L_{p_0}[|(W + f)_{z^4}| + |(\overline{W} + \overline{f})_{z^4}|, \bar D] \leq M_4 L_{p_0}[\rho, \bar D], \end{cases} \qquad (3.16)$$

where $\beta = 1 - 2/p_0$, $M_j = M_j(p_0), j = 2,3,4$, $2 < p_0 < \min(p, 1/(1 - \mu))$.

Theorem 3.3 *Suppose that the nonlinear complex equation (1.32) satisfies Condition C'.*

(1) *Under some conditions, if any solution of Problem B for* (1.32) *satisfies*

$$L_{p_0}[\rho, \bar{D}] < M_5, \quad \rho(z) = w_{z^2 \bar{z}^2}, \tag{3.17}$$

where M_5 is a positive constant, then Problem B for (1.32) *is solvable, and the solution $w(z)$ can be expressed as in* (3.13).

(2) *If the constants q_0, q_0', q_0'', q_0''' and k_1 are sufficiently small, such that*

$$(q_0' + q_0'')M_3 + (q_0 + q_0''')M_4 + k_1 \sum_{j+k=0}^{3} (M_{jk} + \tilde{M}_{jk}) < 1, \tag{3.18}$$

in which M_3, M_4 are the constants in (3.16), *M_{jk} and $\tilde{M}_{jk}$ are the least constants satisfying the inequalities*

$$\begin{cases} C[(W+f)_{z^j \bar{z}^k}, \bar{D}] \le M_{jk} L_{p_0}[\rho, \bar{D}], \\ C[(\overline{W}+\overline{f})_{z^j \bar{z}^k}, \bar{D}] \le \tilde{M}_{jk} L_{p_0}[\rho, \bar{D}], \end{cases} \tag{3.19}$$

where $\rho(z) \in L_{p_0}(\bar{D})$, then Problem B for (1.32) *is solvable.*

Proof (1) We first assume that the coefficients of the complex equation (1.32) are equal to zero in the neighborhood of the boundary Γ, and denote by B_M a bounded open set in the Banach space $L_{p_0}(\bar{D})$ $(2 < p_0 < \min(p, 1/(1-\mu)))$, the elements of which are measurable complex function $\rho(z)$ satisfying (3.17). We choose an arbitrary function $\rho(z) \in \overline{B_M}$ and form a double integral $f(z)$ as in (3.1). Moreover, we find two biharmonic complex functions $W_1(z)$ and $W(z)$ as stated in (3.13) and (3.14), such that $w(z) = W_1(z) + W(z) + f(z)$ satisfy the boundary condition (3.10) and the corresponding integral condition. Substituting $w(z), W_1(z)+W(z)$ into the appropriate positions of (1.32), and consider the following integral equation with the parameter $t \in [0,1]$:

$$\rho^*(z) = tF[z, w, \cdots, (W_1 + W)_{z^4} + \tilde{\Pi}\rho^*, (W_1 + W)_{z^3 \bar{z}}$$
$$+ \Pi\rho^*, (\overline{W_1} + \overline{W})_{z^3 \bar{z}} + \Pi\bar{\rho}^*, (\overline{W_1} + \overline{W})_{z^4} + \tilde{\Pi}\bar{\rho}^*]. \tag{3.20}$$

By Condition C and the principle of contracting mappings, from (3.20) we can find a unique solution $\rho^*(z) \in L_{p_0}(\bar{D})$. Denote by $\rho^*(z) = S(\rho, t)$ the mapping from $\rho(z) \in B_M$ onto $\rho^*(z)$. Similarly to the proof of Theorem 2.4, Chapter 5 in [140], it can be verified that $\rho^* = S(\rho, t)(0 \le t \le 1)$ satisfies the three conditions of the Leray-Schauder theorem (see [78]). Hence when $t = 1$, the integral equation (3.20) has a solution $\rho^*(z) = \rho(z) \in B_M$. Let $\rho(z)$ be substituted into (3.13) and (3.14), the solution $w(z) = W_1(z) + W(z) + f(z)$ of Problem B for (1.32) is found. Furthermore, according to the method of the proof of Theorem 2.4, Chapter 5 in [140], we can cancel the hypothesis that the coefficients equal zero in the neighborhhod of the boundary Γ.

(2) As in the proof of Theorem 2.5, we can prove the second result.

In addition, from the above theorem, the solvability conditions of Problem A for (1.32) can be derived (see [140]).

In the case of a multiply connected circular domain, the result of solvability on the Riemann–Hilbert problem for (1.32) was discussed in [95].

4   Nonlinear Elliptic Equations and Sytems of 2nth Order

Under some conditions the nonlinear elliptic equation and system of $2n$th$(n \geq 3)$ order can be transformed into the complex forms by using the method as stated in Section 1. In the future, we shall only discuss the nonlinear elliptic equations in the complex form.

4.1   Nonlinear elliptic equations of 2nth order

We consider the nonlinear equation of $2n$th order

$$u_{z^n \bar{z}^n} = F(z, u, u_z, \cdots, u_{z^{2n}}, u_{z^{2n-1}\bar{z}}, \cdots, u_{z^{n+1}\bar{z}^{n-1}}), \tag{4.1}$$

and suppose that (4.1) satisfies the uniform ellipticity condition, i.e. the following inequality holds:

$$|F(z, u, u_z, \cdots, V_1^{(1)}, V_1^{(2)}, \cdots, V_1^{(n)}) - F(z, u, u_z, \cdots, V_2^{(1)},$$
$$V_2^{(2)}, \cdots, V_2^{(n)})| \leq \sum_{j=1}^{n} q_0^{(j)} |V_1^{(j)} - V_2^{(j)}|, \quad \sum_{j=1}^{n} q_0^{(j)} < 1, \tag{4.2}$$

for almost every point $z \in D$, any $u \in \mathbb{R}$ and $u_z, \cdots, V_1^{(j)}, V_2^{(j)} \in \mathbb{C} (j = 1, \cdots, n)$, where D is a bounded domain, $q_0^{(j)} (j = 1, \cdots, n)$ are all nonnegative constants. The equation (4.1) can be written in the form

$$u_{z^n \bar{z}^n} = F, \quad F = \sum_{j=0}^{n-1} \text{Re}[Q_j u_{z^{2n-j}\bar{z}^j}] + \sum_{j+k=0}^{2n-1} A_{jk} u_{z^j \bar{z}^k} + A_0, \tag{4.3}$$

where

$$Q_j = Q_j(z, u, u_z, \cdots, u_{z^{2n}}, \cdots, u_{z^{n+1}\bar{z}^{n-1}}), \quad 0 \leq j \leq n - 1,$$
$$A_{jk} = A_{jk}(z, u, u_z, \cdots, u_{z^{n+1}\bar{z}^{n-2}}), \quad 0 \leq j + k \leq 2n - 1,$$
$$A_0 = A_0(z, u, u_z, \cdots, u_{z^{n+1}\bar{z}^{n-2}}), \quad j, k \geq 0.$$

We assume that (4.3) satisfies Condition C, which is similar to the case of (1.29) with the main conditions (4.2) and

$$L_p[A_0, \bar{D}] \leq k_0, \quad L_p[A_{jk}, \bar{D}] \leq k_1, \quad j, k \geq 0, \quad 0 \leq j + k \leq 2n - 1, \tag{4.4}$$

in which $k_0, k_1, p(> 2)$ are nonnegative constants.

To give the representation and the existence theorem of solutions for (4.3), we introduce an n–harmonic function except for a point $z(\neq \zeta) \in D$:

$$Y(z, \zeta) = \frac{2}{[(n-1)!]^2} |z - \zeta|^{2(n-1)} \ln |z - \zeta|, \tag{4.5}$$

which is a solution of the n–harmonic equation

$$\Delta^n u = 0, \text{ i.e. } u_{z^n \bar{z}^n} = 0, \tag{4.6}$$

and construct a double integral

$$f(z) = \frac{1}{\pi} \iint_D Y(z, \zeta) \rho(\zeta) d\sigma_\zeta, \ \rho(z) \in L_p(\bar{D}), \ p > 1. \tag{4.7}$$

The integral possesses some similar properties as stated in Lemma 2.1, for instance

$$\begin{cases} f_{z^n \bar{z}^n} = \rho(z), \\ f_{z^{2n}} = \dfrac{n(-1)^n}{\pi} \iint_D \dfrac{(\bar{z} - \bar{\zeta})^{n-1}}{(z - \zeta)^{n+1}} \rho(\zeta) d\sigma_\zeta, \\ f_{z^{n+1} \bar{z}^{n-1}} = \dfrac{-1}{\pi} \iint_D \dfrac{\rho(\zeta)}{(\zeta - z)^2} d\sigma_\zeta = \Pi\rho. \end{cases} \tag{4.8}$$

Let $u(z)$ be a solution of the equation (4.3) in D and $u(z) \in W_{p_0}^{2n}(D)(2 < p_0 \leq p)$. Then $u(z)$ can be expressed as

$$u(z) = U(z) + f(z) = U(z) + \frac{1}{\pi} \iint_D Y(z, \zeta) \rho(\zeta) d\sigma_\zeta, \tag{4.9}$$

where $\rho(z) = u_{z^n \bar{z}^n}$, $U(z)$ is a n–harmonic function.

If the constant k_1 in (4.4) is sufficiently small, and $U(z) \in C^{2n-1}(\bar{D}) \cap W_{P_0}^{2n}(D)$, then the nonlinear uniformly elliptic equation (4.3) satisfying some conditions has a solution of the form (4.9).

The foregoing results can be proved by using a similar methods as in the proofs of Lemma 2.2 and Theorem 2.3.

In this section, we mainly discuss the Dirichlet boundary value problem D for the nonlinear equation (4.3) of $2n$th order in the unit disk D with the boundary condition

$$u(z) = r_1(z), \frac{\partial u}{\partial \nu} = r_2(z), \cdots, \frac{\partial^{(n-1)} u}{\partial \nu^{(n-1)}} = r_n(z), z \in \Gamma = \{|z| = 1\}, \tag{4.10}$$

in which ν is the outward normal to Γ, and $r_j(z) \in C_\mu^{2n-j}(\Gamma)$, $j = 1, \cdots, n, 1/2 < \mu < 1$.

From Chapter 5 in [131]3), we see that there exists an n–harmonic function $H(z)$ satisfying the boundary condition (4.10). Thus $U(z) = u(z) - H(z)$ is the solution of the equation

$$\begin{cases} U_{z^n \bar{z}^n} = F(z, U + H, U_z + H_z, \cdots, U_{z^{2n}} + H_{z^{2n}}, U_{z^{2n-1}\bar{z}}, +H_{z^{2n-1}\bar{z}}, \cdots, \\[2mm] U_{z^{n+1}\bar{z}^{n-1}} + H_{z^{n+1}\bar{z}^{n-1}}) = f(z, U, U_z, \cdots, U_{z^{2n}}, U_{z^{2n-1}\bar{z}}, \cdots, U_{z^{n+1}\bar{z}^{n-1}}) \end{cases} \quad (4.11)$$

in D satisfying the homogeneous boundary condition

$$U(z) = 0, \; \frac{\partial U}{\partial \nu} = 0, \cdots, \frac{\partial^{(n-1)}U}{\partial \nu^{(n-1)}} = 0, \; z \in \Gamma. \quad (4.12)$$

The boundary value problem $(4.11), (4.12)$ is called Problem D_0.

We first introduce a double integral

$$\begin{cases} g(z) = \dfrac{1}{\pi} \iint_D X(z, \zeta)\rho(\zeta)d\sigma_\zeta = V(z) + W(z), \\[3mm] V(z) = \dfrac{1}{\pi} \iint_D Y(z, \zeta)\rho(\zeta)d\sigma_\zeta, \; W(z) = \dfrac{1}{\pi} \iint_D Z(z, \zeta)\rho(\zeta)d\sigma_\zeta, \end{cases} \quad (4.13)$$

where $\rho(z)$ is a real measurable function, $\rho(z) \in L_p(\bar{D})$, $p > 2$, and

$$\begin{cases} X(z, \zeta) = Y(z, \zeta) + Z(z, \zeta), \\[2mm] Y(z, \zeta) = \dfrac{2}{[(n-1)!]^2}|z - \zeta|^{2(n-1)} \ln |z - \zeta|, \\[2mm] Z(z, \zeta) = -\dfrac{|z - \zeta|^{2(n-1)}}{[(n-1)!]^2}[\ln |1 - \zeta z|^2 \\[2mm] \quad + \displaystyle\sum_{m=1}^{n-1} \dfrac{(-1)^m}{m}|z - \zeta|^{2(n-m-1)}(1 - z\bar{z})^m(1 - \zeta\bar{\zeta})^m]. \end{cases} \quad (4.14)$$

Lemma 4.1 *Suppose that $\rho(z) \in L_p(\bar{D})$, $p > 2$.*

(1) If $0 \le j + k \le 2n - 1$, $j, k \ge 0$, then

$$C_\beta[g_{z^j \bar{z}^k}, \bar{D}] \le M_1 L_p[\rho, \bar{D}], \quad (4.15)$$

where $\beta = \min(\mu, 1 - 2/p)$, $M_1 = M_1(p)$.

(2) If $j + k = 2n$, $j, k \ge 0$, then there exist the generalized derivatives $V_{z^j \bar{z}^k}$, $W_{z^j \bar{z}^k}$,

$$\begin{cases} g_{z^n \bar{z}^n} = \rho(z), \; g_{z^{n+1}\bar{z}^{n-1}} = \Pi\rho + W_{z^{n+1}\bar{z}^{n-1}}, \\[2mm] g_{z^{n+m}\bar{z}^{n-m}} = V_{z^{n+m}\bar{z}^{n-m}} + W_{z^{n+m}\bar{z}^{n-m}}, \; m = 2, \cdots, n, \\[2mm] V_{z^{n+m}\bar{z}^{n-m}} = \dfrac{m(-1)^m}{\pi} \iint_D \dfrac{(\bar{\zeta} - \bar{z})^{m-1}}{(\zeta - z)^{m+1}}\rho(\zeta)d\sigma_\zeta, \end{cases} \quad (4.16)$$

and

$$L_p[V_{z^{n+m}\bar{z}^{n-m}}, \bar{D}] \le M_2 L_p[\rho, \bar{D}], \tag{4.17}$$

where $M_2 = M_2(p)$.

(3) *The function $g(z)$ in (4.13) satisfies the boundary condition (4.12).*

Proof   There is no harm in assuming that $\| \rho \| = L_p[\rho, \bar{D}] \ne 0$. Setting $\rho^*(z) = \rho(z) / \| \rho \|$, and using a similar methods as in [140], we can obtain that for $0 < j + k \le 2n - 1, j, k \ge 0$,

$$C_\beta[V_{z^j \bar{z}^k}, \bar{D}] \le M_3, \ C_\beta[W_{z^j \bar{z}^k}, \bar{D}] \le M_4, \ 0 \le j + k \le 2n - 1, \tag{4.18}$$

herein $M_j = M_j(p)$, $j = 3, 4$. Consquently (4.15) is true. Moreover, applying the results in [91]1) and [139]13), it can be seen that the integrals in (4.16) exist in the sense of principle value, and the following inequalities hold:

$$L_p[V_{z^{n+m}\bar{z}^{n-m}}, \bar{D}] \le M_5 L_p[\rho, \bar{D}], \ L_p[W_{z^{n+m}\bar{z}^{n-m}}, \bar{D}] \le M_6 L_p[\rho, \bar{D}], \tag{4.19}$$

where M_5, M_6 are nonnegative constants. Hence we have (4.17). In addition, from (4.14), we can see that the function (4.13) satifies the homogeneous boundary condition (4.12).

In order to discuss the solvability of Problem D_0 for the nonlinear equation (4.11) or (4.3), we need some estimates of solutions $g(z)$ of Problem D_0 for (4.11) or (4.3).

Lemma 4.2   *If the nonlinear equation (4.3) satisfies Condition C and the constants $q_0^{(j)}(j = 1, \cdots, n)$ and k_1 are small enough, then the solution $g(z)$ in the form (4.13) of (4.11) satisfies the estimates*

$$C_\beta^{2n-1}[g, \bar{D}] < M_7, \ L_{p_0}[\rho, \bar{D}] < M_8, \tag{4.20}$$

in which $\beta = \min(\mu, 1-2/p)$, $2 < p_0 \le \min(p, 1/(1-\mu))$, $M_j = M_j(p_0, k_0, H)$, $j = 7, 8$.

Proof   Let the solution $g(z)$ of Problem D_0 be substituted into equation (4.11), which can be written as

$$\begin{cases} \rho(z) = \displaystyle\sum_{j=0}^{n-1} \mathrm{Re}[Q_j g_{z^{2n-j}\bar{z}^j}] + \sum_{j+k=0}^{2n-1} A_{jk} g_{z^j \bar{z}^k} + A, \\[3mm] A = \displaystyle\sum_{j=0}^{n-1} \mathrm{Re}[Q_j H_{z^{2n-j}\bar{z}^j}] + \sum_{j+k=0}^{2n-1} A_{jk} H_{z^j \bar{z}^k} + A_0. \end{cases} \tag{4.21}$$

According to (4.15) and (4.17), from (4.21) we can obtain

$$\| \rho \| = L_{p_0}[\rho, \bar{D}] \le \sum_{j=1}^{n} q_0^{(j)} M_2 \| \rho \| + \sum_{j+k=0}^{2n-1} k_1 M_1 \| \rho \| + \| A \|, \tag{4.22}$$

where $\| A \| \leq k_0 + M_9$, $M_9 = M_9(k_0, H)$. Provided that $q_0^{(j)}(j = 1, \cdots, n)$ and k_1 are sufficiently small such that

$$1 - [\sum_{j=1}^{n} q_0^{(j)} M_2 + \sum_{j+k=0}^{2n-1} k_1 M_1] \geq \frac{1}{2},$$

we can derive

$$\| \rho \| < M_8 = (1 + k_0 + M_9)/[1 - \sum_{j=1}^{n} q_0^{(j)} M_2 - \sum_{j+k=0}^{2n-1} k_1 M_1]. \qquad (4.23)$$

By (4.15), the first estimate in (4.20) can be concluded.

Theorem 4.3   *Under the same hypotheses as in Lemma 4.2, Problem D_0 for (4.11) or (4.3) has a solution.*

Proof   Let us introduce a bounded and open set B_M in the Banach space $B = L_{p_0}(\bar{D})$. Any element in B_M satisfies the inequality

$$L_{p_0}[\rho, \bar{D}] < M_8, \qquad (4.24)$$

where the constant M_8 is as stated in (4.20). Choosing an arbitrary function $\rho(z) \in \overline{B_M}$ and substituting the double integral $g(z)$ from (4.13) into the appropriate positions of equation (4.11), by the principle of contracting mappings we can find a solution $\rho^*(z) \in B$ of the following integral equation with the parameter $t \in [0, 1]$

$$\rho^*(z) = t\{\sum_{j=0}^{n-1} \mathrm{Re}[Q_j g^*_{z^{2n-j}\bar{z}^j}] + \sum_{j+k=0}^{2n-1} A_{jk} g_{z^j \bar{z}^k} + A\}, 0 \leq t \leq 1, \qquad (4.25)$$

where

$$Q_j = Q_j(z, g + H, \cdots, g^*_{z^{2n}} + H_{z^{2n}}, \cdots, g^*_{z^{2n+1}\bar{z}^{2n-1}} + H_{z^{2n+1}\bar{z}^{2n-1}}), 0 \leq j \leq n - 1,$$

$$A_{jk} = A_{jk}(z, g + H, \cdots, g_{z^{n+1}\bar{z}^{n-2}} + H_{z^{n+1}\bar{z}^{n-2}}), j, k \geq 0, 0 \leq j + k \leq 2n - 1,$$

Denote by $\rho^*(z) = S(\rho, t)(0 \leq t \leq 1)$ the mapping from $\rho(z)$ onto $\rho^*(z)$. We can verify that $\rho^* = S(\rho, t)$ satisfies the conditions of the Leray–Schauder theorem. Hence there exists a function $\rho(z) \in B_M$, so that $\rho = S(\rho, t)$. Especially when $t = 1$, the corresponding function $g(z)$ is just a solution of Problem D_0 for (4.11). Obviously the function $u(z) = H(z) + g(z)$ is a solution of Problem D for (4.3) (see [143]).

When the equation (4.3) is linear, i.e. $Q_j = Q_j(z), j = 0, 1, \cdots, n - 1, A_{jk} = A_{jk}(z), j, k \geq 0, 0 \leq j + k \leq 2n - 1$, we may not assume that the constant k_1 in (4.4) is sufficiently small. Similarly to Theorem 2.6, the solvability result of Problem D for the linear equation (4.3) can be obtained by using the Fredholm theorem for the corresponding integral equations.

Finally, we mention that some oblique derivative boundary value problems for (4.3) remains to be discussed.

4.2 Nonlinear elliptic complex equations of 2nth order

We consider the following nonlinear complex equation of $2n$th order

$$
\begin{cases}
w_{z^n \bar{z}^n} = F(z, w, \cdots, w_{z^{2n}}, \cdots, w_{z^{n+1}\bar{z}^{n-1}}, \bar{w}_{z^{n+1}\bar{z}^{n-1}}, \cdots, \bar{w}_{\bar{z}^{2n}}), \\[2mm]
F = \displaystyle\sum_{\substack{j+k=2n \\ (j,k)\neq(n,n)}} (Q_{jk} w_{z^j \bar{z}^k} + \tilde{Q}_{jk} \bar{w}_{z^j \bar{z}^k}) \\[2mm]
\quad + \displaystyle\sum_{j+k=0}^{2n-1} (A_{jk} w_{z^j \bar{z}^k} + \tilde{A}_{jk} \bar{w}_{z^j \bar{z}^k}) + A_0,
\end{cases}
\tag{4.26}
$$

where $Q_{jk}, \tilde{Q}_{jk}$ are functions of $z, w, \cdots, w_{z^{2n}}, \cdots, \bar{w}_{\bar{z}^{2n}}$, and $A_{jk}, \tilde{A}_{jk}, A_0$ are functions of $z, w, w_z, \cdots, \bar{w}_{\bar{z}^{2n-1}}$, and assume that (4.26) satisfies Condition C', the main conditions of which are as follows:

$$
|F(z, w, \cdots, V_1^{(1)}, \cdots, V_1^{(2n)}) - F(z, w, \cdots, V_2^{(1)}, \cdots, V_2^{(2n)})|
$$

$$
\leq \sum_{j=1}^{2n} q_0^{(j)} |V_1^{(j)} - V_2^{(j)}|, \quad \sum_{j=1}^{2n} q_0^{(j)} < 1,
\tag{4.27}
$$

for almost every point $z \in D$, any $w, \cdots, V_1^{(j)}, V_2^{(j)} \in \mathcal{C}\, (j = 1, \cdots, 2n)$, where D is a bounded domain, $q_0^{(j)}\, (j = 1, \cdots, 2n)$ are nonnegative constants, and

$$
\begin{cases}
L_p[A_{jk}, \bar{D}] \leq k_1, \ L_p[\tilde{A}_{jk}, \bar{D}] \leq k_1, \ j, k \geq 0, \\[2mm]
0 \leq j + k \leq 2n - 1, \ L_p[A_0, \bar{D}] \leq k_0,
\end{cases}
\tag{4.28}
$$

in which $k_0, k_1, p(> 2)$ are nonnegative constants.

We introduce a double integral

$$
f(z) = \frac{1}{\pi} \iint_D Y(z, \zeta) \rho(\zeta) d\sigma_\zeta, \ \rho(z) \in L_p(\bar{D}), \ p > 1,
\tag{4.29}
$$

where $\rho(z), f(z)$ are complex functions in D, $Y(z, \zeta)$ is the same as in (4.5).

If the complex equation (4.26) satisfies Condition C' in D, the constant k_1 in (4.28) is sufficiently small, and $W(z)$ is an n–harmonic complex function, $W(z) \in C^{2n-1}(\bar{D}) \cap W_{p_0}^{2n}(D)(2 < p_0 \leq p)$, then (4.26) has a solution of the following form

$$
w(z) = W(z) + f(z) = W(z) = \frac{1}{\pi} \iint_D Y(z, \zeta) \rho(\zeta) d\sigma_\zeta,
\tag{4.30}
$$

where $\rho(z) \in L_{p_0}(\bar{D})$.

Besides, we can discuss the solvability of the Dirichlet boundary value problem for the complex equation (4.26) (see [143]). The solvability of the Riemann–Hilbert boundary value problem for (4.26) remains to be investigated.

VII   Applications of Elliptic Boundary Value Problems in Continuum Mechanics

In this chapter, we shall discuss some free boundary value problems in gas dynamics, filtration theory and elastico–plastic mechanics, where some parts of the boundaries of the domains are unknown. These are called free boundaries. By using the method of quasiconformal mappings or hodograph and other methods, the free boundary value problems can be reduced to discontinuous boundary value problems for linear or non-linear elliptic complex equations in fixed domains. According to the result of Chapter 4 in [140] and Chapter 1, we know about the solvability of discontinuous boundary value problems. Thus the above free boundary value problems are solved. In the following, sometimes a free boundary value problem will shortly be called free boundary problem. Besides, we also discuss some mixed–contact boundary value problem in orthotropic elasticity.

1   Free Boundary Problems in Planar and Axisymmetric Filtration

1.1   The planar filtration of earth dam with homogeneous medium

The so–called filtration is a fluid flowing through a porous medium. For the case of steady planar filtration of a fluid, its density $\rho(z)$ is a real constant. Let $V = V_x + iV_y$ denote the velocity of a planar filtration being irrotational and having no source in the filtration domain D. Then V_x and V_y satisfy the Cauchy–Riemann system

$$\frac{\partial V_x}{\partial x} + \frac{\partial V_y}{\partial y} = 0, \quad \frac{\partial V_y}{\partial x} - \frac{\partial V_x}{\partial y} = 0 \text{ in } D, \tag{1.1}$$

and the stream function $\psi(x,y)$ and potential function $\phi(x,y)$ satisfy

$$\begin{cases} d\psi = \dfrac{\partial \psi}{\partial x}dx + \dfrac{\partial \psi}{\partial y}dy = -V_y dx + V_x dy, \\[3mm] d\phi = \dfrac{\partial \phi}{\partial x}dx + \dfrac{\partial \phi}{\partial y}dy = V_x dx + V_y dy, \end{cases} \tag{1.2}$$

hence

$$\frac{\partial \phi}{\partial x} = \frac{\partial \psi}{\partial y} = V_x, \quad \frac{\partial \phi}{\partial y} = -\frac{\partial \psi}{\partial x} = V_y, \tag{1.3}$$

the function $f(z) = \phi(x,y) + i\psi(x,y) = \phi(z) + i\psi(z)$ is called the complex potential of filtration, which is an analytic function in any simply connected subdomain of D. The function

$$\overline{V} = f'(z) = \frac{\partial \phi}{\partial x} - i\frac{\partial \phi}{\partial y} \tag{1.4}$$

is called complex velocity, which is also an analytic function in D.

For any point $z \in D$, the value

$$h = y + \frac{p}{\gamma} = y + \frac{p}{\rho g} \tag{1.5}$$

will be called the head of water, where p is the pressure at $z \in D$, g is the acceleration of gravity. According to Darcy's law,

$$V = -k \operatorname{grad} h = -k\left(\frac{\partial h}{\partial x} + i\frac{\partial h}{\partial y}\right) = \overline{f'(z)}, \tag{1.6}$$

where k is the coefficient of filtration being a constant. From (1.3) and (1.6), we have

$$V_x = -k\frac{\partial h}{\partial x} = \frac{\partial \phi}{\partial x}, \quad V_y = -k\frac{\partial h}{\partial y} = \frac{\partial \phi}{\partial y}. \tag{1.7}$$

Therefore if we discard a constant, then

$$\phi(x,y) = \phi(z) = -kh = -k\left(y + \frac{p}{\gamma}\right). \tag{1.8}$$

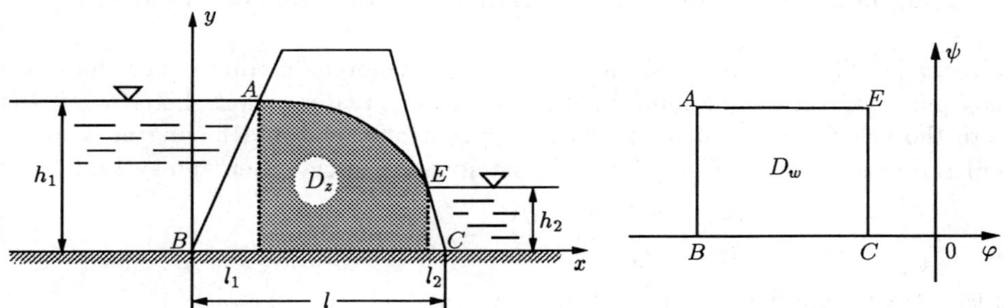

Figure 1.1   Figure 1.2

Now, we introduce a planar filtration problem associated with a homogeneous dam drawn as in Figure 1.1, where the coefficient of filtration is a constant k. It is clear that the complex potential

$$f(z) = \phi(z) + i\psi(z) = \int \overline{V(z)}\, dz$$

is an analytic function in the seepage domain D_z. Noting that the line segments AB, CE are permeable paragraphs, the line segment BC is an impermeable paragraph, and the curve EA is an infiltration paragraph, which is a free boundary, we can write down the boundary condition of the analytic function $w = f(z) = \phi(z) + i\psi(z)$ as

$$\begin{cases} \operatorname{Re} w = \phi(z) = \begin{cases} \phi_1 = -k(p/\gamma + y) = -k(p_0/\gamma + h_1) \text{ on } AB, \\ \phi_2 = -k(p/\gamma + y) = -k(p_0/\gamma + h_2) \text{ on } CE, \end{cases} \\ \operatorname{Im} w = \psi(z) = \begin{cases} \psi_1 \text{ on } BC, \\ \psi_0 \text{ on } EA, \end{cases} \end{cases} \tag{1.9}$$

$$\operatorname{Re} w = \phi(z) = -kh = -k\,(p_0/\gamma + y) = \phi_0 - ky \text{ on } EA, \tag{1.10}$$

where

$$p = \begin{cases} p_0 + \gamma(h_1 - y) \text{ on } AB, \\ p_0 + \gamma(h_2 - y) \text{ on } CE, \end{cases}$$

h_1, h_2 are the heights of water levels in the upper reach and the lower reach respectively, p_0 is the pressure of air, $\gamma = \rho g$, ρ is the density of the fluid, g is the acceleration of gravity. Because BC, AE are stream lines and AB, CE are isopotential lines, $\psi_0, \psi_1, \phi_0, \phi_1$ and ϕ_2 in (1.9),(1.10) are real constants. There is no harm in assuming $\psi_1 = 0$. We only discuss the cases where are no leaks and evaporation, and the pressure of capillary $p = 0$ on the free boundary EA. It is not difficult to see that the domain of complex potential is a rectangle D_w as shown in Figure 1.2. The above free boundary problem can be solved by using a complex analytic method, namely we can find out a bounded analytic function $w = f(z)$ in D satisfying the boundary conditions (1.9) and (1.10), but the constant ϕ_0 needs to be chosen appropriately (see [57]1)).

1.2 The planar filtration of earth dam with nonhomogeneous medium

Next, we consider the case when the soil of the dam is a nonhomogeneous medium different in nature in each direction, and $k = \{k_{ij}(x, y, h, q)\}(k_{12} = k_{21})$ is the tensor of filtration, where $h = y + p/\rho g$ is the head of water, p is the pressure at z, ρ is the density of fluid, g is the acceleration of gravity, $q = |V| = \sqrt{V_x^2 + V_y^2}$ is the magnitude of the velocity of the fluid flow. We assume that the velocity V and the pressure p are continuously differentiable in the domain D of filtration, and V, p comply with the continuity equation and Darcy's law,

$$\begin{cases} \operatorname{div} V = \dfrac{\partial V_x}{\partial x} + \dfrac{\partial V_y}{\partial y} = 0, \\[2mm] V = -\dfrac{k}{\mu}\operatorname{grad}(p + \rho g y), \end{cases} \tag{1.11}$$

in which μ is the viscidity of the fluid. From (1.11), we obtain the stream function $\psi(x, y)$ and the potential function $\phi(x, y)$:

$$\begin{cases} \psi(x, y) = \int [-V_y dx + V_x dy], \\ \phi(x, y) = -k_0 h, \ k_0 = \dfrac{\rho g}{\mu} = \text{constant}, \end{cases} \tag{1.12}$$

and

$$\begin{cases} \psi_x = -V_y, \ \psi_y = V_x, \\ V = \{k_{ij}\} \, \text{grad} \, \phi. \end{cases} \tag{1.13}$$

Hence $\psi(x, y)$ and $\phi(x, y)$ satisfy the following nonlinear system of first order partial differential equations

$$\begin{cases} \psi_y = k_{11}\phi_x + k_{12}\phi_y, \\ -\psi_x = k_{21}\phi_x + k_{22}\phi_y. \end{cases} \tag{1.14}$$

We suppose that the cosfficients $\{k_{ij}\}$ are continuous in $(x, y) \in D$, h and $q(\geq 0) \in \mathbb{R}$, and continuously differentiable with respect to q, and satisfy the conditions

$$|k_{ij}| \leq M_0 < \infty, \ k_{11}k_{22} - k_{12}^2 \geq \delta_0 > 0, \tag{1.15}$$

where M_0, δ_0 are positive constants.

Note that

$$\begin{cases} \Phi_1(\phi_x, \phi_y, \psi_x, \psi_y) = -\psi_y + k_{11}\phi_x + k_{12}\phi_y, \\ \Phi_2(\phi_x, \phi_y, \psi_x, \psi_y) = \psi_x + k_{21}\phi_x + k_{22}\phi_y. \end{cases} \tag{1.16}$$

It can be seen that $\{k_{ij}\}$ are continuously differentiable with respect to $\phi_x, \phi_y, \psi_x, \psi_y$. From (1.34) of Chapter 2 in [140], we have

$$I = 4 \frac{D(\Phi_1, \Phi_2)}{D(\phi_x, \psi_x)} \frac{D(\Phi_1, \Phi_2)}{D(\phi_y, \psi_y)} - \left(\frac{D(\Phi_1, \Phi_2)}{D(\phi_y, \psi_x)} + \frac{D(\Phi_1, \Phi_2)}{D(\phi_x, \psi_y)} \right)^2$$

$$= 4 \frac{D(\Phi_1, \Phi_2)}{D(\phi_x, \phi_y)} \frac{D(\Phi_1, \Phi_2)}{D(\psi_x, \psi_y)} - \left(\frac{D(\Phi_1, \Phi_2)}{D(\phi_y, \psi_x)} - \frac{D(\Phi_1, \Phi_2)}{D(\phi_x, \psi_y)} \right)^2 = 4 \begin{vmatrix} k_{11} & k_{12} \\ k_{21} & k_{22} \end{vmatrix}$$

$$\times \left\{ \begin{vmatrix} \dfrac{\partial k_{11}}{\partial q}\dfrac{\partial q}{\partial \psi_x}\phi_x + \dfrac{\partial k_{12}}{\partial q}\dfrac{\partial q}{\partial \psi_x}\phi_y & -1 + \dfrac{\partial k_{11}}{\partial q}\dfrac{\partial q}{\partial \psi_y}\phi_x + \dfrac{\partial k_{12}}{\partial q}\dfrac{\partial q}{\partial \psi_y}\phi_y \\ 1 + \dfrac{\partial k_{21}}{\partial q}\dfrac{\partial q}{\partial \psi_x}\phi_x + \dfrac{\partial k_{22}}{\partial q}\dfrac{\partial q}{\partial \psi_x}\phi_y & \dfrac{\partial k_{21}}{\partial q}\dfrac{\partial q}{\partial \psi_y}\phi_x + \dfrac{\partial k_{22}}{\partial q}\dfrac{\partial q}{\partial \psi_y}\phi_y \end{vmatrix} \right.$$

$$
-\left|\begin{array}{cc} k_{11} & k_{12} \\ k_{21} & k_{22} \end{array}\right|^{-1}\left[\left|\begin{array}{cc} k_{12} & \dfrac{\partial k_{11}}{\partial q}\dfrac{\partial q}{\partial \psi_x}\phi_x + \dfrac{\partial k_{12}}{\partial q}\dfrac{\partial q}{\partial \psi_x}\phi_y \\ k_{22} & 1 + \dfrac{\partial k_{21}}{\partial q}\dfrac{\partial q}{\partial \psi_x}\phi_x + \dfrac{\partial k_{22}}{\partial q}\dfrac{\partial q}{\psi_x}\phi_y \end{array}\right|\right.
$$

$$
\left.-\left|\begin{array}{cc} k_{11} & -1 + \dfrac{\partial k_{11}}{\partial q}\dfrac{\partial q}{\partial \psi_y}\phi_x + \dfrac{\partial k_{12}}{\partial q}\dfrac{\partial q}{\partial \psi_y}\phi_y \\ k_{21} & \dfrac{\partial k_{21}}{\partial q}\dfrac{\partial q}{\partial \psi_y}\phi_x + \dfrac{\partial k_{22}}{\partial q}\dfrac{\partial q}{\partial \psi_y}\phi_y \end{array}\right|\right]^{2}\Bigg\} = 4(k_{11}k_{22}-k_{12}^2)
$$

$$
\times\left[\left(\dfrac{\partial k_{11}}{\partial q}\dfrac{\partial q}{\partial \psi_x}\phi_x + \dfrac{\partial k_{12}}{\partial q}\dfrac{\partial q}{\partial \psi_x}\phi_y\right)\left(\dfrac{\partial k_{21}}{\partial q}\dfrac{\partial q}{\partial \psi_y}\phi_x + \dfrac{\partial k_{22}}{\partial q}\dfrac{\partial q}{\partial \psi_y}\phi_y\right)\right.
$$

$$
\left.-\left(1 + \dfrac{\partial k_{21}}{\partial q}\dfrac{\partial q}{\partial \psi_x}\phi_x + \dfrac{\partial k_{22}}{\partial q}\dfrac{\partial q}{\partial \psi_x}\phi_y\right)\left(-1 + \dfrac{\partial k_{11}}{\partial q}\dfrac{\partial q}{\partial \psi_y}\phi_x + \dfrac{\partial k_{12}}{\partial q}\dfrac{\partial q}{\partial \psi_y}\phi_y\right)\right]
$$

$$
-\left[\dfrac{\partial q}{\partial \psi_x}\phi_x\left|\begin{array}{cc} k_{11} & \dfrac{\partial k_{11}}{\partial q} \\ k_{22} & \dfrac{\partial k_{21}}{\partial q}\end{array}\right| + \dfrac{\partial q}{\partial \psi_x}\phi_y\left|\begin{array}{cc} k_{12} & \dfrac{\partial k_{12}}{\partial q} \\ k_{22} & \dfrac{\partial k_{22}}{\partial q}\end{array}\right| - \dfrac{\partial q}{\partial \psi_y}\phi_x\right.
$$

$$
\left.\times\left|\begin{array}{cc} k_{11} & \dfrac{\partial k_{11}}{\partial q} \\ k_{21} & \dfrac{\partial k_{21}}{\partial q}\end{array}\right| - \dfrac{\partial q}{\partial \psi_y}\phi_y\left|\begin{array}{cc} k_{11} & \dfrac{\partial k_{12}}{\partial q} \\ k_{21} & \dfrac{\partial k_{22}}{\partial q}\end{array}\right|\right]^{2} = 4(k_{11}k_{22}-k_{12}^2)
$$

$$
\times\left(1 - \dfrac{\partial k_{11}}{\partial q}\dfrac{\partial q}{\partial \psi_y}\phi_x - \dfrac{\partial k_{12}}{\partial q}\dfrac{\partial q}{\partial \psi_y}\phi_y + \dfrac{\partial k_{21}}{\partial q}\dfrac{\partial q}{\partial \psi_x}\phi_x + \dfrac{\partial k_{22}}{\partial q}\dfrac{\partial q}{\partial \psi_x}\phi_y\right)
$$

$$
-\left[\dfrac{\partial q}{\partial \psi_x}\phi_x\left(k_{12}\dfrac{\partial k_{21}}{\partial q} - k_{22}\dfrac{\partial k_{11}}{\partial q}\right) + \dfrac{\partial q}{\partial \psi_x}\phi_y\left(k_{12}\dfrac{\partial k_{22}}{\partial q} - k_{22}\dfrac{\partial k_{12}}{\partial q}\right)\right.
$$

$$
\left.- \dfrac{\partial q}{\partial \psi_y}\phi_x\left(k_{11}\dfrac{\partial k_{21}}{\partial q} - k_{21}\dfrac{\partial k_{11}}{\partial q}\right) - \dfrac{\partial q}{\partial \psi_y}\phi_y\left(k_{11}\dfrac{\partial k_{22}}{\partial q} - k_{21}\dfrac{\partial k_{12}}{\partial q}\right)\right]^{2} \tag{1.17}
$$

$$
= 4(k_{11}k_{22}-k_{12}^2)\left[1 - \dfrac{1}{q}\dfrac{\partial k_{11}}{\partial q}\phi_x\psi_y - \dfrac{1}{q}\dfrac{\partial k_{12}}{\partial q}(\phi_y\psi_y - \phi_x\psi_x)\right.
$$

$$
\left.+ \dfrac{1}{q}\dfrac{\partial k_{22}}{\partial q}\phi_y\psi_x\right] - \left[\dfrac{1}{q}\dfrac{\partial k_{11}}{\partial q}\phi_x(k_{21}\psi_y - k_{22}\psi_x) + \dfrac{1}{q}\dfrac{\partial k_{12}}{\partial q}\psi_x(k_{12}\phi_x\right.
$$

$$
\left.- k_{22}\phi_y) - \psi_y(k_{11}\phi_x - k_{12}\phi_y) + \dfrac{1}{q}\dfrac{\partial k_{22}}{\partial q}\phi_y(k_{12}\psi_x - k_{11}\psi_y)\right]^{2},
$$

in which we use the relations

$$
k_{12} = k_{21} \text{ and } \dfrac{\partial q}{\partial \psi_x} = \dfrac{\psi_x}{q},\ \dfrac{\partial q}{\partial \psi_y} = \dfrac{\psi_y}{q}.
$$

Next, we discuss the uniform ellipticity condition, i.e.

$$I \geq I_0 = \text{constant} > 0. \tag{1.18}$$

There are three cases:

(1) In the linear and quasilinear case, $\{k_{ij}\}$ are independent of q. Consequently

$$I = 4(k_{11}k_{22} - k_{12}^2) \geq I_0 > 0. \tag{1.19}$$

(2) If the soil of dam is nonhomogeneous and isotropic, i.e. $k_{11} = k_{22}$, $k_{12} = k_{21} = 0$, then

$$I = 4k_{11}^3 \frac{\partial}{\partial q}\left(\frac{q}{k_{11}}\right) \geq I_0 > 0. \tag{1.20}$$

(3) In the general case, let $[\phi, \psi]$ be a solution of system (1.14). It is clear that

$$\begin{cases} \phi_x = \dfrac{1}{k_{11}k_{22} - k_{12}^2}(k_{22}\psi_y + k_{12}\psi_x), \\[2mm] \phi_y = \dfrac{1}{k_{11}k_{22} - k_{12}^2}(-k_{21}\psi_y - k_{11}\psi_x). \end{cases} \tag{1.21}$$

Substituting (1.21) into (1.18), we have

$$\begin{aligned}
I = {}& 4(k_{11}k_{22} - k_{12}^2) - \frac{4}{q}\frac{\partial k_{11}}{\partial q}(k_{22}\psi_y + k_{12}\psi_x)\psi_y \\[2mm]
& + \frac{4}{q}\frac{\partial k_{12}}{\partial q}\left(k_{12}\psi_x^2 + k_{21}\psi_y^2 + (k_{11} + k_{22})\psi_x\psi_y\right) \\[2mm]
& - \frac{4}{q}\frac{\partial k_{22}}{\partial q}(k_{21}\psi_y + k_{11}\psi_x)\psi_x - \left(\frac{1}{q}\frac{1}{k_{11}k_{22} - k_{12}^2}\right)^2 \\[2mm]
& \times \left[\frac{\partial k_{11}}{\partial q}\left(-k_{12}k_{22}\psi_x^2 + k_{21}k_{22}\psi_y^2 + (k_{12}^2 - k_{22}^2)\psi_x\psi_y\right)\right. \\[2mm]
& + \frac{\partial k_{12}}{\partial q}\left((k_{12}^2 + k_{11}k_{22})\psi_x^2 - (k_{11}k_{22} + k_{12}^2)\psi_y^2\right. \\[2mm]
& + 2k_{12}(k_{22} - k_{11})\psi_x\psi_y\Big) + \frac{\partial k_{22}}{\partial q}(-k_{11}k_{22}\psi_x^2 \\[2mm]
& + k_{11}k_{21}\psi_y^2 + (k_{11}^2 - k_{12}^2)\psi_x\psi_y)\Big]^2 \geq 4(k_{11}k_{22} - k_{12}^2) \\[2mm]
& - 16qM\max\left(\left|\frac{\partial k_{11}}{\partial q}\right|, \left|\frac{\partial k_{12}}{\partial q}\right|, \left|\frac{\partial k_{22}}{\partial q}\right|\right) - \frac{1}{(k_{11}k_{22} - k_{12}^2)^2} \\[2mm]
& \times 64q^2M_1^2M^4 \geq 4\delta_0 - 16qM_0M_1 - 64q^2M_0^4M_1^2/\delta_0^2,
\end{aligned} \tag{1.22}$$

where

$$M_1 = \max\left(\left|\frac{\partial k_{11}}{\partial q}\right|, \left|\frac{\partial k_{12}}{\partial q}\right|, \left|\frac{\partial k_{22}}{\partial q}\right| \right).$$

If

$$qM_1 \leq (\delta_0/4M_0)^2, \ M_0 \geq 1, \ \delta_0 \leq 1, \tag{1.23}$$

then $16qM_0M_1 \leq \delta_0$ and $64q^2M_0^4M_1^2/\delta_0^2 \leq \delta_0$. From (1.15) and (1.23), it follows that (1.18) holds.

1.3 Planar filtration problems associated with nonhomogeneous and anisotropic medium

Here, we consider a planar filtration problem associated with a nonhomogeneous medium different in nature in each direction as described in Figure 1.3. In this case, the stream function $\psi(x,y)$ and the potential function $\phi(x,y)$ satisfy system (1.14), and the boundary condition

$$\begin{cases} y = \dfrac{h_1}{l_1}x, \ \phi = -k_0(\dfrac{p_0}{\gamma} + h_1) = \phi_1, \ z \in AB, \\[2mm] y = 0, \ \psi = \psi_1 = 0, \ z \in BC, \\[2mm] y = -\dfrac{h_2}{l_2}(x - l), \ \phi = -k_0(\dfrac{p_0}{\gamma} + h_2) = \phi_2, \ z \in CE, \\[2mm] y = -\dfrac{h_2}{l_2}(x - l), \ \psi = -k_0(\dfrac{p_0}{\gamma} + y), \ z \in EF, \\[2mm] \phi = -k_0 y + \phi_0, \ \psi = \psi_0, \ z \in FA, \end{cases} \tag{1.24}$$

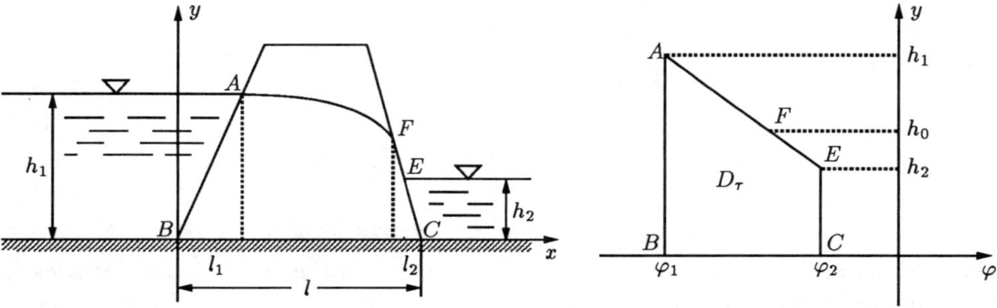

Figure 1.3   Figure 1.4

where FA is an infiltration paragraph (a free boundary). To solve the preceding free boundary problem, we introduce a new complex function $\omega = x + i\psi = \omega(\tau)$, $\tau = \phi + iy$. It is not difficult to see that the domain D_τ in the τ-plane corresponding to D_z is a fixed domain which is a ladder–shape (see Figure 1.4). It can be seen that y is a decreasing function of x on the free boundary AE, hence the pressure p is also a decreasing function of x in D_z, i.e. $p_x < 0$, $\phi_x > 0$. Thus from $\phi = \phi(x,y)$ we can solve $x = x(\phi,y)$, and substitue it into $\psi = \psi(x,y)$ and $\phi = \phi(x,y)$. We can obtain a system of equations

$$\begin{cases} \dfrac{\partial x}{\partial \phi}\dfrac{\partial \psi}{\partial y} - \dfrac{\partial \psi}{\partial \phi}\dfrac{\partial x}{\partial y} - k_{11} + k_{12}\dfrac{\partial x}{\partial y} = 0, \\[4mm] \dfrac{\partial \psi}{\partial \phi} + k_{21} - k_{22}\dfrac{\partial x}{\partial y} = 0. \end{cases} \tag{1.25}$$

Denoting

$$\begin{cases} \Phi_1^*\left(\dfrac{\partial x}{\partial \phi}, \dfrac{\partial x}{\partial y}, \dfrac{\partial \psi}{\partial \phi}, \dfrac{\partial \psi}{\partial y}\right) = \dfrac{\partial x}{\partial \phi}\dfrac{\partial \psi}{\partial y} - \dfrac{\partial \psi}{\partial \phi}\dfrac{\partial x}{\partial y} - k_{11} + k_{12}\dfrac{\partial x}{\partial y}, \\[4mm] \Phi_2^*\left(\dfrac{\partial x}{\partial \phi}, \dfrac{\partial x}{\partial y}, \dfrac{\partial \psi}{\partial \phi}, \dfrac{\partial \psi}{\partial y}\right) = \dfrac{\partial \psi}{\partial \phi} + k_{21} - k_{22}\dfrac{\partial x}{\partial y}, \end{cases} \tag{1.26}$$

and $\partial x/\partial \phi$, $\partial x/\partial y$, $\partial \psi/\partial \phi$, $\partial \psi/\partial y$, ϕ_x, ϕ_y, ψ_x, ψ_y by v_1, v_2, v_3, v_4, u_1, u_2, u_3, u_4 respectively, we can prove that

$$\begin{aligned} I &= 4\frac{D(\Phi_1, \Phi_2)}{D(u_1, u_3)}\frac{D(\Phi_1, \Phi_2)}{D(u_2, u_4)} - \left(\frac{D(\Phi_1, \Phi_2)}{D(u_1, u_4)} + \frac{D(\Phi_1, \Phi_2)}{D(u_2, u_3)}\right)^2 = \\[3mm] &= 4\frac{D(\Phi_1^*, \Phi_2^*)}{D(v_1, v_2)}\frac{D(\Phi_1^*, \Phi_2^*)}{D(v_2, v_4)} - \left(\frac{D(\Phi_1^*, \Phi_2^*)}{D(v_1, v_4)} + \frac{D(\Phi_1^*, \Phi_2^*)}{D(v_2, v_3)}\right)^2. \end{aligned} \tag{1.27}$$

Thereby under the conditions (1.15),(1.23), the systems (1.25) is uniformly elliptic, and $x(\phi,y), \psi(\phi,y)$ satisfy the following boundary condition

$$\begin{cases} x = \dfrac{l_1}{h_1}y, \text{ on } AB, \\[3mm] \psi = 0, \text{ on } BC, \\[3mm] x = -\dfrac{l_2}{h_2}y + l, \text{ on } CEF, \\[3mm] \psi = \psi_0, \text{ on } FA, \end{cases} \tag{1.28}$$

where we assume the position of point E is known, and ψ_0 is an unknown constant to be determined appropriately. Due to system (1.25) is a system of second class of elliptic equations, it can be reduced to the complex equation in the form

$$\omega_\tau = G^*(\tau, \omega, \omega_{\bar\tau}), \quad \tau \in D_\tau, \tag{1.29}$$

and the uniform ellipticity condition can be written as

$$|G^*(\tau,\omega,V_1) - G^*(\tau,\omega,V_2)| \le q_0|V_1 - V_2| \tag{1.30}$$

for $\tau \in D_\tau$, any $\omega, V_1, V_2 \in \mathbb{C}$, where $q_0(< 1)$ is a real constant (see [140]). Setting $\sigma = \phi - iy$, it is obvious that the corresponding domain D_σ in the σ–plane is as indicated in Figure 1.5. The function $\omega(\sigma)$ satisfies a system of first class of elliptic equations, whose complex form is

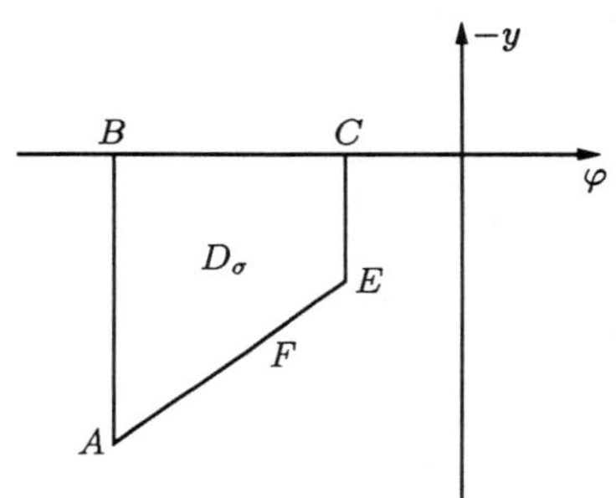

Figure 1.5

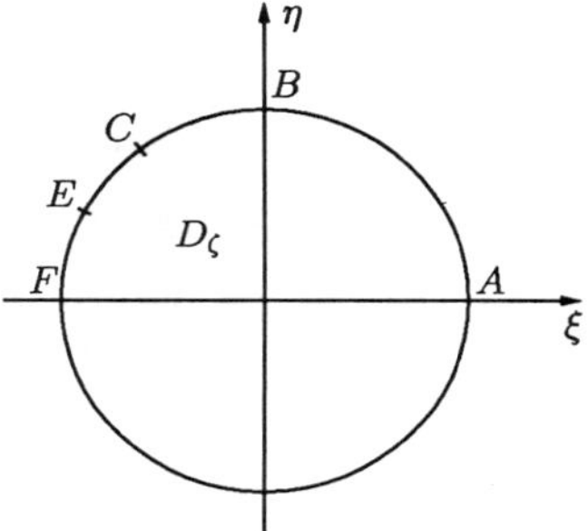

Figure 1.6

$$\omega_{\bar\sigma} = G(\sigma,\omega,\omega_\sigma), \quad \sigma \in D_\tau, \tag{1.31}$$

with uniform ellipticity condition

$$|G(\sigma,\omega,U_1) - G(\sigma,\omega,U_2)| \le q_0|U_1 - U_2|, \tag{1.32}$$

for any $\sigma \in D_\sigma$, $\omega, U_1, U_2 \in \mathbb{C}$. Moreover, $\omega(\sigma)$ satisfies the boundary condition

$$\begin{cases} \mathrm{Re}\,\omega = -\dfrac{l_1}{h_1}\mathrm{Im}\,\sigma, \quad \text{on } AB, \\[2mm] \mathrm{Im}\,\omega = 0, \quad \text{on } BC, \\[2mm] \mathrm{Re}\,\omega = \dfrac{l_2}{h_2}\mathrm{Im}\,\sigma + l, \quad \text{on } CEF, \\[2mm] \mathrm{Im}\,\omega = \psi_0, \quad \text{on } FA. \end{cases} \tag{1.33}$$

We introduce a conformal mapping $\zeta = \zeta(\sigma)$ from D_σ onto the unit disk $D_\zeta = \{|\zeta| < 1\}$, and denote by $\sigma = \sigma(\zeta)$ the inverse function of $\zeta(\sigma)$. It is clear that $\omega = \omega[\sigma(\zeta)]$ satisfies the complex equation

$$\omega_{\bar\zeta} = F(\zeta,\omega,\omega_\zeta), \quad F = \overline{\sigma'(\zeta)}G[\sigma(\zeta),\omega,\omega_\zeta/\sigma'(\zeta)] \tag{1.34}$$

216   *Heinrich Begehr and Guo Chun Wen*

with the uniform ellipticity condition

$$|F(\zeta,\omega,U_1) - F(\zeta,\omega,U_2)| \le q_0|U_1 - U_2| \qquad (1.35)$$

for any $\zeta \in D_\zeta$ and $\omega, U_1, U_2 \in \mathbb{C}$. The boundary condition (1.33) can be transformed into

$$\begin{cases} \mathrm{Re}\,\omega = -\dfrac{l_1}{h_1}\mathrm{Im}\,\sigma(\zeta), \ \ \text{on } AB, \\[2mm] \mathrm{Im}\,\omega = 0, \ \ \text{on } BC, \\[2mm] \mathrm{Re}\,\omega = \dfrac{l_2}{h_2}\mathrm{Im}\,\sigma(\zeta) + l, \ \ \text{on } CEF, \\[2mm] \mathrm{Im}\,\omega = \psi_0, \ \ \text{on } FA. \end{cases} \qquad (1.36)$$

We may write the boundary condition (1.36) in complex form as

$$\mathrm{Re}\,[\overline{\lambda(\zeta)}\omega(\zeta)] = r(\zeta) + h(\zeta), \ \ \text{for } |\zeta| = 1, \qquad (1.37)$$

in which

$$\lambda(\zeta) = e^{i\theta(\zeta)}, \ \theta(\zeta) = \begin{cases} 0, \ \zeta \in AB, \\[2mm] \dfrac{\pi}{2}, \ \zeta \in BC, \\[2mm] 0, \ \zeta \in CEF, \\[2mm] \dfrac{\pi}{2}, \ \zeta \in FA, \end{cases}$$

$$r(\zeta) = \begin{cases} -\dfrac{l_1}{h_1}\mathrm{Im}\,\sigma(\zeta), \ \zeta \in AB, \\[2mm] 0, \ \zeta \in BC, \\[2mm] \dfrac{l_2}{h_2}\mathrm{Im}\,\sigma(\zeta) + l, \ \zeta \in CEF, \\[2mm] 0, \ \zeta \in FA, \end{cases}$$

$$h(\zeta) = \begin{cases} \psi_0, \ \zeta \in FA, \\[2mm] 0, \ \zeta \in ABCEF, \end{cases}$$

where ψ_0 is an unknown real constant to be determined appropriately. Setting $\zeta_1 = A$, $\zeta_2 = B$, $\zeta_3 = C$, $\zeta_4 = F$, and noting that

$$\frac{\lambda(\zeta_j - 0)}{\lambda(\zeta_j + 0)} = e^{i\phi_j}, \ j = 1,2,3,4, \ \phi_1 = \phi_3 = -\frac{\pi}{2}, \ \phi_2 = \phi_4 = \frac{\pi}{2},$$

we require $0 < \gamma_j < \phi_j/\pi - K_j < 1$, herein $K_j(j = 1,2,3,4)$ are integers, hence $K_1 = K_3 = 0$, $K_2 = K_4 = -1$, it shows that the index of above boundary value problem is equal to $K = (K_1 + K_2 + K_3 + K_4)/2 = -1$. By Theorem 3.5 of Chapter 1, if there is $-2K - 1 = 1$ undetermined constant ψ_0, then the preceding discontinuous boundary value problem has a bounded solution $\omega(\zeta)$ in $\overline{D_\zeta}$. Moreover, if $F(\zeta, \omega, U)$ satisfies the condition

$$|F(\zeta, \omega_1, U) - F(\zeta, \omega_2, U)| \le R(\zeta)|\omega_1 - \omega_2|, \tag{1.38}$$

for $\zeta \in D_\zeta$, $\omega_1, \omega_2, U \in \mathbb{C}$, and $R(\zeta) \in L_p(\overline{D_\zeta})$, $p > 2$, then we can prove that the solution of the problem $(1.34), (1.37)$ is unique.

Setting $\omega(\sigma) = \omega[\zeta(\sigma)]$, $\omega(\tau) = \omega[\sigma(\tau)] = \omega(\phi - iy) = x + i\psi$, from $x = x(\phi, y)$, we can find out $\phi = \phi(x, y)$ and substituting it into $\psi = \psi(x, y) = \psi[\phi(x, y), y]$. Thus, we obtain $w(z) = \phi(z) + i\psi(z)$, $z = x + iy$, which is just a solution of the free boundary problem for the planar filtration as described in Figure 1.3. When we find out the complex potential $w(z)$, the quantity ψ_0 of filtration and the equation $\psi = \psi_0$, $\phi = -k_0 y + \phi_0$ of the free boundary FA are obtained. The above result is collected in the following theorem.

Theorem 1.1 *Under the condition (1.15) and (1.23), the preceding free boundary problem associated with the nonhomogeneous and anisotropic earth dam has a solution. Moreover, if condition (1.38) is satisfied, then the solution of the above free boundary problem is unique (see $[162],[57]1)$).*

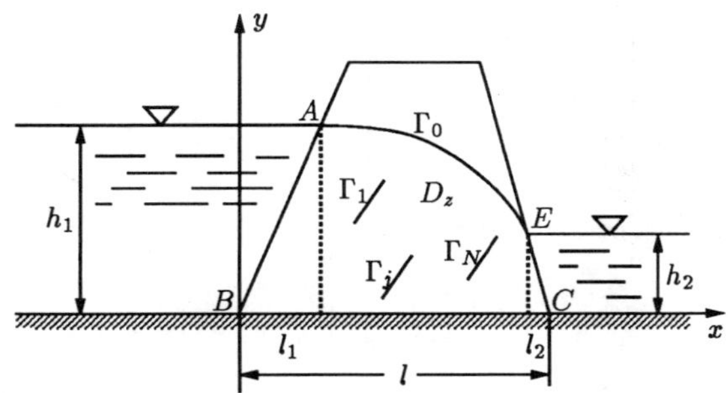

Figure 1.7

Secondly, we discuss the planar filtration of an earth dam with nonhomogeneous medium different in nature of each direction as described in Figure 1.7, where the planar seepage domain is a multiply connected domain D_z, the boundary of which consists of $N+1$ curves: $\Gamma_0 = ABCEA, \Gamma_1, \cdots, \Gamma_N$, where the line segments $\Gamma_1, \cdots, \Gamma_N$ are impermeable paragraphs and others are the same as before. The stream function $\psi(x, y)$ and the potential function $\phi(x, y)$ satisfy the uniformly elliptic system (1.14), in which $k = \{k_{ij}(x, y, h, q)\}(k_{12} = k_{21})$ is the tensor of filtration. The boundary condition of the free boundary problem is

$$
\begin{cases}
y = \dfrac{h_1}{l_1}, \ \phi = -k_0\left(\dfrac{p}{\gamma} + h_1\right) = \phi_1, \ \text{ on } AB, \\[2ex]
y = 0, \ \psi = \psi_* = 0, \ \text{ on } BC, \\[2ex]
y = -\dfrac{h_2}{l_2}(x - l), \ \phi = -k_0\left(\dfrac{p}{\gamma} + h_2\right) = \phi_2, \ \text{ on } CE, \\[2ex]
\phi = -k_0 y + \phi_0, \ \psi = \psi_0, \ \text{ on } EA, \\[2ex]
y = k_j x + l_j, \ \psi = \psi_j, \ \text{ on } \Gamma_j, \ j = 1, \cdots, N.
\end{cases}
\tag{1.39}
$$

By using that the inverse function of a homeomorphic solution of (1.14) satisfies a similar uniformly elliptic system of first order, we can prove the following result.

Theorem 1.2   *Let system (1.14) satisfy the same condition as stated in Theorem 1.1, the above free boundary problem associated with a nonhomogeneous earth dam in an $(N+1)$–connected seepage domain D_z has $N+1$ solvability conditions. If the condition (1.35) holds, then the solution of the free boundary problem is unique* (see [57]1)).

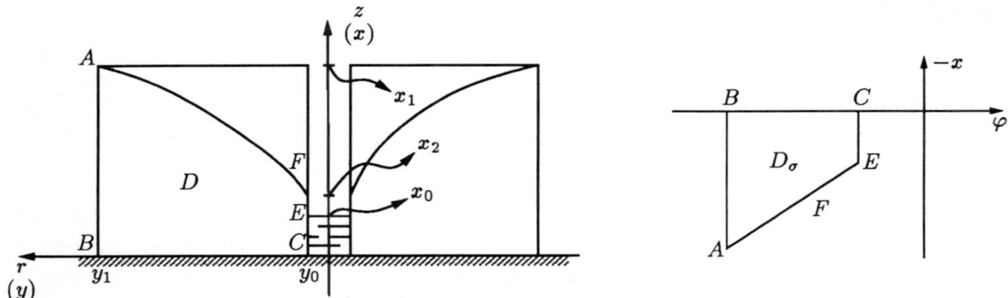

Figure 1.8   Figure 1.9

1.4   The axisymmetric filtration in nonhomogeneous soil

Finally, we consider a free boundary problem of axisymmetric filtration of water in nonhomogeneous and anisotropic soil (see Figure 1.8), where D is the filtration domain, $k = \{k_{ij}(r, z, h, q)\}$ is the symmetric filtration tensor dependent on the spatial variables r, z, the piezometric head $h(r, z) = p/\rho g + z$ and the filtration speed $q(r, z) = |V| = \sqrt{V_r^2 + V_x^2}$, in which p is the pressure at (r, z), ρ, g and V are as stated before. There is no harm in assuming that ρ is the constant 1. For convenience, the variables z, r are replaced by x, y respectively. Introducing the stream function $\psi(x, y)$ and the generalized potential $\phi(x,y) = -k_0 h(x,y)$, where k_0 is a real constant and $k^0 = k/k_0 = \{k_{ij}/k_0\}$ is the reduced filtration tensor, it can be derived that $[\phi, \psi]$ is a solution of the nonlinear system of first order equations

$$\begin{cases} \psi_y = y k_{11}^0 \phi_x + y k_{12}^0 \phi_y, \\ -\psi_x = y k_{21}^0 \phi_x + y k_{22}^0 \phi_y, \end{cases} \quad \text{in } D. \tag{1.40}$$

We suppose that $\{k_{ij}\}(k_{12} = k_{21})$ are continuous in $(x,y) \in D$, h and $q\,(\geq 0) \in \mathbb{R}$, and being continuously differentiable with respect to $q = \sqrt{\psi_x^2 + \psi_y^2}$, satisfying the conditions

$$|k_{ij}^0| \leq M_0 < \infty, \quad k_{11}^0 k_{22}^0 - k_{12}^{02} \geq \delta_0 > 0, \tag{1.41}$$

where M_0, δ_0 are nonnegative constants. Similarly to (1.23), if

$$qM_1 = q \max\left(\left| \frac{\partial k_{11}^0}{\partial q} \right|, \left| \frac{\partial k_{12}^0}{\partial q} \right|, \left| \frac{\partial k_{22}^0}{\partial q} \right| \right) \leq |y| \delta_0^2 / 16 M_0^2, \tag{1.42}$$

then the system (1.40) is uniformly elliptic.

In Figure 1.8, the boundary of the domain D is the closed curve $ABCEFA$, where the line segment BC is an impermeable paragraph and is the stream line, the line segments AB, CE are permeable paragraphs and are equipotential lines, the curve AFE is a stream line, and AF is an infiltration paragraph, which is a free boundary. Thus, the boundary conditions can be written as follows:

$$\psi = \begin{cases} \psi_1 \text{ on } BC, \\ \psi_0 \text{ on } FA, \end{cases}$$

$$\phi = \begin{cases} \phi_1 \text{ on } AB, \\ \phi_2 \text{ on } CE, \\ -k_0(x + p/\gamma) \text{ on } EFA, \end{cases} \tag{1.43}$$

in which $\psi_0, \psi_1, \phi_1, \phi_2, k_0$ are real constants, and we may assume that $\psi_1 = 0$. The above free boundary problem (1.40),(1.43) will be called Problem F. Noting that

$$AB = \{y = y_1, 0 \leq x \leq x_1\},$$

$$BC = \{x = 0, y_1 \leq y \leq y_0\},$$

$$CF = \{y = y_0, 0 \leq x \leq x_2\},$$

and setting $\Omega = y + i\psi$, $\tau = \phi - ix$, it is evident that the domain G in the τ-plane is corresponding to D as indicated in Figure 1.9. The function $\Omega(\tau)$ in G satisfies the nonlinear uniformly elliptic complex equation of first order

$$\Omega_{\bar{\tau}} = G(\tau, \Omega, \Omega_\tau) \tag{1.44}$$

and the boundary conditions

$$\begin{cases} \operatorname{Re}\Omega = y = y_1 \text{ on } AB, \\ \operatorname{Im}\Omega = \psi = 0 \text{ on } BC, \\ \operatorname{Re}\Omega = y = y_0 \text{ on } CEF, \\ \operatorname{Im}\Omega = \psi = \psi_0 \text{ on } FA. \end{cases} \tag{1.45}$$

Denote by $\tau = \tau(\zeta)$ the conformal mapping from the unit disk $\Delta = \{|\zeta| < 1\}$ onto the domain G, it can be derived that the function $\Omega = \Omega[\tau(\zeta)]$ satisfies the complex equation

$$\Omega_{\bar\zeta} = F(\zeta, \Omega, \Omega_\zeta), \quad F = \overline{\tau'(\zeta)}G[\tau(\zeta), \Omega, \Omega_\zeta/\tau'(\zeta)], \tag{1.46}$$

and the boundary conditions

$$\operatorname{Re}[\overline{\lambda(\zeta)}\Omega(\zeta)] = r(\zeta) + h(\zeta), \quad |\zeta| = 1, \tag{1.47}$$

where

$$\lambda(\zeta) = \begin{cases} 1 \text{ on } AB, \\ i \text{ on } BC, \\ 1 \text{ on } CEF, \\ i \text{ on } FA, \end{cases} \qquad r(\zeta) = \begin{cases} y_1 \text{ on } AB, \\ 0 \text{ on } BC, \\ y_0 \text{ on } CEF, \\ 0 \text{ on } FA, \end{cases}$$

$$h(\zeta) = \begin{cases} 0 \text{ on } ABCEF, \\ \psi_0 \text{ on } FA, \end{cases}$$

in which the index of $\lambda(\zeta)$ is $K = -1$, and ψ_0 is an unknown real constant. Provided that we appropriately choose the constsant ψ_0, the discontinuous boundary value problem (1.46),(1.47) has a bounded solution $\Omega(\zeta)$ in Δ. If $F(\zeta, \Omega, V)$ satisfies a condition similar to (1.38), then the above solution is unique (see Section 3 of Chapter 1). Let $\Omega(\zeta)$ be a solution of the discontinuous boundary value problem (1.46),(1.47), and $\zeta(\tau)$ be the inverse function of $\tau(\zeta)$. Then $\Omega(\zeta) = \Omega[\zeta(\tau)]$ is just a solution of the boundary value problem (1.44),(1.45). We can solve equation $y = y(\phi, x)$ for $\phi = \phi(x, y)$, and then $\psi(\phi, x) = \psi[\phi(x, y), y] = \psi(x, y)$ is obtained. Hence $w(z) = \phi(z) + i\psi(z)$, $z = x + iy$, the constant ψ_0 and the free boundary FA are found. The foregoing result is formulated as follows.

Theorem 1.3   *Under the conditions* (1.41) *and* (1.42), *the above free boundary problem of axisymmetric filtration possesses a bounded solution* $w(z) = \phi(z) + i\psi(z)$ (see [141]).

2   Some Free Boundary Problems in Gas Dynamics

2.1   The system of planar gas dynamics equations and related equations

Let V be the velocity of gas flow and S be the velocity of voice. If $S/2 < |V| < S$, we may assume that the gas is compressible, and its density ρ is not a constant. In this section, we consider the planar subsonic steady flow of the compressible gas in the z-plane, $z = x + iy$. It is clear that in the flow domain D of no source, we have

$$\operatorname{div}(\rho V) = 0, \ \text{ i.e. } \frac{\partial(\rho V_x)}{\partial x} = \frac{\partial(-\rho V_y)}{\partial y}, \tag{2.1}$$

and the stream function

$$\psi(x, y) = \int [-\rho V_y dx + \rho V_x dy] \tag{2.2}$$

is single–valued in any simply connected subdomain Ω of D, and

$$\frac{\partial \psi}{\partial x} = -\rho V_y = -\rho v, \ \frac{\partial \psi}{\partial y} = \rho V_x = \rho u, \tag{2.3}$$

where $u = V_x$, $v = V_y$. On the other hand, for the domain of absence of vertices,

$$\operatorname{rot} V = 0, \ \text{ i.e. } \frac{\partial V_x}{\partial y} = -\frac{\partial(-V_y)}{\partial x}, \tag{2.4}$$

and the potential function

$$\phi(x, y) = \int [V_x dx + V_y dy] \tag{2.5}$$

is single–valued in any simply connected domain $\Omega \subset D$, and

$$\frac{\partial \phi}{\partial x} = V_x, \ \frac{\partial \phi}{\partial y} = V_y. \tag{2.6}$$

From (2.3) and (2.6), we obtain the system of gas dynamics equations

$$\rho \frac{\partial \phi}{\partial x} = \frac{\partial \psi}{\partial y}, \ -\rho \frac{\partial \phi}{\partial y} = \frac{\partial \psi}{\partial x}. \tag{2.7}$$

According to the Bernoulli formula, it can be seen that the density ρ of the gas is a function of $q = |V| = (V_x^2 + V_y^2)^{1/2}$, namely

$$\rho = \rho(q), \ q = \left[\left(\frac{\partial \phi}{\partial x}\right)^2 + \left(\frac{\partial \phi}{\partial y}\right)^2 \right]^{1/2}. \tag{2.8}$$

Hence, (2.7) is a nonlinear system of equations.

If we take $V = V_x + iV_y = qe^{i\theta}$ as a complex variables, where $q = (V_x^2 + V_y^2)^{1/2}$, $\theta = \text{tg}^{-1}(V_y/V_x)$, then system (2.7) can be linearized. In fact,

$$
\begin{aligned}
d\phi + \frac{i}{\rho}d\psi &= \frac{\partial\phi}{\partial x}dx + \frac{\partial\phi}{\partial y}dy + \frac{i}{\rho}\left(\frac{\partial\psi}{\partial x}dx + \frac{\partial\psi}{\partial y}dy\right) \\
&= V_x dx + V_y dy + i(-V_y dx + V_x dy) = \\
&= (V_x - iV_y)(dx + idy) = qe^{-i\theta}dz.
\end{aligned}
\tag{2.9}
$$

Suppose that $\rho q \in C_\alpha^1 (0 < \alpha < 1)$ and ϕ, ψ possess continuous partial derivatives up to second order with respect to q, θ, then we have

$$
\frac{\partial z}{\partial\theta} = \frac{e^{i\theta}}{q}\left(\frac{\partial\phi}{\partial\theta} + \frac{i}{\rho}\frac{\partial\psi}{\partial\theta}\right), \; \frac{\partial z}{\partial q} = \frac{e^{i\theta}}{q}\left(\frac{\partial\phi}{\partial q} + \frac{i}{\rho}\frac{\psi}{\partial q}\right),
\tag{2.10}
$$

$$
\begin{aligned}
\frac{\partial^2 z}{\partial\theta\partial q} &= e^{i\theta}\left(\frac{-1}{q^2}\frac{\partial\phi}{\partial\theta} + i\frac{\partial(1/\rho q)}{\partial q}\frac{\partial\psi}{\partial\theta}\right) + \frac{e^{i\theta}}{q}\left(\frac{\partial^2\theta}{\partial\theta\partial q} + \frac{i}{\rho}\frac{\partial^2\psi}{\partial\theta\partial q}\right) \\
&= \frac{\partial^2 z}{\partial q\partial\theta} = \frac{ie^{i\theta}}{q}\left(\frac{\partial\phi}{\partial q} + \frac{i}{\rho}\frac{\partial\psi}{\partial q}\right) + \frac{e^{i\theta}}{q}\left(\frac{\partial^2\phi}{\partial q\partial\theta} + \frac{i}{\rho}\frac{\partial^2\psi}{\partial q\partial\theta}\right).
\end{aligned}
\tag{2.11}
$$

From (2.11), the system of linear equations

$$
\frac{\partial\phi}{\partial q} = q\frac{\partial(1/\rho q)}{\partial q}\frac{\partial\psi}{\partial\theta}, \; \frac{\partial\phi}{\partial\theta} = \frac{q}{\rho}\frac{\partial\psi}{\partial q},
\tag{2.12}
$$

follows, where

$$
\frac{q}{\rho} \; \text{ and } \; q\frac{\partial(1/\rho q)}{\partial q} = -\frac{1}{\rho^2 q}\frac{\partial\rho q}{\partial q}
$$

are functions of q.

Denote by M the score $|V|/S$, which is called the Mach number. If

$$
M < 1, \; \text{ and } \; 1 - M^2 = \frac{1}{\rho}\frac{\partial\rho q}{\partial q} > 0,
$$

then the flow is said to be subsonic, and if $1 - M^2 \geq \delta > 0$, then the flow is called uniformly subsonic. Under this condition, (2.12) can be written in the form

$$
\frac{\partial\phi}{\partial\theta} = \frac{q}{\rho}\frac{\partial\psi}{\partial q}, \; \frac{\partial\phi}{\partial q} = -\frac{1 - M^2}{\rho q}\frac{\partial\psi}{\partial\theta}.
\tag{2.13}
$$

Introducing a fictitious speed q^* and a fictitious density $\rho^*(q^*)$ by

$$
q^* = \int_1^q \frac{\sqrt{1 - M^2(q)}}{q}dq, \; \rho^*(q^*) = \frac{\rho(q)}{\sqrt{1 - M^2(q)}},
\tag{2.14}
$$

it is clear that $q^*(q)$ has a logarithmic singularity for $q = 0$ and $q = \infty$. Due to

$$\frac{dq^*}{dq} = \frac{\sqrt{1-M^2}}{q}, \quad \frac{\partial\phi}{\partial q} = \frac{\partial\phi}{\partial q^*}\frac{\sqrt{1-M^2}}{q}, \quad \frac{\partial\psi}{\partial q} = \frac{\partial\psi}{\partial q^*}\frac{\sqrt{1-M^2}}{q},$$

system (2.13) can be written as

$$\rho^*\frac{\partial\phi}{\partial\theta} = \frac{\partial\psi}{\partial q^*}, \quad -\rho^*\frac{\partial\phi}{\partial q^*} = \frac{\partial\psi}{\partial\theta}. \tag{2.15}$$

Applying the formulae of partial derivatives for the inverse functions $q^* = q^*(\phi,\psi)$, $\theta = \theta(\phi,\psi)$ of $\phi = \phi(q^*,\theta)$, $\psi = \psi(q^*,\theta)$, we obtain the system of first order equations

$$\rho^*\frac{\partial q^*}{\partial\psi} = \frac{\partial\theta}{\partial\phi}, \quad -\rho^*\frac{\partial\theta}{\partial\psi} = \frac{\partial q^*}{\partial\phi}. \tag{2.16}$$

Setting $\omega = q^* - i\theta$, $w = \phi + i\psi$, where ω is called the generalized Joukowsky function and w is the complex potential, and by using the method as stated in Chapter 2, [140], system (2.16) can be transformed in complex form

$$\frac{\partial\omega}{\partial\bar{w}} = q(\omega)\frac{\partial\omega}{\partial w}, \quad \text{i.e. } \omega_{\bar{w}} - q(\omega)\omega_w = 0, \tag{2.17}$$

where $q(\omega) = (\rho^* - 1)/(\rho^* + 1)$.

Similarly, system (2.7) can be written in complex form as

$$w_{\bar{z}} + \frac{\rho - 1}{\rho + 1}\overline{w}_{\bar{z}} = 0, \tag{2.18}$$

and the inverse functions $x = x(\phi,\psi)$, $y = y(\phi,\psi)$ of $\phi = \phi(x,y)$, $\psi = \psi(x,y)$ satisfy the system

$$\rho\frac{\partial y}{\partial\psi} = \frac{\partial x}{\partial\phi}, \quad \rho\frac{\partial x}{\partial\psi} = -\frac{\partial y}{\partial\phi}, \tag{2.19}$$

the complex form of which is

$$z_{\overline{w}} - \frac{\rho - 1}{\rho + 1}z_w = 0. \tag{2.20}$$

Combining (2.17) and (2.20), the complex equation

$$\omega_{\bar{z}} - \tilde{q}(\omega)\omega_z = 0, \tag{2.21}$$

can be derived, where

$$\tilde{q}(\omega) = e^{2i\theta}\frac{1 - \sqrt{1-M^2}}{1 + \sqrt{1-M^2}}, \quad \theta = \text{tg}^{-1}\left(\frac{\partial\phi}{\partial y}\bigg/\frac{\partial\phi}{\partial x}\right).$$

In fact, inserting

$$q(\omega) = \frac{\rho^* - 1}{\rho^* + 1} = \frac{\rho/\sqrt{1 - M^2} - 1}{\rho/\sqrt{1 - M^2} + 1} = \frac{\rho - \sqrt{1 - M^2}}{\rho + \sqrt{1 - M^2}},$$

$$\omega_{\overline{w}} = \omega_{\bar{z}}\bar{z}_{\overline{w}} + \omega_z z_{\overline{w}} = \omega_{\bar{z}}\bar{z}_{\overline{w}} + \omega_z \frac{\rho - 1}{\rho + 1} z_w,$$

$$q(\omega)\omega_w = q(\omega)(\omega_{\bar{z}}\bar{z}_w + \omega_z z_w) = q(\omega)(\omega_{\bar{z}}\frac{\rho - 1}{\rho + 1}\bar{z}_{\overline{w}} + \omega_z z_w),$$

in (2.17) leads to

$$\omega_{\bar{z}} = \frac{q(\omega) - (\rho - 1)/(\rho + 1) \cdot z_w}{1 - q(\omega)(\rho - 1)/(\rho + 1) \cdot \bar{z}_{\overline{w}}}\omega_z = e^{2i\theta}\frac{1 - \sqrt{1 - M^2}}{1 + \sqrt{1 - M^2}}\omega_z = \tilde{q}(\omega)\omega_z.$$

Moreover, denoting $\omega = q^* - i\theta$, $\tau = x + i\psi$, we can prove that $\omega = \omega(\tau)$ satisfies the complex equation

$$\omega_{\overline{\tau}} - Q(\omega)\omega_\tau = 0, \quad Q(\omega) = \frac{\rho q^2 - u\sqrt{1 - M^2} + iv}{\rho q^2 + u\sqrt{1 - M^2} - iv}. \tag{2.22}$$

If the flow is uniformly subsonic, i.e.

$$1 - M^2 = \frac{1}{\rho}\frac{d}{dq}(\rho q) \geq \delta > 0,$$

then $M \leq \sqrt{1 - \delta} < 1$ and $\rho^*(q^*) = \rho(q)/\sqrt{1 - M^2}$ is bounded, and $\rho^* \geq \delta_0 > 0$. Thus the coefficients $q(\omega)$ of (2.17) and $\tilde{q}(\omega)$ of (2.21) satisfy

$$|q(\omega) \leq \left|\frac{\rho^* - 1}{\rho^* + 1}\right| \leq q_0 < 1,$$

$$|\tilde{q}(\omega)| \leq \left|\frac{1 - \sqrt{1 - M^2}}{1 + \sqrt{1 - M^2}}\right| \leq q_0' < 1,$$

respectively. These estimates show that (2.17) and (2.21) are uniformly elliptic. If q^* and θ satisfy the conditions

$$|q^*| \leq N < \infty, \ |\theta| \leq \frac{\pi}{2} - \eta, \ 0 < \eta < \frac{\pi}{2}, \tag{2.23}$$

then system (2.22) is uniformly elliptic. In this case, the coefficient $Q(\omega)$ of (2.22) satisfies

$$|Q(\omega)| \leq Q_0(N, \eta) < 1, \tag{2.24}$$

where N and η are positive constants and $Q_0(N, \eta)$ is a nonnegative constant dependent on N and η.

2.2 Two free boundary problems in planar subsonic steady flow

Now, we discuss two free boundary problems in the subsonic steady flow of planar multiply connected domains.

Problem I There are given N profiles with boundaries $\Gamma_1, \Gamma_2, \cdots, \Gamma_N$. The problem of subsonic flow of a gas jet past the profiles is to find a continuous solution to (2.21), when in a certain distance from the profiles the jet breaks up into several jets going off to infinity (see Figure 2.1). The coordinates of the sharp points $R_1, R_3, \cdots, R_{2N-1}$, the abscissas of the sharp points $R_2, R_4, \cdots, R_{2N}, R'_1, \cdots, R'_n$, and the quantities $S_1, S_2, \cdots$ of the flow between the boundaries are known, and

$$\begin{cases} q^*(x) = q_0(x) \text{ on the free boundary } \Gamma_0, \\ \theta(x) = \theta_0(x) \text{ on the known boundary } \Gamma\backslash\Gamma_0, \end{cases} \tag{2.25}$$

where $q_0(x)$, $\theta_0(x)$ are known functions, $q_0(x) \in C_\alpha(\Gamma_0)$, $\theta_0(x) \in C_\alpha(\Gamma\backslash\Gamma_0)$, $0 < \alpha < 1$. Find a continuous solution $\omega(z) = q^*(z) - i\theta(z)$ of (2.21) in D_z, which satisfies the boundary condition (2.25).

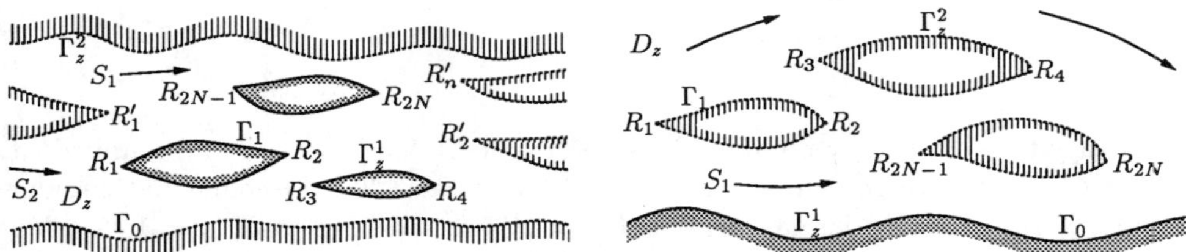

Figure 2.1   Figure 2.2

Problem II There are N free boundaries $\Gamma_1, \cdots, \Gamma_N$ in a planar subsonic steady flow, the bottom of which is uneven (see Figure 2.2). The coordinates of the sharp points $R_1, R_3, \cdots, R_{2N-1}$, the abscissas of the sharp points $R_2, R_4, \cdots, R_{2N}$ of the free boundary and the quantities $S_1, S_2, \cdots$ of the flow between the boundaries are known. Find a continuous solution $\omega(z) = q^*(z) - i\theta(z)$ of the complex equation (2.21) in D_z satisfying the boundary conditions

$$\begin{cases} q^*(x) = q_0(x) \text{ on the free boundary } \Gamma\backslash\Gamma_0, \\ \theta(x) = \theta_0(x) \text{ on the known boundary } \Gamma_0, \end{cases} \tag{2.26}$$

where $q_0(x) \in C_\alpha(\Gamma\backslash\Gamma_0)$, $\theta_0(x) \in C_\alpha(\Gamma_0)$, $0 < \alpha < 1$.

For the above two problems, the flow regions D_z are $(N+1)$–connected domains surrounded by boundary components: $\Gamma_0, \Gamma_1, \cdots, \Gamma_N$. In any flow domain D_z, we discuss the function $\omega = \omega(z) = q^* - i\theta$, where

$$q^* = \int_1^q \frac{\sqrt{1-M^2}}{q}\,dq, \quad q = \sqrt{V_x^2 + V_y^2}, \quad \theta = \mathrm{tg}^{-1}\frac{V_y}{V_x}. \tag{2.27}$$

From the preceding discussion, we see that the function $\omega(\tau)(\tau = x + i\psi)$ satisfies the quasilinear elliptic complex equation (2.22), in which $Q(\omega)$ is required a continuous function of ω. For Problem I and Problem II, the domain D_τ in the τ-plane, $\tau = x + i\psi$, corresponding to D_z are described in Figure 2.3 and Figure 2.4 respectively. There is no harm in assuming the stream function $\psi = 0$ on a part of Γ_0.

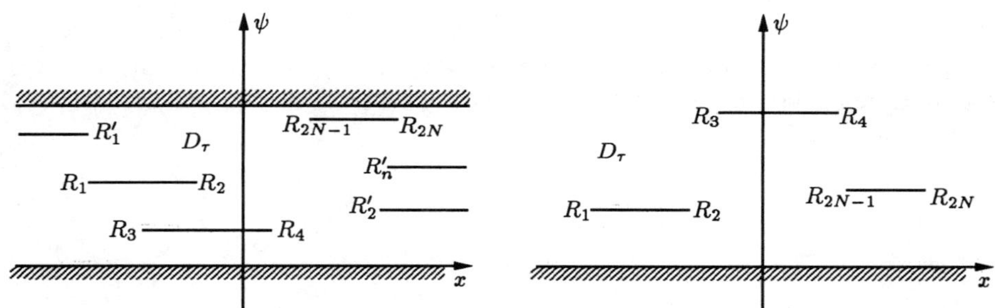

Figure 2.3   Figure 2.4

It is easy to see that any domain D_τ is determined. Hence we can find a univalent function $\zeta = \zeta(\tau)$, which conformally maps D_τ onto an $(N+1)$–connected circular domain D_ζ, such that Γ_0 is mapped onto $L_0 = \{|\zeta| = 1\}$, Γ_j onto L_j $(j = 1, \cdots, N)$, and $\zeta = 0 \in D_\zeta$. Let $L = \cup_{j=0}^{N} L_j$, $L_j = \{|\zeta - \zeta_j| = \rho_j\}, j = 0, 1, \cdots, N$, and $\tau = \tau(\zeta)$ be the inverse function of $\zeta(\tau)$. Then $\omega = \omega[\tau(\zeta)]$ satisfies the quasilinear elliptic complex equation

$$\omega_{\bar\zeta} - \tilde{Q}(\zeta, \omega)\omega_\zeta = 0, \quad \tilde{Q} = Q(\omega)\overline{\tau'(\zeta)}/\tau'(\zeta) \ \text{ in } D_\zeta \tag{2.28}$$

and the corresponding boundary condition

$$\mathrm{Re}\left[\overline{\lambda(\zeta)}\Omega(\zeta)\right] = r(\zeta), \quad \zeta \in L, \tag{2.29}$$

where

$$\lambda(\zeta) = \begin{cases} \lambda_1(\zeta) = \begin{cases} 1, & \zeta \in L_0, \\ -i, & \zeta \in L\backslash L_0, \end{cases} & \text{for Problem I,} \\[3ex] \lambda_2(\zeta) = \begin{cases} 1, & \zeta \in L\backslash L_0, \\ -i, & \zeta \in L_0, \end{cases} & \text{for Problem II,} \end{cases}$$

$$
r(\zeta) = \begin{cases} r_1(\zeta) = \begin{cases} q_0[\tau(\zeta)], & \zeta \in L_0, \\ \theta_0[\tau(\zeta)], & \zeta \in L \backslash L_0, \end{cases} & \text{for Problem I,} \\[3ex] r_2(\zeta) = \begin{cases} q_0[\tau(\zeta)], & \zeta \in L \backslash L_0, \\ \theta_0[\tau(\zeta)], & \zeta \in L_0, \end{cases} & \text{for Problem II,} \end{cases}
$$

in which we require $r(\zeta) \in C_\alpha(L)$. We have to find a bounded solution $\omega(\zeta)$ of the complex equation (2.28) with the boundary condition (2.29). This problem is called Problem R. It is clear that the index of $\lambda(\zeta)$ is $K = -1/2$.

Supposing that (2.23) holds,

$$
|\tilde{Q}(\zeta, \omega)| \le \tilde{Q}_0(N, \eta) < 1 \tag{2.30}
$$

can be derived. If N and η are fixed constants, then (2.28) is a uniformly elliptic complex equation. In order to discuss the solvability of Problem R, we introduce the corresponding modified Problem S with the boundary condition

$$
\text{Re}\,[\overline{\lambda(\zeta)}\omega(\zeta)] = r(\zeta) + h(\zeta), \quad \zeta \in L, \tag{2.31}
$$

where

$$
h(\zeta) = \begin{cases} 0, & \zeta \in L_0 \cup L_1, \\ h_j, & \zeta \in L_j,\ j = 2, \cdots, N, \end{cases} \tag{2.32}
$$

herein $h_j\,(j = 2, \cdots, N)$ are unknow constants to be determined appropriately. In analogy to the method in Section 3, Chapter 1, we first give an a priori estimate of the solution of problem S for the uniformly elliptic complex equation (2.28), and then use the Schauder fixed–point theorem. The following theorem can be proved.

Theorem 2.1 *If $N(< \infty)$, $\eta(0 < \eta < \pi/2)$ are fixed constants, then Problem S for system (2.28) has a bounded solution. Under $N - 1$ conditions, Problem R for (2.28) is solvable. Moreover, under $N - 1$ conditions, the preceding Problem I and Problem II in planar subsonic steady flow for corresponding $(N + 1)$–connected domains are solvable* (see [147]4)).

3 Free Boundary Problems in Elastico-Plastic Mechanics

3.1 Free boundary problem in the case of simply connected domains

We consider a sheet of metal, the boundary of which is a convex polygon L. Due to the affect of outside force, the neighborhood of some parts of the boundary L_0 looses its elasticity and becomes a plastic domain D'(see Figure 3.1). When the sheet of metal attains a new elasticity equilibrium, the boundary of the elastic domain consists of a convex polygonal broken–line L_z^1 and of an unknown arc L_z^2, which is the common boundary of the elastic domain D_z and the plastic domain D', and is called the free boundary. In order to determine the free boundary L_z^2 and the stress in the domain D_z, we need to propose the system of equations in D_z and the boundary condition on the boundary ∂D_z for this situation.

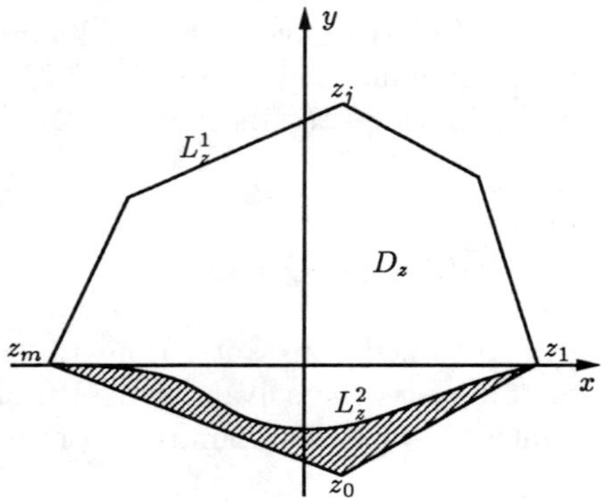

Figure 3.1

The above elasticity equilibrium problem of a sheet of metal can be considered as a plane problem. When the volume force is neglected, in the case of statics the condition of elasticity equilibrium is the system of equations

$$\frac{\partial X_x}{\partial x} + \frac{\partial X_y}{\partial y} = 0, \quad \frac{\partial Y_x}{\partial x} + \frac{\partial Y_y}{\partial y} = 0, \tag{3.1}$$

where X_x, Y_x and X_y, Y_y are the components of stresses F_x, F_y vertical to the x-axis and the y-axis repectively, and $X_y = Y_x$. Hence, in any simply connected domain $\tilde{D} \subset D$, there exists two single–valued functions $A(x,y)$ and $B(x,y)$, such that

$$\begin{cases} dA = Y_y dx - Y_x dy, \\ dB = -X_y dx + X_x dy, \end{cases} \tag{3.2}$$

and then

$$\frac{\partial A}{\partial y} = -Y_x = -X_y = \frac{\partial B}{\partial x}, \quad X_x = \frac{\partial B}{\partial y}, \quad Y_y = \frac{\partial A}{\partial x}. \tag{3.3}$$

Similarly there exists a single–valued function $U(x,y)$, so that

$$dU = A dx + B dy. \tag{3.4}$$

U is called the stress function, which is a biharmonic function in $\tilde{D}$, i.e. $\Delta^2 U = 0$. Because a biharmonic function $U(x,y) = U(z)$ can be expressed by two analytic functions $\phi(z)$ and $\psi(z)$, namely (see [139]17))

$$2U(z) = \bar{z}\phi(z) + z\overline{\phi(z)} + \psi(z) + \overline{\psi(z)}, \tag{3.5}$$

we can give the following representation

$$\begin{cases} X_x + Y_y = 4\mathrm{Re}\,\Phi(z), \\ Y_y - X_x + 2iX_y = 2[\bar{z}\Phi'(z) + \Psi(z)] = 2W(z), \end{cases} \tag{3.6}$$

where $\Phi(z) = \phi'(z)$, $\Psi(z) = \psi''(z)$. From the second formula of (3.6), it follows that

$$W_{\bar{z}} = \Phi'(z) \ \text{in} \ D_z. \tag{3.7}$$

This is a simple elliptic complex equation.

Next, we shall establish the boundary condition. There is no harm in assuming that the intersection of L_z^1 and L_z^2 is located on the x–axis. If the stress and its partial derivatives satisfy the boundary conditions

$$\begin{cases} \dfrac{\partial E}{\partial x} = u(\tau), \\[2mm] \dfrac{\partial E}{\partial y} = v(\tau), \qquad\qquad z \in L_z^2, \\[2mm] \dfrac{1}{4}(Y_y - X_x)^2 + X_y^2 = k^2(E), \end{cases} \tag{3.8}$$

$$\begin{cases} f\left(\dfrac{\partial E}{\partial x}, \dfrac{\partial E}{\partial y}\right) = 0, \\[2mm] \qquad\qquad\qquad z \in L_z^1, \\[2mm] F_s = 0, \end{cases} \tag{3.9}$$

where $u(\tau)$, $v(\tau)$, $k(E) \in C_\alpha^1\,(0 < \alpha < 1)$, $k(E) \geq k_0 > 0$, and $E = (X_x + Y_y)/4$, F_s is the tangent component of the outside stress F, and τ is one of the parameters $x = \mathrm{Re}\,z$, $r = |z|$, the arc length, $\theta = \arg\partial z/\partial s$ and so on. Our problem is how to determine the free boundary L_z^2 and the stress in D_z from conditions (3.8) and (3.9). This free boundary problem is simply called Problem F.

Let F be the outside force at the point A and F_s, F_n be the tangent component and normal component of F, F_x, F_y be the components in x–axis direction and y–axis direction of F respectively. We choose an arbitrary small triangle ABC in D_z as a differential unit, and denote $\phi = (n, x)$. It can be obtained that

$$\begin{cases} X_n BC - X_x BC \cos\phi - X_y BC \sin\phi = 0, \\ Y_n BC - Y_x BC \cos\phi - Y_y BC \sin\phi = 0, \end{cases} \tag{3.10}$$

consequently,

$$\begin{cases} X_n = X_x \cos\phi + X_y \sin\phi, \\ Y_n = Y_y \cos\phi + Y_y \sin\phi. \end{cases} \tag{3.11}$$

If we use $\theta = \arg dz/ds$, and taking $\theta = \phi + \pi/2$ into account, then from (3.11) it follows that

$$\begin{cases} X_n = X_x \sin\theta - X_y \cos\theta, \\ Y_n = Y_y \sin\theta - Y_y \cos\theta. \end{cases} \qquad (3.12)$$

Hence

$$F_s = Y_n \cos\phi - X_n \sin\phi = Y_n \sin\theta + X_n \cos\theta$$

$$= (Y_x \sin\theta - Y_y \cos\theta)\sin\theta + (X_x \sin\theta - X_y \cos\theta)\cos\theta \qquad (3.13)$$

$$= \frac{1}{2}(X_x - Y_y)\sin 2\theta - X_y \cos 2\theta.$$

Noting (3.6), the equality (3.13) can be written in the complex form

$$\operatorname{Re}\left[\overline{\lambda(z)}W(z)\right] = F_s, \quad \lambda(z) = -ie^{-2i\theta}. \qquad (3.14)$$

Thereby the second boundary conditions in (3.8) and (3.9) can be written as

$$\begin{cases} \operatorname{Re}\left[\overline{\lambda(z)}W(z)\right] = 0, \ z \in L_z^1, \\ |W| = k(E), \ z \in L_z^2. \end{cases} \qquad (3.15)$$

In the following, we shall find a bounded solution $W(z)$ of (3.7) in $\bar{D}_z$ satisfying the boundary condition (3.15). The preceding free boundary problem will be called Problem H. We can break up Problem F into two boundary value problems, namely Problem H and the boundary value problem (Problem G) for analytic functions in D_z with the boundary conditions

$$\begin{cases} \operatorname{Re} G = u(\tau), \ \operatorname{Im} G = -v(\tau), \ z \in L_z^2, \\ f(\operatorname{Re} G, \operatorname{Im} G) = 0, \ z \in L_z^1. \end{cases} \qquad (3.16)$$

In order to do so introduce the function

$$G(z) = U(z) - iV(z), \ U = \frac{\partial E}{\partial x}, \ V = \frac{\partial E}{\partial y}. \qquad (3.17)$$

Noting that $\Phi(z)$ in (3.6) is an analytic function in D_z, it is clear that

$$G(z) = \frac{\partial E}{\partial x} - i\frac{\partial E}{\partial y} = \frac{\partial \operatorname{Re}\Phi}{\partial x} - i\frac{\partial \operatorname{Re}\Phi}{\partial y} = \Phi'(z).$$

Hence $G(z)$ is also analytic in D. From the boundary conditions (3.8) and (3.9), we have

$$\begin{cases} U = u(\tau), \ V = v(\tau), \ z \in L_z^2, \\ f(U, V) = 0, \ z \in L_z^1. \end{cases} \qquad (3.18)$$

This shows that the boundary L_G^1 and L_G^2 in the G–plane corresponding to L_G^1 and L_G^2 are known arcs, and $L_G^1 \cup L_G^2$ is a simple closed contour, which bounds a domain D_G. Without loss of generality, we take the parameter $\tau = \mathrm{Re}\, z = x$.

Let $W = W(\zeta)$ be a univalent function which conformally maps the unit disk D_ζ onto the domain D_G such that $G(\zeta_1) = G_1$, $G(\zeta_m) = G_m$ and $G(\zeta_0) = G_0$, where G_1 and G_m are the points of intersection of L_G^1 and L_G^2, $G_0 \in L_G^2$, and $\zeta_1, \zeta_0, \zeta_m$ are three arbitrary points on $|\zeta| = 1$, $\zeta_0 \in (\zeta_m \zeta_1)$ (see Figure 3.2 and Figure 3.3). Comparing (3.18) and the function $G(\zeta)$, the equality and corresponding relation

$$G(\zeta) = u(x) + iv(x), \ x = x(\zeta), \ \zeta \in (\zeta_m \zeta_1) \tag{3.19}$$

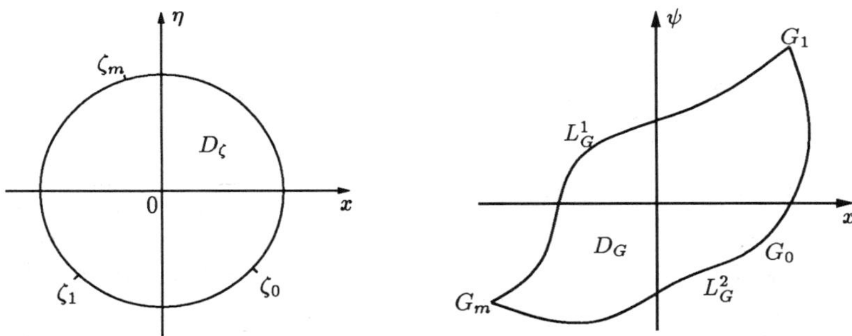

Figure 3.2   Figure 3.3

can be obtained. Moreover, the equation of each straight line l_k on L^1 is

$$l_k : \ a_k x - y = b_k, \ k = 2, \cdots, m. \tag{3.20}$$

Let $(\zeta_{k-1} \zeta_k)$ denote the arc corresponding on l_k between ζ_{k-1} and ζ_k $(k = 2, \cdots, m)$. The boundary condition (3.19),(3.20) can be written in complex form as

$$\mathrm{Re}\left[\overline{\Lambda(\zeta)} z(\zeta)\right] = R(\zeta), \ |\zeta| = 1. \tag{3.21}$$

This is a discontinuous boundary value problem (Problem G^*). Using the result from Section 3, Chapter 1, we can obtain the solvability result of Problem G^*.

Theorem 3.1 *Problem G^* and Problem G for analytic functions have unique solutions.*

Proof We can write the coefficients of the boundary condition (3.21) as follows

$$\Lambda(\zeta) = \begin{cases} 1, \ \zeta \in (\zeta_m \zeta_0 \zeta_1) = (\zeta_m \zeta_0) \cup (\zeta_0 \zeta_1), \\[2mm] \dfrac{a_k - i}{\sqrt{1 + a_k^2}}, \ \zeta \in (\zeta_{k-1} \zeta_k), \end{cases}$$

$$R(\zeta) = \begin{cases} x(\zeta), \ \zeta \in (\zeta_m \zeta_0 \zeta_1), \\[2mm] \dfrac{b_k}{\sqrt{1 + a_k^2}}, \ \zeta \in (\zeta_{k-1} \zeta_k), \ k = 2, \cdots, m. \end{cases}$$

Suppose that L_z^2 has tangents at the points z_1 and z_m and denote

$$\gamma_k = \frac{1}{\pi i} \ln \frac{\Lambda(\zeta_k - 0)}{\Lambda(\zeta_k + 0)} = \frac{\phi_k}{\pi} - K_k, \ \frac{\Lambda(\zeta_k - 0)}{\Lambda(\zeta_k + 0)} = e^{i\phi_k}, \ k = 1, \cdots, m. \tag{3.22}$$

If we take γ_k such that $0 < \gamma_k < 1$, then $K_k = 0$, $k = 2, \cdots, m$, and take

$$\gamma_1 = \frac{1}{\pi i} \ln \frac{\Lambda(\zeta_1 - 0)}{\Lambda(\zeta_1 + 0)} = \frac{\phi_1}{\pi} - K_1 > 0, \ -\pi < \phi_1 < 0,$$

then $K_1 = -1$. It is clear that the index of $\Lambda(\zeta)$ is equal to

$$K = \frac{1}{2} \sum_{k=1}^{m} K_k = -\frac{1}{2}.$$

According to Theorem 3.5 in Chapter 1, Problem G^* has a unique bounded solution $z(\zeta)$. Taking into account that $x = \operatorname{Re} z(\zeta)$ is decreasing on $(\zeta_1 \zeta_m)$ and increasing on $\zeta_m \zeta_0 \zeta_1$ and using Theorem 6, Section 3, Chapter 2 in [101]2), we known that the function $z(\zeta)$ is univalent, and the free boundary L_z^2 is also determined. Denoting by $\zeta = \zeta(z)$ the inverse function of $z = z(\zeta)$, then the function $G = G(\zeta) = G[\zeta(z)]$ is a unique solution of Problem G for analytic functions.

Secondly, we shall prove that Problem H is solvable. Let $z = z(\zeta)$ be a solution of Problem G^* and $W = W[z(\zeta)]$. Then the complex equation (3.7) can be reduced to

$$W_{\bar\zeta} = f(\zeta), \ f(\zeta) = \overline{z'(\zeta)} \Phi'[z(\zeta)]. \tag{3.23}$$

Moveover, the boundary condition (3.15) can be transformed into

$$\begin{cases} |W(\zeta)| = k[E(\zeta)] = g(\zeta), \ \zeta \in (\zeta_m \zeta_0 \zeta_1), \\[2mm] \operatorname{Re}[\overline{\Lambda(\zeta)} W(\zeta)] = 0, \ \zeta \in (\zeta_1 \zeta_m), \end{cases} \tag{3.24}$$

in which $\Lambda(\zeta) = -ie^{-2\theta(\zeta)i}$. We assume that

$$\begin{cases} [(u(x))^2 + (v(x))^2]^{1/2} \leq \dfrac{k_0}{2M} - \varepsilon, \ z \in L_z^2, \\[3mm] \left[(\dfrac{\partial E}{\partial x})^2 + (\dfrac{\partial E}{\partial y})^2 \right]^{1/2} \leq \dfrac{k_0}{2M} - \varepsilon, \ z \in L_z^1, \\[3mm] k[E(\zeta)] \geq k_0, \ z \in L, \end{cases} \tag{3.25}$$

where k_0, $M = \max_{z \in \bar{D}} |z|$ and ε are positive constants and ε is sufficiently small. From (3.25) and (3.6), and by using the maximum modulus principle, we see that

$$|W(\zeta)| = |\overline{z(\zeta)}\Phi'[z(\zeta)] + \Psi[z(\zeta)]| \geq |\Psi[z(\zeta)]| - |\overline{z(\zeta)}\Phi'[z(\zeta)]|$$

$$\geq \frac{k_0}{2} + M\varepsilon - (\frac{k_0}{2} - M\varepsilon) = 2M\varepsilon > 0, \ \zeta \in L.$$

This shows that $W(\zeta)$ has no zero on $\overline{D_\zeta}$. Setting $V(\zeta) = \ln W(\zeta)$, the complex equation (3.23) can be reduced to the following quasilinear elliptic complex equation

$$V_{\bar{\zeta}} = f(\zeta)e^{-V(\zeta)}, \tag{3.26}$$

and the boundary condition can be rewritten in the form

$$\begin{cases} \operatorname{Re} V(\zeta) = \ln g(\zeta), \ \zeta \in (\zeta_1\zeta_m), \\ \operatorname{Im} V(\zeta) = -2\theta(\zeta), \ \zeta \in (\zeta_m\zeta_0\zeta_1), \end{cases} \tag{3.27}$$

where

$$g(\zeta) = k[E(\zeta)], \ E(\zeta) = \operatorname{Re}\left[\int G[z(\zeta)]z'(\zeta)d\zeta\right] = \frac{1}{4}(X_x + Y_y).$$

The above boundary value problem (3.26) and (3.27) is called Problem H^*. The boundary condition (3.27) can be written in the complex form

$$\operatorname{Re}[\overline{\lambda(\zeta)}V(\zeta)] = r(\zeta), \ |\zeta| = 1, \tag{3.28}$$

in which

$$\lambda(\zeta) = \begin{cases} 1, \ \zeta \in (\zeta_1\zeta_m), \\ i, \ \zeta \in (\zeta_m\zeta_0\zeta_1), \end{cases} \quad r(\zeta) = \begin{cases} \ln g(\zeta), \ \zeta \in (\zeta_1\zeta_m), \\ -2\theta(\zeta), \ \zeta \in (\zeta_m\zeta_0\zeta_1). \end{cases}$$

Similarly to Section 3 of Chapter 1, we may arrange that the index of Problem H^* is $K = -1/2$.

Theorem 3.2 *Under the condition* (3.25), *the above Problem H^* has a unique bounded solution $V(\zeta)$, and $W(\zeta) = e^{V[\zeta(z)]}$ is a bounded solution of Problem H for* (3.7).

Proof Because the total number of solvability condition of Problem H^* equals $2|K| - 1 = 0$, and $\gamma_j > 0$, $j = 1, \cdots, m$, on the basis of Theorem 3.5 in Chapter 1, it can be shown that Problem H^* possesses a bounded solution $V(\zeta)$ on $|\zeta| \leq 1$.

Furthermore, applying the representation of the solution for the complex equation and the method used in Section 3 of Chapter 1, we can prove the uniqueness of solutions as stated in this theorem.

234   *Heinrich Begehr and Guo Chun Wen*

Now, we prove the result of solvability of the free boundary problem F for the elastico–plastic mechanics of a sheet metal.

Theorem 3.3   *If condition* (3.25) *holds, then Problem F for the above elastico–plastic mechanics has a unique bounded solution, and the free boundary L is determined.*

Proof   According to Theorem 3.1, we can find out $G(z) = \Phi'(z)$ and determine the free boundary L^2. From Theorem 3.2, the solution $V(\zeta)$ of Problem H^* can be obtained. Hence the function $W(z) = e^{V[\zeta(z)]}$ is given. Taking $\Phi(z) = \int G(z)dz$ into account, from (3.6), we can find the functions X_x, Y_y and X_y. This shows that Problem F has a unique bounded solution.

3.2   Free boundary problem in the case of a multiply connected domain

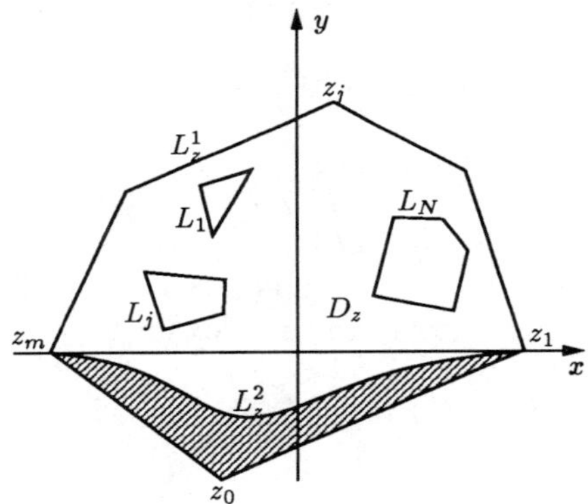

Figure 3.4

Finally, we discuss the case of a multiply connected elastic domain D_z, i.e. the boundary of D_z consists of $L_z^1 \cup L_z^2$ and N convex polygonal lines $L_1, \cdots, L_N$, where L_z^1, L_z^2 are as stated before, $L_1, \cdots, L_N$ are situated inside $L_z^1 \cup L_z^2$ (see Figure 3.4). Similarly to (3.8) and (3.9), if $E = (X_x + Y_y)/4$ and its partial derivatives satisfy the boundary condition

$$\frac{\partial E}{\partial x} = u(x), \quad \frac{\partial E}{\partial y} = v(x), \quad \frac{1}{4}(Y_y - X_x)^2 + X_y^2 + X_y^2 = k(x), \quad z \in L^2, \tag{3.29}$$

$$f(\frac{\partial E}{\partial x}, \frac{\partial E}{\partial y}) = 0, \quad F_s = 0, \quad z \in L^1 \cup L_1 \cup \cdots \cup L_n, \tag{3.30}$$

where $u(x), v(x), k(E) \in C_\alpha^1(0 < \alpha < 1)$, $k(E) \geq k_0 > 0$, we require to find the stress in D_z and to determine the free boundary L_z^2. This free boundary problem is called Problem F'.

According to Problem F, we decompose Problem F' into the corresponding Problem G' for analytic functions $G(z)$ and Problem H' for the complex equation $W_{\bar{z}} = \Phi'(z)$ in D_z, where the boundary condition of Problem G' is

$$\begin{cases} \operatorname{Re} G = u(x), \ \operatorname{Im} G = -v(x), \ z \in L_z^2, \\ f(\operatorname{Re} G, -\operatorname{Im} G) = 0, \ z \in L_z^1 \cup L_1 \cup \cdots \cup L_n, \end{cases} \tag{3.31}$$

and the boundary condition of Problem H' is

$$\begin{cases} |W| = k(x), \ z \in L_z^2, \\ \operatorname{Re}[\overline{\lambda(z)}W(z)] = 0, \ z \in L_z^1. \end{cases} \tag{3.32}$$

Now, the domain D_G is an $(N+1)$–connected domain in the G–plane, D_ζ is an $(N+1)$–connected circular domain in the ζ–plane, and the polygonal lines L_j can be written as

$$a_p^j x - y = b_p^j, \ p = 1, \cdots, m_j, \ j = 1, \cdots, N. \tag{3.33}$$

Hence the boundary condition (3.21) must be replaced by

$$\operatorname{Re}[\overline{\Lambda(\zeta)}z(\zeta)] = R(\zeta), \ \zeta \in \partial\Omega, \tag{3.34}$$

in which

$$\Lambda(\zeta) = \begin{cases} 1, \ \zeta \in (\zeta_m\zeta_0\zeta_1), \\ \dfrac{a_k - i}{\sqrt{1 + a_k^2}}, \ \zeta \in (\zeta_{k-1}\zeta_k), \ k = 2, \cdots, m, \\ \dfrac{a_p^j - i}{\sqrt{1 + a_p^{j2}}}, \ \zeta \in (\zeta_{p-1}^j\zeta_p^j), \ p = 1, \cdots, m_j, \ j = 1, \cdots, N, \end{cases}$$

$$R(\zeta) = \begin{cases} x(\zeta), \ \zeta \in (\zeta_m\zeta_0\zeta_1), \\ \dfrac{b_k}{\sqrt{1 + a_k^2}}, \ \zeta \in (\zeta_{k-1}\zeta_k), \ k = 2, \cdots, m, \\ \dfrac{b_p^j}{\sqrt{1 + a_p^{j2}}}, \ \zeta \in (\zeta_{p-1}^j\zeta_p^j), \ p = 1, \cdots, m_j, \ j = 1, \cdots, N. \end{cases}$$

It is not difficult to compute that the index of $\Lambda(\zeta)$ on $\zeta(L_j)(j = 1, \cdots, N)$ are equal to 1, consequently the index of $\Lambda(\zeta)$ on $\partial\Omega$ is $K = N - 1/2$. Using the result of Section 3, Chapter 1 and the Schauder theorem, we can obtain the result of solvability for Problem F' (see [151]1)).

4 A Mixed-Contact Boundary Value Problem for the Orthotropic Elasticity

In this section, a mixed contact boundary value problem for orthotropic elasticity is considered. First we pose this problem and prove a uniqueness theorem. Then by means of the theory of bi–analytic functions representations of solutions to the basic problems for orthotropic elasticity are obtained. Finally we reduce this mixed contact problem to boundary value problems for analytic functions which will be solved in virtue of a function theoretic method (see [14]).

The theory of so–called bi–analytic functions and its applications in elasticity has been developed by Hua, Loo-keng et al (see [68]). Here we shall utilize the methods of bi–analytic functions to solve the mixed contact problem for orthotropic elasticity (see [14]).

4.1 Formulation of the mixed contact problem

For the situation of plane strain of an orthotropic elastic body the equilibrium equations are

$$\begin{cases} \dfrac{\partial X_x}{\partial x} + \dfrac{\partial X_y}{\partial y} + F_1 = 0, \\[2mm] \dfrac{\partial X_y}{\partial x} + \dfrac{\partial Y_y}{\partial y} + F_2 = 0, \end{cases} \tag{4.1}$$

where X_x, X_y, Y_y are the components of stress as stated in Section 3, $F = (F_1, F_2)$ is the body force vector. In the case of small deformations, the generalized Hooke's law is given by the system of equations

$$\begin{cases} X_x = \dfrac{1}{1 - \nu_{12}\nu_{21}}(E_{11}\dfrac{\partial u}{\partial x} + \nu_{12}E_{22}\dfrac{\partial v}{\partial y}), \\[3mm] Y_y = \dfrac{1}{1 - \nu_{12}\nu_{21}}(\nu_{21}E_{11}\dfrac{\partial u}{\partial x} + E_{22}\dfrac{\partial v}{\partial y}), \\[3mm] X_y = G_{12}(\dfrac{\partial u}{\partial y} + \dfrac{\partial v}{\partial x}), \end{cases} \tag{4.2}$$

where G_{12} is the modulus of motion, ν_{12} and ν_{21} are Poisson's ratios, E_{11} and E_{22} are Young's moduli; they are related by the equation

$$\nu_{12}E_{22} = \nu_{21}E_{11}. \tag{4.3}$$

Substituting (4.2) into (4.1), we obtain

$$\left[A^{11} + \frac{\partial^2}{\partial x^2} + A^{12}\frac{\partial^2}{\partial x \partial y} + A^{21}\frac{\partial^2}{\partial y \partial x} + A^{22}\frac{\partial^2}{\partial y^2}\right]\begin{pmatrix} u \\ v \end{pmatrix} = -\begin{pmatrix} F_1 \\ F_2 \end{pmatrix}, \tag{4.4}$$

where

$$A^{11} = \begin{pmatrix} \dfrac{E_{11}}{1 - \nu_{12}\nu_{21}} & 0 \\ 0 & G_{12} \end{pmatrix}, \quad A^{12} = \begin{pmatrix} 0 & \dfrac{\nu_{12}E_{22}}{1 - \nu_{12}\nu_{21}} \\ G_{12} & 0 \end{pmatrix},$$

$$A^{21} = \begin{pmatrix} 0 & G_{12} \\ \dfrac{\nu_{21}E_{11}}{1 - \nu_{12}\nu_{21}} & 0 \end{pmatrix}, \quad A^{22} = \begin{pmatrix} G_{12} & 0 \\ 0 & \dfrac{E_{22}}{1 - \nu_{12}\nu_{21}} \end{pmatrix}. \tag{4.5}$$

If u, v are continuously differentiable up to the second order, then system (4.4) may be written into

$$L\begin{pmatrix} u \\ v \end{pmatrix} = \left[A^{11}\frac{\partial^2}{\partial x^2} + (A^{12} + A^{21})\frac{\partial^2}{\partial x \partial y} + A^{22}\frac{\partial^2}{\partial y^2} \right] \begin{pmatrix} u \\ v \end{pmatrix} = -\begin{pmatrix} F_1 \\ F_2 \end{pmatrix}. \tag{4.6}$$

The displacement boundary value problem is to look for the solution u, v to system (4.4) in the domain Ω, which the elastic body occupies, if the displacement u, v are specified at all points of the boundary Γ of Ω. If the stresses are specified at all points of Γ, then this is the stress boundary value problem. Assume n to be the exterior unit normal to Γ, and X_n and Y_n to be the components of the external stress along the axes, then we may write the boundary conditions for this problem similar to (3.11), i.e.

$$[X_x \cos(n, x) + X_y \cos(n, y)]\,|_\Gamma = X_n,$$

$$[Y_x \cos(n, x) + Y_y \cos(n, y)]\,|_\Gamma = Y_n. \tag{4.7}$$

Noticing (4.2) and using (4.5), we get the matrix form

$$\left[\cos(n, x)\left(A^{11}\frac{\partial}{\partial x} + A^{12}\frac{\partial}{\partial y} \right) + \cos(n, y)\left(A^{21}\frac{\partial}{\partial x} + A^{22}\frac{\partial}{\partial y} \right) \right] \begin{pmatrix} u \\ v \end{pmatrix} \Big|_\Gamma = \begin{pmatrix} X_n \\ Y_n \end{pmatrix}. \tag{4.8}$$

We now start to formulate the problem. Suppose the closed curves Γ_0, Γ_1, Γ_2 of class C^3 are given in the plane $I\!\!R^2$ such that Γ_0 is surrounded by Γ_1 and Γ_2 by Γ_0. By Ω_1 denote the doubly–connected domain with the boundary $\partial\Omega_1 = \Gamma_0 \cup \Gamma_1$, and by Ω_2 the domain with the boundary $\partial\Omega_2 = \Gamma_0 \cup \Gamma_2$. Suppose that Ω_1 and Ω_2 are occupied by two distinct orthotropic elastic bodies respectively. Obviously, they obey the system of equilibrium equations respectively:

$$L_j\begin{pmatrix} u^j \\ v^j \end{pmatrix} = \left[\left(A_j^{11}\frac{\partial}{\partial x} + A_j^{12}\frac{\partial}{\partial y} \right)\frac{\partial}{\partial x} + \left(A_j^{21}\frac{\partial}{\partial x} \right. \right.$$

$$\left. \left. + A_j^{22}\frac{\partial}{\partial y} \right)\frac{\partial}{\partial y} \right] \begin{pmatrix} u^j \\ v^j \end{pmatrix} = -\begin{pmatrix} F_1^j \\ F_2^j \end{pmatrix}, \quad (x, y) \in \Omega_j, \; j = 1, 2, \tag{4.9}$$

where $(u^j, v^j)^T$ is the unknown displacement vector, and the (2×2)–matrices A_j^{ik} take the forms

$$
A_j^{11} = \begin{pmatrix} \dfrac{E_{11}^j}{1 - \nu_{12}^j \nu_{21}^j} & 0 \\ 0 & G_{12}^j \end{pmatrix}, \quad
A_j^{12} = \begin{pmatrix} 0 & \dfrac{\nu_{12}^j E_{22}^j}{1 - \nu_{12}^j \nu_{21}^j} \\ G_{12}^j & 0 \end{pmatrix},
$$

$$
A_j^{21} = \begin{pmatrix} 0 & G_{12}^j \\ \dfrac{\nu_{21}^j E_{11}^j}{1 - \nu_{12}^j \nu_{21}^j} & 0 \end{pmatrix}, \quad
A_j^{22} = \begin{pmatrix} G_{12}^j & 0 \\ 0 & \dfrac{E_{22}^j}{1 - \nu_{12}^j \nu_{21}^j} \end{pmatrix},
\qquad j = 1, 2. \qquad (4.10)
$$

Here, as in the following, T denotes matrix transposition. We divide the curves $\Gamma_j\,(j = 1, 2)$ into arcs Γ_{jl} by points $\zeta_l^j \in \Gamma_j$, $l = 1, 2$, moreover, by rotating Γ_j in the counterclockwise direction, the beginning of the arc Γ_{jl} will be assumed to be the points ζ_l^j, $j, l = 1, 2$. By n_j denote the exterior unit normal to Γ_j, $j = 0, 1, 2$. The mixed—contact boundary problem is to seek out the displacement vector

$$
(u, v) = \begin{cases} (u^1(x, y), v^1(x, y)), & (x, y) \in \Omega_1, \\ (u^2(x, y), v^2(x, y)), & (x, y) \in \Omega_2, \end{cases}
$$

such that $u^j, v^j \in C^2(\Omega_j)$ satisfy system (4.9) in the domain Ω_j and fulfil the boundary conditions

$$
(u^j, v^j) = 0, \ (x, y) \in \Gamma_{j1},
$$

$$
\left[\cos(n_j, x)\left(A_j^{11} \frac{\partial}{\partial x} + A_j^{12} \frac{\partial}{\partial y}\right) + \cos(n_j, y)\left(A_j^{21} \frac{\partial}{\partial x} + A_j^{22} \frac{\partial}{\partial y}\right)\right] \begin{pmatrix} u^j \\ v^j \end{pmatrix} = 0, \qquad (4.11)
$$

$$
(x, y) \in \Gamma_{j2}, \ j = 1, 2,
$$

and the contact conditions

$$
\left[\cos(n_0, x)\left(A_1^{11} \frac{\partial}{\partial x} + A_j^{12} \frac{\partial}{\partial y}\right) + \cos(n_0, y)\left(A_1^{21} \frac{\partial}{\partial x} + A_1^{22} \frac{\partial}{\partial y}\right)\right] \begin{pmatrix} u^1 \\ v^1 \end{pmatrix}
$$

$$
= \left[\cos(n_0, x)\left(A_2^{11} \frac{\partial}{\partial x} + A_2^{12} \frac{\partial}{\partial y}\right) + \cos(n_0, y)\left(A_2^{21} \frac{\partial}{\partial x} + A_2^{22} \frac{\partial}{\partial y}\right)\right] \begin{pmatrix} u^2 \\ v^2 \end{pmatrix}, \qquad (4.12)
$$

$$
(u^1, v^1) = (u^2, v^2), \ (x, y) \in \Gamma_0.
$$

Before we start to discuss the uniqueness and the existence of solutions to this mixed contact problem, we introduce some function classes. We refer a function W on Γ_j to belong to the class $H_0(\Gamma_j)$ if it is Hölder continuous on the closed arcs Γ_{jl}, $l = 1, 2$. A function W is said to belong to the class $H_*(\Gamma_j)$, if on each closed subarc of Γ_j not involving the points $\zeta_l^j \in \Gamma_j$ it is Hölder continuous, while near the points ζ_l^j it can be represented in the form $W(t) = W_*(t)/|t - \zeta_l^j|^\alpha$ $(0 < \alpha < 1)$, where $W_* \in H_0(\Gamma_j)$. We also refer a function $W(z) = W(x, y)$, $z = x + iy$, defined in the domain Ω_j to belong to $H_*(\overline{\Omega_j})$ if it is Hölder continuous in Ω_j up to the boundary $\partial\Omega_j$ everywhere except possibly at the point ζ_l^j near which the inequality $|W(z)| \le M/|z - \zeta_l^j|^\alpha$ holds, where $M > 0$ is a constant and $0 \le \alpha < 1$. Moreover, its limit value $W(t)$ is assumed to exist on $\partial\Omega_j$ everywhere with the possible exception of the points ζ_l^j and belongs to the class $H_*(\partial\Omega_j)$. Similarly, $H_*^1(\overline{\Omega_j})$ denotes the class of functions the first order partial derivatives of which belong to $H_*(\overline{\Omega_j})$.

4.2 Uniqueness theorem of the mixed contact problem

Now, we prove the uniqueness of solutions to the mixed contact problem for orthotropic elasticity.

Lemma 4.1 *If $w = (u, v)^T \in C^2(\Omega) \cap H_*^1(\overline{\Omega})$, and Ω^+ is a subdomain of Ω, then*

$$
\int_{\partial\Omega^+} (u, v)\left\{\left[\left(A^{11}\frac{\partial}{\partial x} + A^{12}\frac{\partial}{\partial y}\right)\begin{pmatrix} u \\ v \end{pmatrix}\right]\cos(n, x) + \left[\left(A^{21}\frac{\partial}{\partial x} + A^{22}\frac{\partial}{\partial y}\right)\right.\right.
$$
$$
\left.\left.\times \begin{pmatrix} u \\ v \end{pmatrix}\right]\cos(n, y)\right\}ds = \iint_{\Omega^+} [((u, v)L(w) + Q(w)]dxdy \tag{4.13}
$$

holds, where A^{ij} and $L(w)$ are defined in (4.5) and (4.6) respectively, and $Q(w)$ is the quadratic form

$$
Q(w) = \frac{\nu_{12}E_{22}}{(1 - \nu_{12}\nu_{21})}\left[\left(\frac{\partial u}{\partial x}\right)^2 + 2\nu_{21}\left(\frac{\partial u}{\partial x}\right)\left(\frac{\partial v}{\partial y}\right) + \frac{\nu_{21}}{\nu_{12}}\left(\frac{\partial v}{\partial y}\right)^2\right]
$$
$$
+ G_{21}\left(\frac{\partial v}{\partial x} + \frac{\partial u}{\partial y}\right)^2. \tag{4.14}
$$

Proof It is easy to verify the identity

$$
\frac{\partial}{\partial x}\left[(u, v)\left(\begin{pmatrix} \dfrac{E_{11}}{1 - \nu_{12}\nu_{21}} & 0 \\ 0 & G_{12} \end{pmatrix}\frac{\partial}{\partial x} + \begin{pmatrix} 0 & \dfrac{\nu_{12}E_{22}}{1 - \nu_{12}\nu_{21}} \\ G_{12} & 0 \end{pmatrix}\frac{\partial}{\partial y}\right)\begin{pmatrix} u \\ v \end{pmatrix}\right]
$$
$$
+ \frac{\partial}{\partial y}\left[(u, v)\left(\begin{pmatrix} 0 & G_{12} \\ \dfrac{\nu_{21}E_{11}}{1 - \nu_{12}\nu_{21}} & 0 \end{pmatrix}\frac{\partial}{\partial x} + \begin{pmatrix} G_{12} & 0 \\ 0 & \dfrac{E_{22}}{1 - \nu_{12}\nu_{21}} \end{pmatrix}\frac{\partial}{\partial y}\right)\begin{pmatrix} u \\ v \end{pmatrix}\right]
$$
$$
= (u, v)L\begin{pmatrix} u \\ v \end{pmatrix} + Q(w). \tag{4.15}
$$

Applying Green's formula, from (4.15) we obtain the desired formula (4.13).

Theorem 4.2 *The mixed contact problem* (4.9), (4.11) *and* (4.12) *has at most one solution in the function class* $C^2(\Omega_j) \cap H^1_*(\overline{\Omega}_j)$, $j = 1, 2$.

Proof Suppose there exist two solution $w^* = (u^*, v^*)^T$ and $w^{**} = (u^{**}, v^{**})^T$ to this problem. Then

$$w(x, y) = w^{**}(x, y) - w^*(x, y) = (u, v)^T$$

fulfils the homogeneous system

$$L(w) = 0, \tag{4.16}$$

the homogeneous boundary conditions and the corresponding homogeneous contact conditions. From the domain Ω_j we remove the set of points z, belonging to the closed discs $|z - \zeta_l^j| \leq \varepsilon \, (j, l = 1, 2)$ with small enough radius ε, and denote the residual domain by $\Omega_{j,\varepsilon}$. It is easy to see that $w^j = (u^j, v^j)^T$ in $\Omega_{j,\varepsilon} \, (j = 1, 2)$ satisfy the conditions of Lemma 4.1. So we get

$$\int_{\partial\Omega_{j,\varepsilon}} (w^j)^T \Big\{ \Big[\Big(A_j^{11}\frac{\partial}{\partial x} + A_j^{12}\frac{\partial}{\partial y}\Big)w^j\Big]\cos(n_j, x) + \Big[\Big(A_j^{21}\frac{\partial}{\partial x} + A_j^{22}\frac{\partial}{\partial y}\Big)w^j\Big]$$

$$\times \cos(n_j, y)\Big\} ds = \iint_{\Omega_{j,\varepsilon}} [(w^j)^T L_j(w^j) + Q_j(w^j)] dx dy, \quad j = 1, 2,$$

where

$$Q_j(w^j) = \frac{\nu_{12}^j E_{22}^j}{(1 - \nu_{12}^j \nu_{21}^j)\nu_{21}^j}\left[\left(\frac{\partial u^j}{\partial x}\right)^2 + 2\nu_{21}\left(\frac{\partial u^j}{\partial x}\right)\left(\frac{\partial v^j}{\partial y}\right)\right.$$

$$\left. + \frac{\nu_{21}^j}{\nu_{12}^j}\left(\frac{\partial v^j}{\partial y}\right)^2\right] + G_{21}^j\left(\frac{\partial v^j}{\partial x} + \frac{\partial u^j}{\partial y}\right)^2.$$

By letting ε tend to 0, it is easy to show

$$\iint_{\Omega_j} Q_j(w^j) dx dy = 0, \quad j = 1, 2.$$

For the orthotropic elasticity $Q_j(w^j)$ is positive definite. Hence we conclude $u^j = v^j = 0 \, (j = 1, 2)$.

4.3 Representation of general solution and the forms of boundary conditions

1) First we consider the case without body forces in a simply connected domain D. As in [56]2), we introduce new parameters ν, E, δ defined by the equations

$$\nu^2 = \nu_{12}\nu_{21}, \quad E^2 = E_{11}E_{22}, \quad \delta^2 = \sqrt{\frac{E_{11}}{E_{22}}}, \quad k_1 = \frac{E}{2G_{12}} - \nu. \tag{4.17}$$

Introducing $k = k_1 - \sqrt{k_1^2 - 1}$, $\lambda = (1 - \nu^2)k/[2(k_1 + \nu)]$ and making transformations $x = \delta x'$, $y = y'/\sqrt{k}$, and $u = (1 + k\nu)u'/(k + \nu)\sqrt{k}$, $v = -\delta v'$, we reduce (4.9) into (see [68])

$$\left[\begin{pmatrix} k & 0 \\ 0 & -\lambda \end{pmatrix} \frac{\partial}{\partial x'} + \begin{pmatrix} 0 & \lambda \\ k & 0 \end{pmatrix} \frac{\partial}{\partial y'}\right]$$
$$\times \left[\begin{pmatrix} 1/k & 0 \\ 0 & 1/k \end{pmatrix} \frac{\partial}{\partial x'} + \begin{pmatrix} 0 & -1 \\ 1 & 0 \end{pmatrix} \frac{\partial}{\partial y'}\right]\begin{pmatrix} u' \\ v' \end{pmatrix} = 0. \tag{4.18}$$

Putting $f = u_1 + iv_1$, $z_1 = x_1 + iy_1$, from (4.18) we get the general solution:

$$f(z_1) = \alpha\Phi(z_1) + \beta\overline{\Phi(z_1)} + \Psi(\frac{k+1}{2k}z_1 + \frac{k-1}{2k}\bar{z}_1), \tag{4.19}$$

where

$$\alpha = \frac{\lambda - k}{2(1-k)\lambda}, \ \beta = \frac{\lambda + k}{2(1+k)\lambda}, \ \Phi(z_1) = \int_{z_0}^{z_1} \phi(\zeta)d\zeta,$$

herein $\phi(z_1)$ is an analytic function of z_1, and $\Psi(z_2)$ is an arbitrary analytic function of the variable $z_2 = (k+1)z_1/2k + (k-1)\bar{z}_1/2k$. Setting

$$\alpha_3 = \frac{1+k}{1-k}\frac{E}{1-\nu^2}, \ \beta_3 = -\frac{E}{k+\nu}, \ p_1 = -\frac{k+\nu}{2E\sqrt{k}},$$
$$p_2 = -\frac{1+k}{2kE\sqrt{k}}, \ q_1 = -\frac{1+\nu k}{2E_{22}\delta k}i, \ q_2 = -\frac{\nu + k}{2E_{22}\delta k}i,$$

we finally obtain the general solution

$$u = 2\text{Re}[p_1\alpha_3\Phi(z_1) + p_2k\beta_3\Psi(z_2)],$$
$$v = 2\text{Re}[q_1\alpha_3\Phi(z_1) + q_2k\beta_3\Psi(z_2)]. \tag{4.20}$$

Now we simplify the boundary conditions of the stress problem (4.11). It is obvious that

$$\cos(n, x) = \frac{dy}{ds} = \frac{1}{\sqrt{k}}\frac{dy_1}{ds}, \ \cos(n, y) = -\frac{dx}{ds} = -\delta\frac{dx_1}{ds}.$$

Putting (4.20) into (4.11), we obtain

$$X_n ds = d\left[\text{Re}\, i\delta\alpha_3\Phi(x_1 + iy_1)\right] + d\left[\text{Re}\, i\delta\beta_3\Psi(x_1 + \frac{i}{k}y_1)\right].$$

Integrating along the boundary Γ with respect to the arc length s we get

$$\text{Re}[i\alpha_3\Phi(x_1 + y_1) + \frac{i}{k}(k\beta_3)\Psi(x_1 + \frac{iy_1}{k})] = \int_0^s \frac{X_n}{\delta}ds + c_2. \tag{4.21}$$

Similarly,

$$\text{Re}[\alpha_3\Phi(x_1 + iy_1) + (k\beta_3)\Psi(x_1 + \frac{iy_1}{k})] = -\int_0^s \sqrt{k}Y_n ds + c_1. \tag{4.22}$$

2) For a doubly connected domain D, setting

$$\alpha_3\Phi(z_1) = \Psi_1(z_1) + A\ln z_1, \quad k\beta_3\Psi(z_2) = \Psi_2(z_2) + B\ln z_2,$$

where A, B are constants determined by the principal vector of the exterior forces along the boundary, we obtain the following result.

Theorem 4.3   *The representation of the boundary conditions for the stress problem and the displacement problem will be respectively*

$$2\mathrm{Re}[\Psi_1(t_1) + \Psi_2(t_2)] + 2\mathrm{Re}[A\ln t_1 + B\ln t_2] = -2\int_0^s kY_n ds + c_1,$$

$$2\mathrm{Re}[i\Psi_1(t_1) + \frac{i}{k}\Psi_2(t_2)] + 2\mathrm{Re}[iA\ln t_1 + \frac{i}{k}B\ln t_2] = 2\int_0^s \frac{X_n}{\delta}ds + c_2, \tag{4.23}$$

and

$$2\mathrm{Re}[p_1\Psi_1(t_1) + p_2\Psi_2(t_2)] + 2\mathrm{Re}[p_1 A\ln t_1 + p_2 B\ln t_2] = U(t),$$

$$2\mathrm{Re}[q_1\Psi_1(t_1) + q_2\Psi_2(t_2)] + 2\mathrm{Re}[q_1 A\ln t_1 + q_2 B\ln t_2] = V(t), \tag{4.24}$$

where $U(t)$, $V(t)$ are given functions.

3) For the general case when body forces exist, we have to seek out a special solution corresponding to the nonhomogeneous system (4.4). Setting

$$F = -\frac{1}{2}\left(\frac{1-\nu^2}{E}\right)\left(\frac{k+\nu}{1+k\nu}\sqrt{k}F_1 + iF_2\right), \quad \phi_0(z_1) = \frac{1}{\pi}\int\int_{G_1} \frac{F(\zeta_1)}{\zeta_1 - z_1}d\xi_1 d\eta_1,$$

where $\zeta_1 = \xi_1 + i\eta_1$, G_1 is the image of Ω under the mapping

$$z_1 = \frac{1}{2}\left(\frac{1}{\delta} + \sqrt{k}\right)z + \frac{1}{2}\left(\frac{1}{\delta} - \sqrt{k}\right)\bar{z} = c(z).$$

Then

$$f_0(z_2) = u_0 + iv_0 = -\frac{1}{\pi}\int\int_{G_2} \frac{[(\lambda - k)\phi_0(a(\zeta_2)) + (\lambda + k)\overline{\phi_0(a(\zeta_2))}]/4\lambda}{\zeta_2 - z_2}d\xi_2 d\eta_2,$$

where $\zeta_2 = \xi_2 + i\eta_2$, G_2 is the image of G_1 under the mapping

$$z_2 = b(z_1) = \frac{k+1}{2k}z_1 + \frac{k-1}{2k}\bar{z}_1,$$

and

$$z_1 = a(z_2) = \frac{1+k}{2}z_2 + \frac{1-k}{2}\bar{z}_2.$$

Consequently, the system (4.4) has a special solution

$$u_0(z) = \frac{1+k\nu}{(k+\nu)\sqrt{k}}\mathrm{Re}[f_0(b(c(z)))], \quad v_0(z) = -\delta\mathrm{Im}[f_0(b(c(z)))].$$

Finally we obtain the general solution

$$u(z) = 2\text{Re}[p_1\Psi_1(z_1) + p_2\Psi_2(z_2)] + 2\text{Re}[p_1 A\ln z_1 + p_2 B\ln z_2] + u_0(z),$$
$$v(z) = 2\text{Re}[q_1\Psi_1(z_1) + q_2\Psi_2(z_2)] + 2\text{Re}[q_1 A\ln z_1 + q_2 B\ln z_2] + v_0(z). \tag{4.25}$$

4.4 Reduction of the contact problem to a boundary value problem for analytic functions

On the basis of the result from Subsection 4.3, we are able to reduce the contact problem to the boundary value problem for analytic functions $\Psi_1(z_1)$ and $\Psi_2(z_2)$. Let Γ_0' and Γ_0'' be the images of Γ_0 under the affine transformations

$$z_1 = c(z) = x_1 + iy_1 = x/\delta + i\sqrt{k}y,$$

$$z_2 = b(z_1) = x_1 + iy_1/k = x/\delta + iy/\sqrt{k},$$

respectively. Substituting (4.25) into the first equation of the contact condition (4.12), we have

$$2\text{Re}[(p_1^1\Psi_1^1(t_1) - p_1^2\Psi_1^2(t_1)) + (p_2^1\Psi_2^1(t_2) - p_2^2\Psi_2^2(t_2))]$$

$$= -2\text{Re}[(p_1^1 A^1 + p_1^2 A^2)\ln t_1 + (p_2^1 B^1 - p_2^2 B^2)\ln t_2]$$

$$-\text{Re}[\frac{1 + k^1\nu^1}{(k^1 + \nu^1)\sqrt{k^1}} + f_0^1(t) - \frac{1 + k^2\nu^2}{(k^2 + \nu^2)\sqrt{k^2}}f_0^2(t)],$$

$$2\text{Re}[(q_1^1\Psi_1^1(t_1) - q_1^2\Psi_1^2(t_1)) + (q_2^1\Psi_2^1(t_2) - q_2^2\Psi_2^2(t_2))]$$

$$= -2\text{Re}[(q_1^1 A^1 - q_1^2 A^2)\ln t_1 + (q_2^1 B^1 - q_2^2 B^2)\ln t_2]$$

$$+\text{Im}[\delta^1 f_0^1(t) - \delta^2 f_0^2(t)],$$

$$t \in \Gamma_0, \ t_1 \in \Gamma_0', \ t_2 \in \Gamma_0''.$$

Moreover, noting (4.23), from the second equation of (4.12) we obtain

$$2\text{Re}[\frac{1}{\sqrt{k^1}}\Psi_1^1(t_1) - \frac{1}{\sqrt{k^2}}\Psi_1^2(t_1) + \frac{1}{\sqrt{k^1}}\Psi_2^1(t_2) - \frac{1}{\sqrt{k^2}}\Psi_2^2(t_2)]$$

$$= -2\text{Re}[(A^1 - A^2)\ln t_1] + (B^1 - B^2)\ln t_2] + c_1^1 + h_1(t),$$

$$2\text{Re}[i(\delta^1\Psi_1^1(t_1) - \delta^2\Psi_1^2(t_1)) + i(\frac{\delta^1}{\sqrt{k^1}}\Psi_2^1(t_2) - \frac{\delta^2}{\sqrt{k^2}}\Psi_2^2(t_2)]$$

$$= -2\text{Re}[i(A^1 - A^2)\ln t_1 + i(\frac{B^1}{k^1} - \frac{B^2}{k^2})\ln t_2] + c_2^1 + h_2(t),$$

$$t \in \Gamma_0, \ t_1 \in \Gamma_0', \ t_2 \in \Gamma_0'',$$

where $h_1(t)$ and $h_2(t)$ are completely determined by the functions $f_0^1(t)$ and $f_0^2(t)$. Similarly, we can get the boundary conditions from (4.11).

References

[1] Adams, R. *Sobolev spaces.* Academic Press, New York, 1975.

[2] Agmon, S, Douglis, A and Nirenberg, L. *Estimates near the boundary for solutions of elliptic partial differential equations satisfying general boundary conditions* I, II. Comm. Pure Appl. Math. 12(1959), 623-727; 17(1964), 35-92.

[3] Ahlfors, L. *Lectures on quasiconformal mappings.* D. Van. Nostrand Co., New York, 1966.

[4] Andreian Cazacu, C. *Theorie der Funktionen mehrerer komplexer Veränderlicher.* Birkhäuser Verlag, Basel-Stuttgart, 1975.

[5] Avantaggiati, A. *Internal and external first order boundary value problems on C^1 domains.* Instituto per le applicazioni del calcolo "Mauro Picone", Pubblicazioni Serie III, N. 215, Roma, 1982.

[6] Bear, R J, Gilbert, R P and Kendall, H. *On a two-dimensional free-boundary problem.* J. Appl. Math. Phys. 11(1960), 341-355; 12(1960), 69-70.

[7] Begehr, H. 1) *Boundary value problems for mixed kind systems of first order partial differential equations.* 3. Roumanian-Finnish Seminar on Complex Analysis, Bucharest 1976, Lecture Notes in Math. 743, Springer-Verlag, Berlin etc., 1979, 600-614.
2) *An approximation method for the Dirichlet problem of nonlinear elliptic systems in R^r.* Rev. Roumaine Math. Pure Appl. 27(1982), 927-934.
3) *Boundary value problems for analytic and generalized analytic functions.* Complex Analysis-Methods, Trends, and Applications, Akademie-Verlag, Berlin, 1983, 150-165.
4) *Boundary value problems for Beltrami equations.* J. Hebei Normal Univ. Math. Sci. Ed. 3(1987), 37-42 (Chinese).
5) *Entire solutions for linear elliptic equations.* J. Hebei Normal Univ. Math. Sci. Ed. 3(1987), 43-47 (Chinese).
6) *Nonlinear evolution equations.* J. Hebei Normal Univ. Math. Sci. Ed. 3(1987), 48-54 (Chinese).
7) *Hele-Shaw flow and the moment problem.* J. Hebei Normal Univ. Math. Sci. Ed. 3(1987), 55-63 (Chinese).
8) *Randwertaufgaben für elliptische Systeme vom Haack-Vekua Typ.* Sber. Berlin Math. Ges. 1972-1987, 159-170.
9) *A priori estimate for elliptic equations and systems.* Proc. Assiut First International Conference, Assiut Univ., Assiut, Part V, 1990, 37-58.
10) *Moving boundary problems of Hele Shaw type.* Proc. 21.st Annual Iranian Math. Conference, Univ. Isfahan, Isfahan, 1990, 43-63.

11) *Estimates of solutions to linear elliptic systems and equations.* Part. Diff. Equations, Banach Centre Publ., 27, Part 1, Warsaw, 1992, 45-63.
12) *Complex analytic methods for partial differential equations.* World Scientific, Singapore, 1994.

[8] Begehr, H and Dai, Dao-qing. *Initial boundary value problem for nonlinear pseudoparabolic equations.* Complex Variables, Theory Appl. 18(1992), 33-47.

[9] Begehr, H and Gilbert, R P. 1) *Das Randwert-Normproblem für ein fastlineares elliptisches System und eine Anwendung.* Ann. Acad. Sci. Fenn. A.I. 3(1977), 179-184.
2) *Randwertaufgaben ganzzahliger Charakteristik für verallgemeinerte hyperanalytische Funktionen.* Appl. Anal. 6(1977), 189-205.
3) *On Riemann boundary value problems for certain linear elliptic systems in the plane.* J. Diff. Equ. 32(1979), 1-14.
4) *Hele-Shaw flows in R^n.* Nonlinear Analysis, 10(1986), 65-85.
5) *Non-Newtonian Hele-Shaw flows in $n \geq 2$ dimensions.* Nonlinear Analysis 11(1987), 17-47; Chinese transl. J. Hebei Normal Univ. Math. Sci. Ed. 3(1987), 67-96.
6) *Pseudohyperanalytic functions.* Complex Variables, Theory Appl. 9(1988), 343-357.
7) *Transformations, transmutations, and kernel functions* I,II. Longman, Harlow, 1992, 1993.

[10] Begehr, H and Hile, G N. 1) *Nonlinear Riemann boundary problems for a semilinear elliptic system in the plane.* Math. Z. 179(1982), 241-261.
2) *Riemann boundary value problem for nonlinear elliptic systems.* Complex Variables, Theory Appl. 1(1983), 239-261.
3) *Schauder estimates and existence theory for entire solutions of linear elliptic equations.* Proc. Royal Soc. Edinburgh 110A(1988), 101-123.

[11] Begehr, H and Hsiao, G C. 1) *Nonlinear boundary value problems for a class of elliptic systems.* Komplexe Analysis und ihre Anwendungen auf partielle Differentialgleichungen. Martin-Luther-Univ., Halle-Wittenberg. Wiss. Beiträge 1980, 90-102.
2) *On nonlinear boundary value problems of elliptic systems in the plane.* Ord. Part. Diff. Equ., Proc. Dundee 1980. Lecture Notes in Math. 864, Springer-Verlag, Berlin etc. 1981, 55-63.
3) *Nonlinear boundary value problems of Riemann-Hilbert type.* Contemporary Math. 11(1982), 139-153.
4) *The Hilbert boundary value problem for nonlinear elliptic systems.* Proc. Roy. Soc. Edinburgh 94A(1983), 97-112.
5) *A priori estimates for elliptic systems.* Z. Anal. Anw. 6(1987), 1-21.

[12] Begehr, H and Jeffrey, A, ed. 1) *Partial differential equations with complex anbalysis*. Longman, Harlow, 1992.
2) *Partial differentisal equations with real analysis*. Longman, Harlow, 1992.

[13] Begehr, H and Kumar, A. *Bi-analytic functions of several complex variables.* Complex Variables, Theory and Appl. 24(1993), 89-106.

[14] Begehr, H and Lin, Wei. *A mixed-contact boundary problem in orthotropic elasticity*. Partial Differential Equations with Real Analysis, Longman, Harlow, 1992, 219-239.

[15] Begehr, H and Wen, Guo-chun. 1) *The discontinuous oblique derivative problem for nonlinear elliptic systems of first order.* Rev. Roumaine Math. Pures Appl. 33(1988), 7-19.
2) *A priori estimates for the discontinuous oblique derivative problem for elliptic systems.* Math. Nachr. 142(1989), 307-336.

[16] Begehr, H, Wen, Guo-chun and Zhao, Zhen. *An initial and boundary value problem for nonlinear composite type systems of three equations.* Math. Pannonica 2(1991), 49-61.

[17] Begehr, H and Xu, Zhen-yuan. *Nonlinear half-Dirichlet problems for first order elliptic equations in the unit ball of $R^m (m \geq 3)$.* Appl. Anal. 45(1992), 3-18.

[18] Bergman, S. *Integral operators in the theory of linear partial differential equations.* Springer-Verlag, Berlin etc. 1961.

[19] Bers, L. 1) *Theory of pseudoanalytic functions.* New York University, New York, 1953.
2) *Mathematical aspects of subsonic and transonic gas dynamics.* Wiley, New York, 1958.
3) *Quasiconformal mappings, with applications to differential equations, function theory and topology.* Bull. Amer. Math. Soc. 83(1977), 1083-1100.

[20] Bers, L, John, F and Schechter, M. *Partial differential equations.* Intersience Publ., New York etc., 1964.

[21] Bers, L and Nirenberg, L. 1) *On a representation theorem for linear elliptic systems with discontinuous coefficients and its application.* Conv. Intern. Eq. Lin. Derivate Partiali, Trieste, Cremonense, Roma, 1954, 111-140.
2) *On linear and nonlinear elliptic boundary value problems in the plane.* Conv. Intern. Eq. Lin. Derivate Partiali, Trieste, Cremonense, Roma, 1954, 141-167.

[22] Beyer, K. *Nichtlineare Randwertprobleme für elliptische Systeme 1. Ordnung für zwei Funktionen von zwei Variablen.* Beiträge zur Analysis 4(1972), 31-34.

[23] Bicadze, A V. 1) *Boundary value problems for elliptic equations of second order.* Nauka, Moscow, 1966(Russian); Engl. Transl. North Holland Publ. Co., Amsterdam, 1968.
2) *The theory of non-Fredholm elliptic boundary value problems.* Amer. Math. Soc. Transl. (2) 105(1976), 95-103.
3) *Some classes of partial differential equations.* Gordon and Breach, New York etc., 1988.

[24] Bliev, N K. *Generalized analytic functions in fractional space.* Izdatel'stro Nauka Kazakhikai SSR, Alma-Ata, 1985 (Russian).

[25] Bojarski, B. 1)*Generalized solutions of a system of differential equations of first order and elliptic type with discontinuous coefficients.* Mat. Sb. N. S. 43(85)(1957), 451-563 (Russian).
2) *Some boundary value problems for systems of $2m$ equations of elliptic type in the plane.* Dokl. Akad. Nauk SSSR 124(1958), 15-18 (Russian).
3) *Riemann-Hilbert problem for a holomorphic vector.* Dokl. Akad. Nauk SSSR 126(1959), 695-698 (Russian).
4) *Theory of generalized analytic vectors.* Ann. Polon. Math. 17(1966), 281-320.
5) *Subsonic flow of compressible fluid.* Arch. Mech. Stos. 18(1966), 497-520; Mathematical Problems in Fluid Mechanics, Polon. Acad. Sci., Warsaw, 1967, 9-32.

[26] Bojarski, B and Iwaniec, T. *Quasiconformal mappings and non-linear elliptic equations in two variables* I, II. Bull. Acad. Polon. Sci., Sér. Sci. Math. Astr. Phys. 22(1974), 473-478; 479-484.

[27] Brackx, F, Delanghe, R and Sommen, F. *Clifford analysis.* Pitman, London etc., 1982.

[28] Brezis, H. *The dam problem revisited.* Proc. Sem. Montecatini, Pitman, London etc. 1983.

[29] Buchanan, J. *Bers-Vekua equations of two complex variables.* Contemporary Math. 11(1982), 71-88.

[30] Chisholm, J S R and Common, A K. *Clifford algebras and their applications in mathematical physics.* Reidel, Dordrecht, 1986.

[31] Ciechanowicz-Halka, B. *Ein nichtlineares Randproblem der flachen Elastizitätstheorie.* Demonstratio Math. 11(1978), 583-590.

[32] Colton, D. 1) *Partial differential equations in the complex domain.* Pitman, London etc., 1976.
2) *Analytic theory of partial differential equations.* Pitman, London etc., 1980.

[33] Colton, D and Gilbert, R P. *Constructive and computational methods for differential and integral equations.* Letures Notes in Math. 430, Springer-Verlag, Berlin etc., 1974.

[34] Courant, R and Hilbert, H. *Methods of mathematical physics* II. Interscience Publ., New York, 1962.

[35] Crank, J. *Free and moving boundary problems.* Clarendon Press, Oxford, 1984.

[36] Dai, Dao-qing. *On the Riemann-Hilbert boundary value problem for nonlinear elliptic equations in the plane.* Partial Differential Equations with Complex Analysis, Longman, Harlow, 1992, 150-155.

[37] Daniljuk, I I. *Nonregular boundary value problems in the plane.* Izdat. Nauka, Moscow, 1975 (Russian).

[38] Delanghe, R. 1) *On regular-analytic functions with values in a Clifford algebra.* Math. Ann. 185(1970), 91-111.
2) *On the singularities of functions with values in a Clifford algebra.* Math. Ann. 196(1972), 293-319.
3) *On regular functions with values in topological modules over a Clifford algebra.* Bull. Soc. Math. Belg. 25(1973), 131-138.

[39] Delanghe, R and Brackx, F. *Hypercomplex function theory and Hilbert modules with reproducing kernel.* Proc. London Math. Soc. 37(1978), 545-576.

[40] Dzhurajev, A. 1) *Study of partial differential equations by the means of generalized analytical functions.* Function Theoretic Methods for Partial Differential Equations. Lecture Notes in Math. 561, Springer-Verlag, Berlin etc., 1976, 29-38.
2) *Systems of equations of composite type.* Nauka SSSR, Moscow, 1972 (Russian); Engl. transl. Longman, Harlow, 1989.
3) *Methods of singular integral equations.* Nauk SSSR, Moscow, 1987 (Russian); Engl. transl. Longman, Harlow, 1992.
4) *Degenerate and other problems.* Longman, Harlow, 1992.

[41] Ding, Zheng-zhong. *The general boundary value problem for a class of semilinear second order degenerate elliptic equations.* Acta Math. Sinica 27(1984), 177-191(Chinese).

[42] Dong, Guang-chang. 1) *Initial and nonlinear oblique boundary value problems for fully nonlinear parabolic equations.* J. Partial Differential Equations, Series A, 1(1988), no.2, 12-42.
2) *Nonlinear second order partial differential equations.* Amer. Math. Soc., Providence, RI, 1991.

[43] Douglis, A. *A function-theoretic approach to elliptic systems of equations in two variables.* Comm. Pure Appl. Math. 6(1953), 259-289.

[44] Douglis, A and Nirenberg, L. *Interior estimates for elliptic systems of partial differential equations.* Comm. Pure Appl. Math. 8(1953), 503-538.

[45] Duduchava, R and Rodino, L. *The Riemann-Hilbert boundary value problem in a bicylinder.* Bollettino U. M. I. (6) 4-A(1985), 327-336.

[46] Fang, Ai-nong. *Quasiconformal mappings and the theory of functions for systems of nonlinear elliptic partial differential equations of first order.* Acta Math. Sinica 23(1980), 280-292 (Chinese).

[47] Fasano, A and Primicerio, M. *New results on some classical parabolic free boundary problems.* Q. Appl. Math. 38(1980-1981), 439-460.

[48] Friedman A. 1) *Variational principles and free boundary problems.* Wiley, New York, 1982.
2) *Partial differential equations of parabolic type.* Malabar Fla., Kriger, 1983.

[49] Gakhov, F D. *Boundary value problems.* Fizmatgiz, Moscow, 1963 (Russian); Pergamon, Oxford, 1966.

[50] Gilbarg, D and Trudinger, N S. *Elliptic partial differential equations of second order.* Springer-Verlag, Berlin, 1977.

[51] Gilbert, R P. 1) *Function theoretic methods in partial differential equations.* Academic Press, New York, 1969.
2) *Constructive methods for elliptic equations.* Lecture Notes in Math. 365, Springer-Verlag, Berlin etc., 1974.
3) *Nonlinear boundary value problems for elliptic systems in the plane.* Nonlinear Systems and Applications, ed. V. Lakshmikantham. Academic Press, New York, 1977, 97-124.
4) *Verallgemeinerte hyperanalytische Funktionentheorie.* Komplexe Analysis und ihre Anwendungen auf partielle Differentialgleichungen. Martin-Luther-Univ. Halle-Wittenberg, Wiss. Beiträge 1980, 124-145.
5) *Plane ellipticity and related problems.* Amer. Math. Soc., Providence, RI, 1981.

[52] Gilbert, R P and Buchanan, J L. *First order elliptic systems: A function theoretic approach.* Academic Press, New York, 1983.

[53] Gilbert, R P and Hile, G N. 1) *Generalized hypercomplex function theory.* Trans. Amer. Math. Soc. 195(1974), 1-29.
2) *Hypercomplex function theory in the sense of L. Bers.* Math. Nachr. 72(1976), 187-200.
3) *Degenerate elliptic systems whose coefficients matrix has a group inverse.* Complex Variables, Theory Appl. 1(1982), 61-88.

[54] Gilbert, R P and Hsiao, G C. *Constructive function theoretic methods for higher order pseudoparabolic equations.* Lecture Notes in Math. 561, Berlin etc., 1976, 51-67.

[55] Gilbert, R P and Kukral, D K. *A function theoretic method for* $\Delta_4^2 u + Q(x)u = 0$. Ann. Math. Pure Appl. 104(1975), 31-42.

[56] Gilbert, R P and Lin, Wei. 1) *Algorithms for generalized Cauchy kernels.* Complex Variables, Theory Appl. 2(1983), 103-124.
2) *Function theoretic solutions to problems of orthotropic elasticity.* J. Elasticity, 15(1985), 143-154.

[57] Gilbert, R P and Wen, Guo-chun. 1) *Free boundary problems occurring in planar fluid dynamics.* Nonlinear Analysis 13(1989), 285-303.
2) *Two free boundary problems occurring in planar filtrations.* Nonlinear Analysis 21(1993), 859-868.

[58] Gilbert, R P and Wendland, W. *Analytic, generalized hyperanalytic function theory and an application to elasticity.* Proc. Roy. Soc. Edinburgh 73A(1975), 317-331.

[59] Goldschmidt, B. 1) *Verallgemeinerte analytische Vektoren in* R^n. Habilitationsschrift, Martin-Luther-Univ., Halle-Wittenberg, 1980.
2) *Generalized analytic functions in* R^n. Komplexe Analysis und ihre Anwendungen auf partielle Differentialgleichungen. Martin-Luther-Univ., Halle-Wittenberg, Wiss. Beiträge 1980, 175-178.
3) *Regularity properties of generalized analytic vectors in* R^n. Math. Nachr. 103(1981), 245-254.

[60] Gong, Zhi-Guo. *Integral of Cauchy type and Riemann problem in Clifford analysis.* J. Sichuan Normal Univ.(Natur. Sci.), 1994, no.2, 16-18.

[61] Haack, W and Wendland, W. *Lectures on Pfaffian and partial differential equations.* Pergamon Press, Oxford, 1972.

[62] Habetha, K. 1) *Über lineare elliptische Differentialgleichungssysteme mit analytischen Koeffizienten.* Ber. Ges. Math. Datenverab. Bonn 77(1973), 65-89.
2) *On zeros of elliptic systems of first order in the plane.* Function Theoretic Methods in Differential Equations. Pitman, London, 1976, 42-62.

[63] Hile, G N. 1) *Hypercomplex function theory applied to partial differential equations.* Ph. D. Dissertation, Indiana University, Bloomington, 1972.
2) *Function theory for a class of elliptic systems in the plane.* J. Diff. Eq. 32(3)(1979), 369-387.

[64] Hile, G N and Protter, M H. 1) *Maximum principles for a class of first order elliptical systems.* J. Diff. Eq. 24(1)(1977), 136-151.
2) *Properties of overdetermined first order elliptic systems.* Arch. Rat. Mech. Anal. 66(1977), 267-293.

[65] Hou, Zong-yi. *Dirichlet problem for a class of linear elliptic second order equations with parabolic degeneracy on the boundary of the domain.* Sci. Record(N. S.) 2(1958), 244-249 (Chinese).

[66] Hörmander, L. *Linear partial differential operators.* Springer-Verlag, Berlin etc., 1964.

[67] Hsiao, G C and Wendland, W. *A finite element method for some integral equations of the first kind.* J. Math. Anal. Appl. 58(1977), 449-481.

[68] Hua, Loo-keng, Lin, Wei and Wu, Tzu-chien. *Second order systems of partial differential equations in the plane.* Pitman, London, 1985.

[69] Huang, Sha. *A nonlinear boundary value problem with Haseman shift in Clifford analysis.* J. Systems Sci. Math. Sci. 11(1991), 336-345.

[70] Huang, Sha and Li, Sheng-xun. *On oblique derivative problem for generalized regular functions with values in a real Clifford algebra.* Integral Equations and Boundary Value Problems, World Scientific, Singapore, 1991, 80-85.

[71] Iftimie, V. *Functions hypercomplexes.* Bull. Math. Soc. Sci. Math. R. S. Roumanie 9(57)(1965), 279-332.

[72] Iwaniec, T. *Quasiconformal mapping problem for general nonlinear systems of partial differential equations.* Symposia Math. 18, Acad. Press, London, 1976, 501-517.

[73] Iwaniec, T and Mamourian, A. *On the first order nonlinear differential systems with degeneration of ellipticity.* Proc. Second Finnish-Polish Summer School in Complex Analysis, (Jyväskylä 1983), Univ. Jyväskylä, 28(1984), 41-52.

[74] Kakicilev, V A. *Boundary value problems for the linear conjugacy for the functions homomorphic in a bicylindrical domain.* Dokl. Akad. Nauk SSSR 178(1968), 1003-1006 (Russian).

[75] Keldych, M V. *On certain cases of degeneration of equations of elliptic type on the boundary of a domain.* Dokl. Akad. Nauk SSSR(N.S.) 77(1951), 181-183 (Russian).

[76] Kharab, A. *A free boundary value problem for water invading an unsaturated medium.* Computing 38(1987), 185-207.

[77] Kracht, M and Kreyszig, E. *Methods of complex analysis in partial differential equations with applications.* Wiley, New York, 1988.

[78] Krushkal, S L and Kühnau, R. *Quasikonforme Abbildungen. Neue Methoden und Anwendungen.* Teubner-Verlag, Leipzig, 1983.

[79] Kumar, A. A generalized Riemann boundary problem in two variables. Arch. Math. (Basel) 62(1994), 531-538.

[80] Kühnau, R and Tutschke, W. 1) *Boundary value and initial value problems in complex analysis.* Longman, Harlow, 1991.
2) *Geometric function theory and applications of complex analysis in mechanics.* Longman, Harlow, 1991.

[81] Ladyshenskaja, O A and Uraltseva, N N. *Linear and quasilinear elliptic equations.* Academic Press, New York, 1968.

[82] Lanckau, E. *Complex integral operators in mathematical physics.* Johann Ambrosius Barth, Leipzig, 1993.

[83] Lanckau, E and Tutschke, W. *Complex analysis: Methods, trends, and applications.* Akademie-Verlag, Berlin, 1983.

[84] Lavrent'ev, M A. *Certain boundary value problems for systems of elliptic type.* Sibirsk Mat. Z. 3(1962), 715-728 (Russian).

[85] Lehto, O. *Remarks on generalized Beltrami equations and conformal mappings.* Proc. Romanian-Finnish Seminar on Teichmüller Spaces and Quasiconformal Mappings, Brasov, 1969. Sém. Inst. Math. Acad. R. S. Roumanie, Publ. House Acad. S. R. Roumanie, 1971, 203-214.

[86] Leray, J and Schauder, J. *Topologie et équations fonczionelles.* Ann. Sci. École Norm. Sup. 51(1934), 45-78; YMH 1(1946), 71-95 (Russian).

[87] Levinson, N. *Dirichlet problem for $\Delta u = f(x, y, u)$.* J. Math. Mech. 12(1963), 567-575.

[88] Li, Zhong. *The modified Dirichlet problem for a system of quasilinear partial differential equations of elliptic type.* Beijingdaxue Xuebao(Acta Sci. Nat. Univ. Peking), 1964, 319-341 (Chinese).

[89] Li, Zhong and Wen, Guo-chun. *On Riemann-Hilbert boundary value problem of elliptic systems of linear partial differential equations of the first order.* Acta Math. Sinica 15(1965), 599-613 (Chinese); Chinese Math. 7(1965), 323-338.

[90] Li, Ming-de and Qin, Yu-chun. *On boundary value problems for singular elliptic equations.* Hangzhoudaxue Xuebao(Ziran Kexue), 1980., no.2, 1-8.

[91] Li, Ming-zhong. 1) *Hausdorff solvability of the Dirichlet problem for a class of second order elliptic systems.* Chinese Ann. Math. 3(1982), 319-328 (Chinese).
2) *Generalized Riemann-Hilbert problem for a system of first order quasilinear elliptic equations of several complex variables.* Complex Varivables, Theory and Appl. 7(1987), 383-393.

[92] Li, Ming-zhong and Cheng, Jin. 1) *On boundary value problems for overdetermined elliptic systems of two complex variables.* Chinese Ann. Math. 11B(1990), 84-92 (Chinese).
2) *Boundary value problems for complex overdetermined partial differential equations of second order.* Integral Equations and Boundary Value Problems, World Scientific, Singapore, 1991, 11-17.

[93] Lion, J L and Magenes, E. *Non-homogeneous boundary value problems and applications.* Springer-Verlag, Berlin etc., 1972.

[94] Litvinchuk, G S. *Boundary value problems and singular integral equations with shift.* Nauka, Moscow, 1977 (Russian).

[95] Liu, Qi-feng. *The Riemann-Hilbert problem for nonlinear elliptic complex equations in a multiply connected domain.* Sichuan Shiyuan Xuebao(Ziran Kexue), 1984, no.4, 32-38.

[96] Lopatinski, Y B. *On a method of reducing boundary value problems for a system of differential equations of elliptic type to regular equations.* Ukrain. Mat. Z. 5(1953), 123-151 (*Russian*).

[97] Lu, Chien-ke. 1) *On compound boundary problems.* Sci. Sinica 14(1965), 1545-1555.
2) *Methods of complex variables for planar elasticity.* Wuhan University Press, Wuhan, 1986 (Chinese).
3) *Boundary value problems for analytic functions.* World Scientific, Singapore, 1993.

[98] Malonek, H and Müller, B. *Definition and properties of a hypercomplex singular integral operator.* Results Math. 22(1992), 713-724.

[99] Mikhailov, L G. *A new class of singular integral equations and its applications to differential equations with singular coefficients.* Wolters-Noordhoff, Groningen, 1970.

[100] Miranda, C. *Partial differential equations of elliptic type.* Springer-Verlag, Berlin etc., 1970.

[101] Monakhov, V N. 1) *Transformations of multiply connected domains by the solutions of nonlinear L-elliptic systems of equations.* Dokl. Akad. Nauk SSSR

220(1975), 520-523 (Russian).
2) *Boundary value problems with free boundaries for elliptic systems.* Amer. Math. Soc., Providence, RI, 1983.

[102] Morrey, C B. *On the solution of quasilinear elliptic partial differential equations.* Trans. Am. Math. Soc. 43(1938), 126-166.

[103] Mshimba, A S A. *The Hilbert boundary value problem for elliptic differential equations in the plane.* Appl. Anal. 30(1988), 75-86.

[104] Mshimba, A S A and Tutschke W. *Functional analytic methods in complex analysis and applications to partial differential equations.* World Scientific, Singapore, 1990.

[105] Mushelishvili, N I. 1) *Singular integral equations.* Noordhoff, Groningen, 1953.
2) *Some basic problems of the mathematical theory of elasticity.* Nauka, Moscow, 1946(Russian); Noordhoff, Groningen, 1953.

[106] Naas, J and Tutschke, W. 1) *Some probabilistic aspects in partial complex differential equations.* Complex Analysis and its Applications. Akad. Nauk SSSR, Izd. Nauka, Moscow, 1978, 409-412.
2) *On the error in the approximate solution of boundary value problems of nonlinear first order differential equations in the plane.* Appl. Anal. 7(1978), 239-246.

[107] Nirenberg, L. 1) *On nonlinear elliptic partial differential equations and Hölder continuity.* Comm. Pure Appl. Math. 6(1953), 103-156.
2) *An application of generalized degree to a class of nonlinear problems.* Coll. Analyse Fonct. Liége, 1970, Vander, Louvain, 1971, 57-74.

[108] Nitsche, J and Nitsche, J J. *Bemerkungen zum zweiten Randwertproblem der Differentialgleichung* $\Delta \phi = \phi_x^2 + \phi_y^2$. Math. Ann. 126(1953), 69-74.

[109] Oleinik, O A. *On equations of elliptic type degenerating on the boundary of a region.* Dokl. Akad. Nauk SSSR(N.S.) 87(1952), 885-888.

[110] Orton, M. *Hilbert boundary value problems – a distributional approach.* Proc. Roy. Soc. Edinburgh 77A(1977), 193-208.

[111] Oskolkov, A P. *Interior estimates of the first order derivatives for a certain class of quasilinear elliptic systems.* Trudy Mat. Inst. Steklov 110(1970), 102-106 (Russian).

[112] Panejah, B P. *On the theory of solvability of the oblique derivative problem.* Math. Sb.(N.S.) 114(1981), 226-268 (Russian).

[113] Polozhy, G N. *Generalization of the theory of analytic functions of a complex variable. p-analytic and (p,q)-analytic functions and some of their applications.* Izdat. Kiev. Univ., Kiev, 1965 (Russian).

[114] Protter, M H and Weinberger, H F. *Maximum principles in differential equations.* Prentice-Hall, Englewood Cliffs, N.J., 1967.

[115] Rudin, W. *Function theory in the unit ball of C^n.* Springer-Verlag, Berlin, 1980.

[116] Ryan, J. 1) *Complexified Clifford analysis,* Complex Variables, Theory and Appl. 1(1982), 119-149.
2) *Special functions and relation within complex Clifford analysis,* I. Complex Variables, Theory and Appl. 2(1983), 177-198.

[117] Schauder, L. *Der Fixpunktsatz in Funktionalräumen.* Studia Math. 2(1930), 171-181.

[118] Schleiff, M. *Über geschlossene Lösungen des Poincaréschen Randwertproblems mit Hilfe einer komplexen Differentialgleichung.* Beiträge zur Analysis 6(1974), 73-85.

[119] Shi, P. *Invariance property of anisotropic Hele-Shaw flows.* Appl. Anal. 30(1988), 201-231.

[120] Sobolev, S L. *Applications of functional analysis in mathematical physics.* Amer. Math. Soc., Providence, RI, 1963.

[121] Sommen, F. *Some connections between Clifford analysis and complex analysis.* Complex Variables, Theory Appl. 1(1982), 97-118.

[122] Stein, E M. *Singular integrals and differentiability properties of functions.* Princeton Univ. Press, Princeton, N. J., 1970.

[123] Stern, I. 1) *Boundary value problems for generalized Cauchy-Riemann systems in the space.* Boundary Value and Initial Value problems in Complex Analysis, Longman, Harlow, 1991, 159-183.
2) *On the existence of Fredholm boundary value problems for generalized Cauchy-Riemann systems in the space.* Complex Variables, Theory Appl. 21(1993), 19-38.

[124] Tai, Chung-wei. *The oblique derivative problem for various elliptic complex equations in a multiply connected domain.* Appl. Math. Num. Math. 1982, no.1, 69-73 (Chinese).

[125] Teichert-Gaida, A. *On some compound nonlinear boundary value problem.* Demostratio Math. 9(1976), 409-421.

[126] Tian, Mao-ying. 1) *The mixed boundary value problem for a class of nonlinear degenerate elliptic equations of second order in the plane.* Beijingdaxue Xuebao (Ziran Kexue), 1982, no. 5, 1-13 (Chinese).
2) *The general boundary value problem for nonlinear degenerate elliptic equations of second order in the plane.* Integral Equations and Boundary Value Problems, World Scientific, Singapore, 1991, 197-203.

[127] Tjurikov, E V. *The nonlinear Riemann-Hilbert boundary value problem for quasilinear elliptic systems.* Soviet Math. Dokl. 20(1979), 863-866.

[128] Trudinger, N S. *Nonlinear oblique boundary value problems for nonlinear elliptic equations.* Trans. Amer. Math. Soc. 295(1986), 509-546.

[129] Tutschke, W. 1) *Die neuen Methoden der komplexen Analysis und ihre Anwendung auf nichtlineare Differentialgleichungssysteme.* Sitzungsber. Akad. Wiss. DDR 17N(1976).
2) *Partielle komplexe Differentialgleichungen in einer und in mehreren komplexen Variablen.* VEB Deutscher Verlag der Wissenschaften, Berlin, 1977.
3) *The Riemann-Hilbert problem for nonlinear systems of differential equations in the plane.* Complex Anaysis and its Applications. Akad. Nauk SSSR, Izd. Nauka, Moscow, 1978, 537-542.
4) *Boundary value problems for generalized analytic functions for several complex variables.* Ann. Pol. Math., 39(1981), 227-238.
5) *Partielle Differentialgleichungen: Klassische, funktionalanalytische und komplexe Methoden.* Teubner, Leipzig, 1983.

[130] Valent, T. *Boundary value problems of finite elasticity.* Springer-Verlag, Berlin, 1987.

[131] Vekua, I N. 1) *Generalized analytic functions.* Pergamon, Oxford, 1962.
2) *Theory of thin shallow shells of variable thickness.* Akad. Nauk Gruzin. SSR Trudy Tbilisi. Mat. Inst. Razmadze 1965 (Russian).
3) *New methods for solving elliptic equations.* North-Holland Publ., Amsterdam, 1967.
4) *On two ways of constructing the consistent theory of elastic shells.* Mater. perv. vsesoyuz. shk. po. teor. i chisl. met. rasch. obol. i. plast., Metsniereba, Tbilisi, 1974 (Russian).
5) *Shell theory: General methods of construction.* Pitman, London etc., 1985.

[132] Vinogradov, V S. 1) *On some boundary value problems for quasilinear elliptic systems of first order in the plane.* Dokl. Akad. Nauk SSSR 121(1958), 579-581 (Russian).
2) *An analog of an integral of Cauchy type for analytic functions of several complex variables.* Soviet Math. Dokl. 9(1968), 73-76.
3) *A certain boundary value problem for an elliptic system of special form.* Diff.

Uravn. 77(1971), 1226-1234 (Russian).

4) *On an elliptic system which does not possess any Noetherian boundary value problem.* Dokl. Acad. Nauk SSSR 199(1971), 1008-1010 (Russian).

5) *On a method of solution of a boundary value problem for a first order elliptic system on the plane.* Dokl. Akad. Nauk SSSR 201(1971) (Russian); Soviet Math. Dokl. 12(1971), 1699-1703.

6) *Elliptic systems in the plane.* Complex Analysis-Methods, Trends, and Applications. Akademie-Verlag, Berlin, 1983, 166-173.

7) A boundary value problem for pseudo-symmetric systems. Diff. Uravn. 21,1(1985), 161-163 (Russian).

[133] Vvedenskaya, N D. *On a boundary problem for equations of elliptic type degenerating on the boundary of a region.* Dokl. Akad. Nauk SSSR(N.S.) 91(1953), 711-714 (Russian).

[134] Wang, Chung-fang. *Dirichlet problems for singular elliptic equations.* Hangzhoudaxue Xuebao(Ziran Kexue), 1978, no.2, 19-32 (Chinese).

[135] Wang Li-ping. *Riemann-Hilbert boundary value problem for analytic functions in several bicylinder domain.* J. Ocean University of Qingdao 21(1991), 134-142 (Chinese).

[136] Warowna-Dorau, G. *A compound boundary value problem of Hilbert-Vekua type.* Demonstratio Math. 7(1974), 337-352.

[137] Warschawski, W. *On differentiability at the boundary in conformal mapping.* Proc. Amer. Math. Soc. 12(1961), 614-620.

[138] Wegert, E. *Nonlinear boundary value problems for homomorphic functions and singular integral equations.* Akademie-Verlag, Berlin, 1992.

[139] Wen, Guo-chun. 1) *Modified Dirichlet problem and quasiconformal mappings for nonlinear elliptic systems of first order.* Kexue Tongbao (A monthly J. Sci.) 25(1980), 449-453.

2) *The Riemann-Hilbert boundary value problem for nonlinear elliptic systems of first order in the plane.* Acta. Math. Sinica 23(1980), 244-255 (Chinese).

3) *The oblique derivative boundary value problem for nonlinear elliptic systems of second order* I. Hebei Huagong Xueyuan Xuebao (Shuxue Zhuanji), 1980, 119-144 (Chinese).

4) *The singular case of the Riemann-Hilbert boundary value problem.* Beijingdaxue Xuebao (Acta Sci. Natur. Univ. Peki.), 1981, no.4, 1-14 (Chinese).

5) *The Poincaré problem with negative index for linear elliptic systems of second order in a multiply connected domain.* J. Math. Res. Expos., 1981, no.3, 61-76 (Chinese).

6) *The mixed boundary value problem for nonlinear elliptic equations of second*

order in the plane. Proc. 1980 Beijing Sym. Diff. Geom. Diff. Eq., Beijing, 1982, 1543-1557.

7) *On a representation theorem of solutions and mixed boundary value problems for second order nonlinear elliptic equations with unbounded measurable coefficients*. Acta Math. Sinica 26(1983), 533-537 (Chinese).

8) *Oblique derivative boundary value problems for nonlinear elliptic systems of second order*. Scientia Sinica, Ser. A, 26(1983), 113-124.

9) *Some nonlinear boundary value problems for nonlinear elliptic equations of second order in the plane*. Complex Variables, Theory Appl. 4(1985), 189-210.

10) *Nonlinear discontinuous boundary value problems for nonlinear elliptic systems of first order in a multiply connected domain*. Beijingdaxue Xuebao, 1985, no.3, 1-10 (Chinese).

11) *Nonlinear oblique derivative boundary value problems for nonlinear elliptic systems of several second order equations*. Proc. 1982 Changchun Sym. Diff. Geom. Diff. Eq., Changchun, 1986, 623-636.

12) *Applications of complex analysis to nonlinear elliptic systems of partial differential equations*. Analytic Functions of one Complex Variable, Amer. Math. Soc., Providence, RI, 1985, 217-234.

13) *Linear and nonlinear elliptic complex equations*. Shanghai Science Techn. Publ., Shanghai, 1986 (Chinese).

14) *Some boundary value problems for nonlinear degenerate elliptic complex equations*. Lectures on Complex Analysis, World Scientific, Singapore, 1988, 265-281.

15) *Oblique derivative problems in Clifford analysis*. J. Yantai Univ.(Natur. Sci. Eng.), 1990, no.1, 1-6 (Chinese).

16) *Clifford analysis and elliptic systems, hyperbolic systems of first order*. Integral Equations and Boundary Value Problems, World Scientific, Singapore, 1991, 237-243.

17) *Conformal mappings and boundary value problems*. Amer. Math. Soc., Providence, RI, 1992.

18) *Irregular oblique derivative problems for fully nonlinear elliptic equations of second order*. J. Yantai Univ.(Natur. Sci. Eng.), 1992, no.4, 1-8 (Chinese).

19) *Function theoretic methods for partial differential equations and their applications and computations*, Advan. Math. 22(1993), 391-402.

20) *Some discontinuous boundary value problems for nonlinear elliptic equations of second order in multiply connected domains*. J. Yantai Univ.(Natur. Sci. Eng.), 1994, no.1, 6-11 (Chinese).

[140] Wen, Guo-chun and Begehr, H. *Boundary value problems for elliptic equations and systems*. Longman, Harlow, 1990.

[141] Wen, Guo-chun and Chen, Li-cheng. *Complex variable methods for the solution of some filtration problems with free boundaries*. Arab. J. Sci. Eng. 17(1992), 4B:605-609.

[142] Wen, Guo-chun and Fang Ai-nong. *The complex form and some boundary value problems for nonlinear elliptic systems of second order.* Chinese Ann. Math. 2(1981), 201-216 (Chinese).

[143] Wen, Guo-chun and Kang, Shi-xiang. *The Dirichlet problem of elliptic complex equations of 2nth order.* J. Sichuan Normal Univ.(Natur. Sci.), 1990, no.3, 11-17 (Chinese).

[144] Wen, Guo-chun, Li, Hong-zhen and Li, Zi-shi. *Properties of solutions for first order elliptic complex equations in the whole plane.* Hebeidaxue Xuebao (Ziran Kexue), 1981, no.2, 28-35 (Chinese).

[145] Wen, Guo-chun and Li, Ping-qian. *The weak solution of Riemann-Hilbert problems for elliptic complex equations of first order.* Appl. Anal. 45(1992), 209-227.

[146] Wen, Guo-chun, Li, Sheng-xun and Xu, Ke-ming. *On basic theorems of quasiconformal mappings.* Hebei Huagong Xueyuan Xuebao (Shuxue Zhuanji), 1980, 20-40 (Chinese).

[147] Wen, Guo-chun and Liu, Qi-feng. 1) *The Riemann-Hilbert problem for elliptic systems of fourth order.* Sichuan Shiyuan Xuebao (Shuxue Zhuanji), 1981, 90-103 (Chinese).
2) *The oblique derivative boundary value problem for elliptic systems of fourth order.* Sichuan Shiyuan Xuebao (Ziran Kexue), 1983, no.2, 53-64 (Chinese).
3) *Nonlinear Riemann-Hilbert problems for nonlinear elliptic complex equations of first order.* Sichuan Shiyuan Xuebao (Ziran Kexue), 1983, no.4, 36-47 (Chinese).
4) *Some gas-dynamics problems in a multiply connected domain with free boundaries.* Sichuan Shiyuan Xuebao (Ziran Kexue), 1986, no.1, 26-38 (Chinese).

[148] Wen, Guo-chun, Liu, Qi-feng and Kang, Shi-xiang. *The oblique derivative boundary value problem for elliptic system of fourth order.* Sichun Shiyuan Xuebao (Ziran Kexue), 1983, no.2, 53-64 (Chinese).

[149] Wen, Guo-chun and Tai, Chung-wei. *Nonlinear boundary problems for elliptic complex equations of second order.* Hebeishida Xuebao (Ziran Kuxue), 1983, no.1, 6-18 (Chinese).

[150] Wen, Guo-chun and Tain, Mao-ying. 1) *Solutions for elliptic equations of second order in the whole plane.* J. Math. 2(1982), no.1, 23-36(Chinese).
2) *Some boundary value problems for several complex variables.* J. Yantai Univ. (Natur. Sci. Eng.), 1989, no.2, 1-14 (Chinese).

[151] Wen, Guo-chun and Xu, Zuo-liang. 1) *Some free boundary value problems in elasticity.* Geometric Function Theory and Applications of Complex Analysis in Mechanics, Longman, Harlow, 1991, 167-175.

2) *The Poincaré problem for nonlinear elliptic equations of second order with nonsmooth boundaries,* J. Liaoning Inst. of Tech., 1994, no.1, 13-18.(Chinese)

[152] Wen, Guo-chun and Yang, C C. *Some discontinuous boundary value problems for nonlinear elliptic systems of first order in the plane.* Complex Variables, Theory Appl. 25(1994), 217-226.

[153] Wen, Guo-chun and Yang, Guang-wu. *The Poincaré boundary value problem for linear equations of second order.* Advan. Math. 10(1981), 157-160 (Chinese).

[154] Wen, Guo-chun, Yang, Guang-wu and Huang, Sha. *Generalized analytic functions and their generalizations.* Hebei Education Press, Hebei, 1989 (Chinese).

[155] Wen, Guo-chun, Yang, Guang-wu ang Kang Shi-xiang. *Some basic boundary value problem for elliptic complex equations of first order with non-smooth boundaries.* J. Sichuan Normal Univ. (Natur. Sci.), 1994, no.6, 17-25.

[156] Wen, Guo-chun and Zhao, Zhen. *Integral equations and boundary value problems.* World Scientific, Singapore, 1991.

[157] Wendland, W. 1) *On a class of semilinear boundary value problems for certain elliptic systems in the plane.* Complex Analysis and its Applications. Akad. Nauk USSR, Izd. Nauka, Moscow, 1978, 108-119.
2) *On the imbedding method for semilinear first order elliptic systems and related finite element methods.* Continuation methods. Academic Press, New York etc. 1978, 277-336.
3) *Elliptic systems in the plane.* Pitman, London etc., 1979.
4) *Zur Behandlung elliptischer Randwertaufgaben mit Integralgleichungen.* Wiss. Schriftenreihe TH Karl-Marx-Stadt, 1979, 143-153.

[158] Wilson, D G, Soloman, A D and Boggs, P T. *Moving boundary problems.* Academic Press, New York, 1978.

[159] Wloka, J. *Partielle Differentialgleichungen.* Teubner-Verlag, Leipzig, 1982.

[160] Wolfersdorf, L v. 1) *A class of nonlinear Riemann-Hilbert problems for holomorphic functions.* Math. Nachr. 116(1984), 89-107.
2) *On strongly nonlinear Poincaré boundary value problems for harmonic functions.* Z. Anal. Anw. 3(1984), 385-399.

[161] Wolferdorf, L v and Wolska-Bochenek, J. *A compound Riemann-Hilbert problem for homomorphic functions with nonlinear boundary condition.* Demonstratio Math. 17(1984), 545-556.

[162] Wu, Su-ming. *A free boundary problem in filtration theory of an earth dam.* Beijingdaxue Xuebao(Acta Sci. Natur. Univ. Peki.), 1987, no.5, 78-91 (Chinese)

[163] Xu, Zhen-yuan. 1) *Nonlinear Poincaré problem for a system of first order elliptic equations in the plane.* Complex Variables, Theory and Application. 7(1987), 363-381.
2) *Boundary value problems and function theory for spin-invariant differential operators.* Ph.D thesis, State Univ. Ghent, Ghent, 1989.
3) *On linear and nonlinear Riemann-Hilbert problems for regular function with values in a Clifford algebra.* Chinese Ann. Math. Ser. 11B(1990), 349-357.

[164] Xu, Zuo-liang. 1)*A free boundary in axisymmetric nonstationary filtration.* Integral Equations and Boundary Value Problems, World Scientific, Singapore, 1991, 251-256.
2) *Riemann-Hilbert boundary value problem for analytic functions in the unit ball of C^2.* to appear.

[165] Yang, C C, Wen, G C, Li, K Y and Chiang, Y M. *Complex analysis and its applications.* Longman, Harlow, 1993.

[166] Yang, Pi-wen. *Riemann boundary value problems for functions of several complex variables in the generalized polycylinder domain.* J. Sichuan Normal Univ. (Natur. Sci.), 1991, no.2, 117-124 (Chinese).

Index